The Latin American Studies Book Series

The Latin American Studies Book Series promotes quality scientific research focusing on Latin American countries. The series accepts disciplinary and interdisciplinary titles related to geographical, environmental, cultural, economic, political, urban and health-related research dedicated to Latin America. The series publishes comprehensive monographs, edited volumes and textbooks refereed by a region or country expert specialized in Latin American studies.

The series aims to raise the profile of Latin American studies, showcasing important works developed focusing on the region. It is aimed at researchers, students, and everyone interested in Latin American topics.

Submit a proposal: Proposals for the series will be considered by the Series Advisory Board. A book proposal form can be obtained from the Publisher, Andrea Brody-Barre (andrea.brody-barre@springer.com).

Vanda Claudino-Sales · José Falcão Sobrinho
Editors

Geomorphology of the Northeast Region of Brazil

Springer

Editors
Vanda Claudino-Sales
Federal University of Ceara
Fortaleza, Ceará, Brazil

José Falcão Sobrinho
Acarau Valley State University
Sobral, Ceará, Brazil

ISSN 2366-3421 ISSN 2366-343X (electronic)
The Latin American Studies Book Series
ISBN 978-3-031-58873-0 ISBN 978-3-031-58874-7 (eBook)
https://doi.org/10.1007/978-3-031-58874-7

This Springer imprint is published by the registered company Springer Nature Switzerland AG
The registered company address is: Gewerbestrasse 11, 6330 Cham, Switzerland

This book is dedicated to the geomorphologist Tereza Cardoso

Foreword

This work is a praiseworthy and brilliant initiative of the authors, who dealt with the Brazilian Northeast from the perspective of geomorphological knowledge in its state of the art for nine States. Sectoring the distribution of the chapters referring to the States by sub-regions, even though these are not formally and institutionally constituted, seemed to us a reasonable solution to present the contents treated here. The first chapter is the introduction, which contains the theoretical and methodological basis of geomorphology concisely, without reveries, while the nine others contemplate each State. Due to the distributive arrangement, a repetition of information and interpretations happened because many States have the same relief units or compartments, mainly when treated on the medium or regional scale as it happened here.

What could be a problem—the repetition of communication of contents—has revealed as an interesting contribution as the protagonists of each one of the chapters have exercised absolute independence to deal with the geomorphologic theme of their respective States in the way they prefer, that is, on the methodological, technical, and theoretical approach in the manner of presenting the facts and interpretations. Nowadays, the appreciation of differences has been a treasure as it allows creativity and freedom of academic and scientific manifestations without necessarily submitting to the impositions of a single, standardized, rigorously homogenized, and impoverishing method.

Despite the different forms of presenting the units and major landforms over the nine chapters, there is no contradictory information or interpretations among them. It has revealed that, within the differences, there are convergences of knowledge consolidated in the geological, geomorphological, and geographic literature, which appropriately fit and are adjusted when worked on a regional scale.

The chapters concerning the States are not products of a single project but derive from the rescue of knowledge produced over time, some based on systematic geomorphological mapping and others on recognized and diverse literature. Some chapters are erudite and academic, while others are more technical and scientific. Instead of being a problem, it reveals various possibilities to interpret and represent geomorphological facts. Nature is unequal, diverse, and different, even though some elements for analysis are prerequisites for a hegemonic understanding, but not necessarily the same. Science is researching, experimenting, daring, creating, disseminating, and confronting results. Because of the facts, the organizers and authors of each of the chapters have performed their academic and technical-scientific abilities nobly and impartially, gracing the Brazilian geographic and geomorphological community with outstanding products of scientific and educational interest.

Professor Jurandyr Luciano Sanches Ross
São Paulo University (USP)
São Paulo, Brazil

Introduction

The Brazilian Northeast, comprising 1,600,000 km^2 and nine States (Alagoas, Bahia, Ceará, Maranhão, Paraíba, Pernambuco, Piauí, Rio Grande do Norte, Sergipe), has an enormous diversity of geomorphological landscapes. Despite the wealth of relief, the available studies do not present, in detail, the regional geomorphological elements. However, it is necessary to allow the advancement of geographic science and related areas, create appropriate processes for the use and occupation of geographic space, guide land-use planning and environmental planning, and encourage conservationist practices.

Thinking about the significant aspects, we organized and offered to the national and international public a set of works explaining the genesis and evolution of the geomorphologic landscape of each State in the Northeast region. The set undergoes the optics of the book Geomorphology of the Northeast Region of Brazil, which has a chapter for each State.

The chapters, written by the best geomorphological specialists in each segment, present unique, rich, and unpublished material that treats the reality of the geomorphological landscape in the Brazilian Northeast. It starts with a theoretical discussion on advances in the analysis of geomorphology by State, showing an unprecedented and functional geographic division. Indeed, the division of the book was into theoretical issues, Western Northeast, Northern Northeast, Eastern Northeast, and Southern Northeast, following the morphological and genetic characteristics of each geographic segment.

The relief in the Brazilian Northeast reveals episodes of morphotectonic and paleoclimate evolution. The area, which is around paleosurfaces, has several geomorphological compartments derived from significant tectonic events, such as the formation cycle of the supercontinent Atlantis, known as the Transamazon Cycle in older literature, in the Paleoproterozoic, as well as the Brasiliano Cycle, which joined the supercontinent Gondwana in the Neoproterozoic, followed by the separation of Pangea, which divided South America and Africa in the Middle Cretaceous.

This group suffered remobilizations in the Cenozoic and underwent important neotectonic activities, some of them capable of generating reliefs, and the group modeled by climatic conditions is notably semi-arid but with variations to sub-humid and humid in some segments, including the coastal zone, which geologically makes up the Brazilian Marginal Province.

Such geological characteristics, which rebound in a complex way in geomorphological aspects, gave rise to an innovative and functional presentation of the sequence of chapters that make up this book. Here, we divide the Northeast into four segments according to the geological composition, which defines, together with climate and weather, the main geomorphological characteristics of the States.

Thus, we classify the Northeast region, from a structural and geographical point of view, the following segments: (1) Western Northeast, which includes Maranhão and Piauí, the so-called Mid North; (2) Northern Northeast, which includes Ceará, Rio Grande do Norte, and Paraíba; (3) the Eastern Northeast, formed by Pernambuco, Alagoas, and Sergipe; (4) and the Southern Northeast, defined by Bahia.

Concerning the Western Northeast, the main geological configuration of Maranhão is the generation of three sedimentary basins that occupy more than 90% of the state. They are the Parnaíba Gondwanic Basin, the Grajaú—São Luís Post-Gondwanic Basin, resulting from the opening of the Equatorial Atlantic, and the Espigão Mestre Basin, deposited during the Upper Cretaceous. The remaining 10% of the state corresponds to an outcrop of Precambrian crystalline rocks from the São Luís Province.

Piauí is in portions of four different geological provinces, three dating from the Archean to the Mesozoic and one from the Cenozoic. They are (a) Borborema; (b) São Francisco; (c) Parnaíba; and (d) coastal. These structures keep in their regional geomorphological compartmentation elements from tectonic, sedimentological, and flattening events from their genesis and evolution.

The Northern and Eastern Northeast, associated with the Borborema Province, result from the uplift induced by the split of Gondwana in the Cretaceous and from Cenozoic pulses of uplift and climatic fluctuations. The area reflects crustal segments with the diverse post-Cretaceous formation and deformation histories, which effectively allows us to divide it into the Northern Northeast (transform margin) and the Eastern Northeast (passive margin).

The Southern Northeast, characterized by the terrains that compose Bahia, is the domain area of the São Francisco Province, which constitutes a geotectonic macro-unit of South America, with rocks of the Paleoproterozoic, having limits delineated by folded belts associated with the Brazilian Orogeny of the Neoproterozoic. This is the most well-exposed and diversified tectonic unit of the South American platform basement, which impacts the relief extraordinarily.

The geological aspects in the chapters follow a logical and didactic way, tying geology and climate in the genesis of the regional reliefs. We strongly believe the book, concerning its production and presentation, will fill a gap in the academy and society concerning such natural aspects, besides the repressed demand from students, professionals, technicians, planners, administrators, and related professionals.

It is, therefore, with great joy that we release this product to the Brazilian and international society, and we thank the authors for having so kindly accepted our invitation to produce the material. The reviewers also played a significant role in ensuring the quality of the texts, and to them, we extend our thanks.

We expect this material will guide research, discussions, analyses, and debates and improve and strengthen the slow and necessary process of detailed knowledge of the natural aspects of the Brazilian territory, especially in the Northeast. We invite the readers to enjoy it!

Geomorphological greetings,

Vanda Claudino-Sales
José Falcão Sobrinho

Reviewers

Jonas Otaviano Praça de Souza, Federal University of Paraiba, Brazil
Marcelo Martins Moura-Fé, Regional University of Cariri, Brazil
Osvaldo Girão da Silva, Federal University of Pernambuco, Brazil
Rafael Albuquerque Xavier, State University of Rio Grande do Norte, Brazil

Contents

Editors and Contributors

About the Editors

Vanda Claudino-Sales holds a degree in geography from the University of Brasilia (UNB), a specialization in coastal geology at the Federal University of Rio Grande do Sul (UFRGS), a master's degree in physical geography at the University of São Paulo (USP), a Ph.D. in environmental geography at the Université Paris-Sorbonne, and post-doctoral studies in coastal geomorphology at the University of Florida. She is a professor of the Department of Geography of the Federal University of Ceara (UFC) and was a visiting professor at the University of Florida, at the University Paris-Sorbonne, and the geography master's program of the Vale do Acaraú State University (UVA). She is the regional vice president of the American Coastal Research Foundation (CERF) and has experience in structural geomorphology, mega-geomorphology, coastal geomorphology, and environmental analysis, and is an environmentalist, having been in the movement for the preservation of the environment in Ceará State for over 40 years.

José Falcão Sobrinho is a geographer, holding a post-doctorate in geography from the Federal University of Ceará (UFC) and a Ph.D. in physical geography from the University of São Paulo (USP). He is an associate professor of the geography graduate program at the Vale do Acaraú State University (UVA), the leader of the research and extension network of the semi-arid region/CNPq, a member of the editorial team of the Willian Moris Davis – Geomorphology Journal, and a research productivity fellow at CNPq (Level 2). He deals with geomorphology and erosion processes.

Contributors

Hélio Mário de Araújo Federal University of Sergipe, São Cristóvão, Brazil

Renata Nunes Azambuja Federal University of Sergipe, São Cristóvão, Brazil

Helen Nébias Barreto Federal University of Maranhão (UFMA), São Luís, Brazil

Frederico de Holanda Bastos State University of Ceará, Fortaleza, Brazil

Vanda Claudino-Sales Federal University of Ceará, Fortaleza, Brazil

Antônio Carlos de Barros Corrêa Federal University of Pernambuco (UFPE), Recife, Brazil

Marcelo Eduardo Dantas Geological Survey of Brazil SGB/CPRM, Rio de Janeiro, Brazil

Marco Túlio Mendonça Diniz Federal University of Rio Grande do Norte, Caico, Brazil

Bruno Ferreira Federal University of Pernambuco, Recife, Brazil

Iracilde Maria de Moura Fé Lima Federal University of Piauí, Teresina, Brazil

Daniel Rodrigues de Lira Federal University of Pernambuco (UFPE), Recife, Brazil

Rubson Pinheiro Maia Federal University of Ceará, Fortaleza, Brazil

Ronaldo Missura Federal University of Sergipe, São Cristóvão, Brazil

Kleython de Araújo Monteiro Federal University of Alagoas, Maceio, Brazil

George Pereira de Oliveira Federal University of Pernambuco (UFPE), Recife, Brazil

Edilea Dutra Pereira Federal University of Maranhão (UFMA), São Luís, Brazil

Jorge Hamilton Souza dos Santos Federal University of Maranhão (UFMA), São Luís, Brazil

Jémison Mattos dos Santos State University of Feira de Santana, Feira de Santana, Brazil

Drielly Naamma Fonsêca da Silva Metropolitan University of Technology, Santiago, Chile

José Falcão Sobrinho Acaraú Valley State University (UVA), Sobral, Brazil

Bruno de Azevedo Cavalcanti Tavares Federal University of Pernambuco (UFPE), Recife, Brazil

Part I
Theoretical Aspects

Chapter 1
Evolution of Geomorphological Knowledge: Theoretical and Methodological Aspects

José Falcão Sobrinho

Abstract The relief materializes as a stage for human activities. Resulting from endogenous and exogenous activities, its studies over time placed it as an interphase and integrative basis of natural units such as rock, climate, soil, vegetation and water resources, therefore, considering it as dynamic and being an object of study sometimes associated with an element, sometimes in the integrated landscape set. In this way, its theoretical and methodological basis has constituted different perspectives that allow understanding the relief according to the premise that it is approached. Here, we prioritize understanding relief as an integrative basis not only for nature but also for the organization of natural landscapes and, consequently, the actions imposed by society.

Keywords Relief · Methodology · Geomorphological knowledge

1.1 Opening Remarks

To contribute to the methodological basis that I am proposing, I will now outline brief considerations on the evolution of geomorphological knowledge.

Relief, abstract as matter, materializes as the stage for human activities.

The relief—resulting from endogenous and exogenous activities—interacts with the rock, the climate, the soil, the vegetation, and the water resources. This scope results in the constitution of the natural or cultural landscape associated with human activities. Thus, the understanding of relief focuses not only on its structure and shape but also on the unadulterated processes and human actions influencing its modeling. It is directed as a guiding element for decision-making to plan and manage the use of specific environments, whether in a rural or urban area and more than ever in the peri-urban, in progressive occupation of increasingly less natural spaces.

J. F. Sobrinho (✉)
Acaraú Valley State University (UVA), Sobral, Brazil
e-mail: falcao_sobrinho@uvanet.br

V. Claudino-Sales and J. F. Sobrinho (eds.), *Geomorphology of the Northeast Region of Brazil*, The Latin American Studies Book Series,
https://doi.org/10.1007/978-3-031-58874-7_1

The words of Abreu (1993) reflect the evolution of geomorphological knowledge by stating that the study of relief has been a segment of geology and geography. Besides, its object development has happened in a theoretical order or an empirical basis regarding the form of approach.

The focus here is not on a retrospective of the geomorphological evolution. However, it would help to understand the approach that treats relief nowadays and how I intend to place it in the methodological proposal.

The study of relief, like other geography branches, has taken different dimensions through geomorphology, contextualizing itself in the historical moment. According to Marques (1994), James Hutton, during the second half of the eighteenth century, had great importance in the geomorphological context, through the ideas of actualism, for being the first to identify the importance of knowledge of the present to understand the past.

However, Davis (1899) is the primary contributor to geomorphology history. The geographical cycle, idealized by him, was the first set of conceptions that could describe and explain, coherently, the genesis and evolutionary sequence of the existing relief forms, emphasizing three elements, namely the geological structure, fluvial processes, and time. The phases of relief—youth, maturity, and senility—supported this theory, which focuses on the geological structure and is a fruit of the American School. Despite much criticism of such an interpretation, his contribution to relief studies is significant for literature and decision-making attitudes in search of new theories.

In the American School, the works of Stralher (1950) stand out, highlighting the morphometry of relief, and Hack (1960), in which he postulates a systemic conception for the study of relief.

In another approach, Penck (1953), the fruit of the Germanic School, highlights the relief resulting from the morphogenetic and morphodynamic power provided by the hydraulic gradient that acts through the force of gravity. The initial energy would come from tectonics and lithostratigraphy, while the matter output would be provided by external agents, especially climate, through water and solar radiation that act on rocks and soils. From the contrary action of uplift and denudation with different timing, intensities, and rhythms, relief forms would result from this abrasion and weathering. Consequently, the exogenous factors, through climate, would be inserted into the approach to relief. This vision evolves and makes possible more integrated studies among the geography branches.

Through the reading of the denudational forms and the correlative processes, Penck (1953) considers that it is possible to evaluate what is acting with greater relative intensity, the exogenous or endogenous processes, which is fundamental because it is the result of this balance that will command morphogenesis and morphodynamics.

Three aspects that would become principal to geomorphological thinking were attributed to the studies of Penck (1953), constituting the basis for the understanding of morphology, these being:

> 1. the exogenetic processes; 2. the endogenetic processes; 3. the products due to both, which may be called the proper morphological characteristics (Penck 1953, p. 3 E 4 apud Abreu, 2003, p. 59).

Following the German School, there are authors, such as Passarge (1912), valuing integrally the relief, climate, and vegetation, and Troll (1950), who introduced the concept of landscape ecology.

Davis (1899) and Penck (1953) are the precursors of geomorphology in the twentieth century because, in cultural aspects, there is a powerful element that each author provides when interpreting the relief. Davis, who has a geologist background, and Penck, in turn, were influenced by a school of naturalists.

Systematically and based on the evolution of geomorphological knowledge in Brazil, Ab'Saber (1969) proposed a study on relief, where he designated three levels for the interpretation of it:

(a) The first level is topographic compartmentalization, which relates to horizontal analysis through the domain of forms.
(b) The second level, the surface structure survey, refers to altimetry and correlative deposits.
(c) The third level, landscape physiognomy, refers to the morphogenetic processes caused by climatic dynamics, which includes human action. The approach to geomorphological study will receive attention concerning dimensioning and, consequently, scale.

According to Marques (1994), the relief forms participate in the composition of landscapes at different scales. Large reliefs, when observed quickly, show a static and unchanging appearance. However, they are permanently undergoing erosive or depositional processes triggered by the existing climatic conditions.

To this understanding, Ross (1992) postulates a relief taxonomy applied for planning purposes based on theoretical and methodological aspects in the works of Penck (1953), Demek (1967), and Mescerjakov (1968), highlighting the contributions of Calilleux and Tricart (1956), seconded by Tricart (1965).

The geomorphological map corresponds to an analysis of specialized studies. It is a document that contains a great deal of information necessary for environmental planning since it deals with not only the forms but also their materials and processes, giving an understanding of the dynamics of the relief or evolutionary trends.

Tricart (1965) states that Cailleux and Tricart (1956) propose an analysis methodology based on two fundamental principles: the dynamic and the dimensional. The dynamic principle relates to the structural and climatic mechanisms that originate the forms. The dimensional relates to the notion of the scale of space–time dimensions. Following the taxonomic proposal, Cailleux and Tricart (1956) classify the geomorphological facts in seven magnitudes, later increased to eight, according follows:

- We have in ORDER 1: expressed as surface units (Km^2): 10^7—*hundreds of millions*; Follow as CHARACTERISTICS OF THE UNITS (example): Continents, Ocean Basins (Globe Setup); Regarding CLIMATE UNITS: Large seasonal ensembles commanded by astronomical factors: Below are the GENETIC MECHANISMS THAT COMMAND RELIEF: Differentiation of the earth's crust (SIAL and SIMA); And, finally, ORDER OF MAGNITUDE OF TEMPORAL PERMANENCE (years): 10^9—billions of years.

- We have in ORDER II: expressed as surface units (Km^2): 10^6—*tens of millions*; Follow as CHARACTERISTICS OF THE UNITS (example): Large structural assemblies; Regarding CLIMATE UNITS: Great kinds of weather. Interference of geographical influences with astronomical factors; Below are the GENETIC MECHANISMS THAT COMMAND RELIEF: Movements of the earth's crust, such as the formation of geosinclines. Climate influence on dissection; And, finally, ORDER OF MAGNITUDE OF TEMPORAL PERMANENCE (years): 10^8—hundreds of millions.
- We have in ORDER III: expressed as surface units (Km^2): 10^4—*hundreds of thousands*; Follow as CHARACTERISTICS OF THE UNITS (example): Large structural units; Regarding CLIMATE UNITS: Differentiation of climatic types, but of no great importance for dissection; Below are the GENETIC MECHANISMS THAT COMMAND RELIEF: Tectonic units having a connection with paleogeography. Dissection velocities influenced by lithology; And, finally, ORDER OF MAGNITUDE OF TEMPORAL PERMANENCE (years): 10^7—tens of millions.
- We have in ORDER IV: expressed as surface units (Km^2): 10^2—*hundreds*; Follow as CHARACTERISTICS OF THE UNITS (example): Elementary tectonic units. Massifs, trenches, horsts; Regarding CLIMATE UNITS: Regional climates with geographical influences, particularly in mountainous regions; Below are the GENETIC MECHANISMS THAT COMMAND RELIEF: Predominant influence of tectonics and, secondarily, of lithology; And, finally, ORDER OF MAGNITUDE OF TEMPORAL PERMANENCE (years): 10^7—tens of millions.
- We have in ORDER V: expressed as surface units (Km^2): 10—*Dozens*; Follow as CHARACTERISTICS OF THE UNITS (example): Tectonic accidents. Anticlines, synclines, hills, valleys. Tectonic accidents. Anticlines, synclines, hills, valleys; Regarding CLIMATE UNITS: Local climate influenced by relief layout; Below are the GENETIC MECHANISMS THAT COMMAND RELIEF: Predominance of lithology and tectostatics. Classic structural influences; And, finally, ORDER OF MAGNITUDE OF TEMPORAL PERMANENCE (years): 10^8 a 10^7 - millions to tens of millions.
- We have in ORDER VI: expressed as surface units (Km^2): 10^{-2}—*hundreds of thousands of* m^2; Follow as CHARACTERISTICS OF THE UNITS (example): Landforms: ridge, terminal moraine, dejection cone; Regarding CLIMATE UNITS: Mesoclimate directly linked to form. Ex: nivation niche; Below are the GENETIC MECHANISMS THAT COMMAND RELIEF: Predominance of the morphodynamic factor, influenced by lithology; And, finally, ORDER OF MAGNITUDE OF TEMPORAL PERMANENCE (years): 10^6—hundreds of thousands.
- We have in ORDER VII: expressed as surface units (Km^2): 10^{-2}—*tens of* m^2; Follow as CHARACTERISTICS OF THE UNITS (example): Microforms: soliflusion lenses, gullies; Regarding CLIMATE UNITS: Microclimates directly linked to forms by autocatalysis. Ex. Lapies (fluting); Below are the

GENETIC MECHANISMS THAT COMMAND RELIEF: Predominance of the morphodynamic factor, influenced by lithology,ORDER OF MAGNITUDE OF TEMPORAL PERMANENCE (years): 10^2—hundreds.
- We have in ORDER VIII: expressed as surface units (Km^2): *10^n—tenths of m^2*; Follow as CHARACTERISTICS OF THE UNITS (example): Microscopic: details of corrosion, polishing, etc.; Regarding CLIMATE UNITS: Micromedium; Below are the GENETIC MECHANISMS THAT COMMAND RELIEF: Rock dynamics and texture interface; And, finally, ORDER OF MAGNITUDE OF TEMPORAL PERMANENCE (years): current.

Contributing to the advancement of the classification of geomorphological facts on a large scale, Demek (1967), aiming at mapping, presents basic taxonomic units, namely (a) genetically homogeneous surfaces, (b) relief forms, and (c) relief types.

Subsequently, Mescerjacov (1968), based on the concepts of morpho-structure and morpho-sculpture, proposes a methodology for the classification of relief in six taxa. For him, the active participation of tectonic movements and litho-structure are always present, according follows: The taxons would be distributed like this: SURFACE (km^2); MORPHOSTRUCTURAL ELEMENTS (morphotectonics of the relief of continents and oceanic depressions); MORPHOSCULTURAL CATEGORIES RELIEF: (a) Mainland y (b) Bottom of Oceans and Seas

- On the SURFACE (km^2) 10^7 a 10^6 from a MORPHOSTRUCTURAL ELEMENTS: 1ª order morpho-structures—continental masses (including continental shelves); in MORPHOSCULTURAL CATEGORIES RELIEF: (a) Mainland: Morphosculptural zones (morphoclimatic) cryogenic, glacial, fluvial, arid and other morpho-sculpture zones and (b) Bottom of Oceans and Seas: Morphosculptural zones: Circumpolar zones of glacial forms temperate altitude zones and tropical equatorial zones.
- On the SURFACE (km^2) 10^6 a 10^5 from a MORPHOSTRUCTURAL ELEMENTS: 2ª order morpho-structures platform planes, mountainous areas (orogenic); in MORPHOSCULTURAL CATEGORIES RELIEF: (a) Mainland: Morphosculptural zones (morphoclimatic) cryogenic, glacial, fluvial, arid and other morpho-sculpture zones and (b) Bottom of Oceans and Seas: Morphosculptural zones: Circumpolar zones of glacial forms temperate altitude zones and tropical equatorial zones.
- On the SURFACE (km^2) 10^5 a 10^3 from a MORPHOSTRUCTURAL ELEMENTS: 1ª order morpho-structures plateaus, low flat regions, mountain ranges, extensive depressions, etc.; in MORPHOSCULTURAL CATEGORIES RELIEF: (a) Mainland: Morpho-sculpture provinces, Mediterranean-type fluvial morpho-sculpture provinces, glacial erosion provinces, accumulation provinces, etc.
- On the SURFACE (km^2) 10^3 a 10^2 from a MORPHOSTRUCTURAL ELEMENTS: 2ª order morpho-structures—tectonic pumping marked in the relief, tectonic depressions, etc.; in MORPHOSCULTURAL CATEGORIES RELIEF: (a) Mainland: Morphosculptural regions of frontal moraines, regions of karst relief, etc.

- On the SURFACE (km^2) 10^2 a 10 from a MORPHOSTRUCTURAL ELEMENTS: 3ª order morpho-structures—anticlines marked in the relief, recent trenches, ridges, etc.: in MORPHOSCULTURAL CATEGORIES RELIEF: (a) Mainland: Particular landforms determined mainly by exogenetic factors. River valleys, karst forms, etc. and (b) Bottom of Oceans and Seas: Submarine valleys, forms of accumulation of icebergs.
- And finally, On the SURFACE (km^2) 10 a 10^{-1} a 10 from a MORPHOSTRUCTURAL ELEMENTS: Tectonic microrelief, dykes, crevasses; in MORPHOSCULTURAL CATEGORIES RELIEF: (a) Mainland: Small landforms of exogenous origin. Polygonal soils, small hydrolacoliths.

Based on such studies, Ross (1992) advances in the classification of the Brazilian relief, based on the works of Penck (1953) and Mescerjakov (1968), aligning himself to the concepts of morpho-structure and morpho-sculpture. Penck (1953) emphasizes three elements: (a) geostructure, (b) morpho-structure, and (c) morpho-sculpture. Mescejakov (1968) states that morpho-structures and morpho-sculptures have different dimensions.

In this sense, Ross (1992) classifies relief into six taxa (Fig. 1.1), with a hierarchy that starts from the morphostructural unit, higher and older, toward the forms resulting from current processes, smaller and younger.

1st taxon: morphostructural units, the principal dimensions, related to the genetic factors of the relief; 2nd taxon: morphosculptural sites; 3rd taxon: similar morphologies and morphometrics; 4th taxon: individual forms that define the morphological unit or similar pattern; 5th taxon: shapes of the slopes and; 6th taxon: smaller relief shapes along the slopes, including those generated by anthropic induction.

The 1° taxon is characterized by morphostructural units corresponding to large macrostructures, referring to the genetic types of lithological groupings and their structural arrangements that determine landforms (Ross 1992). This is characterized by a larger taxon, namely the basic morpho-structure, such as a sedimentary basin, for example.

The 2° taxon corresponds to morphoscultural units, which are equivalent to compartments generated by climatic action over geological time, with the intervention of tectogenetic processes. These morphoscultural units are inserted into a morphostructural unit, presenting sets of landforms that share the same genetic characteristics of age and similarity in patterned forms (Ross 1992).

In the 3° taxon, morphological units or patterns of similar forms are corresponding to groupings of forms related to landforms, distinguished by differences in topographic roughness or relief dissection index, as well as by the shape of hilltops, slopes, and valleys of each pattern. They are related to morphological units or patterns of relief types, where current morphoclimatic processes begin to be more easily noticed.

The 4° taxon refers to the unit of patterns of similar, individualized forms inserted into the morphological units of the previous taxonomic level, relief. The forms in this category can be of aggradation or denudation (Ross 1992).

The 6° taxon corresponds to types of slopes or sectors of the slopes of each landform. Each typology of slope form is genetically distinct (Ross 1992).

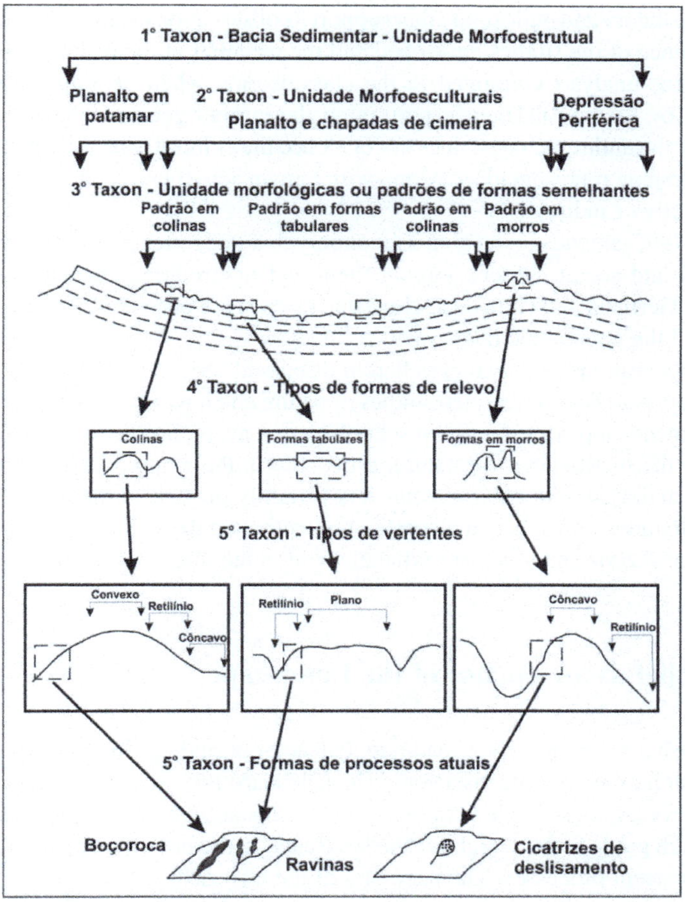

Fig. 1.1 Relief taxonomy. *Source* Ross (1992)

The 7° taxon refers to smaller forms resulting from the action of erosive processes or current deposits (Ross 1992).

Ross (1992), when proposing the classification of landforms, says:

> The qualification bases itself on the physiognomic aspect that each size of relief presents, not being interested in the rigidity of the extension in km^2 but in the morphogenetic meaning and the structural and sculptural influences of the modeling (p. 72).

For the author, his morphogenetic qualification does not have the rigidity of the space–time scale contained in the classifications of Cailleux and Tricart (1956) and of Mescerjakov (1968), in which there is an extension of the taxonomic units of Demek (1967) for purposes of analysis and mapping of relief forms, facilitating its use in the construction of geomorphological maps in several scales, large and small.

From the geomorphological view of Abreu (1982), the taxonomic classification constitutes the central point of the methodological approach in geomorphology and reveals the theory and philosophy that support its observations, allowing us to evaluate the coherence of the studies developed and the meaning of the results achieved.

Thus, the analysis employed in the classification when naming the 6th taxon classified by Ross (1992), even portraying the current geomorphic processes and especially the anthropic ones, does not evidence the cultural aspects, which could be through graphic representation. Along with human activities, a culture accompanies them directly or indirectly.

Ross (1992) intends to develop a taxonomy that integrates and applies geography to natural and social aspects, joining them and interrelating currents such as the Davisian, German, and Humanist. However, he does not place relief as an integrating element of the geographic landscape.

The observations of the authors herein discussed seek the relief of the necessary condition to start from integrated studies of nature and have it as a basis for planning. The treatments approach divergent schools that converge in specific moments.

As for the methodological nature, especially in the mapping, Ross proposed to visualize mainly humid environments and certainly presents difficulties of applicability in dry areas. It leads to a theoretical reflection on the relief and its incidence in the semi-arid zone and as a primordial element in the integrated landscape analysis.

1.2 Relief as an Anchor of the Landscape

To establish a methodological roadmap, Falcão Sobrinho (2006) makes a claim that will appear here with some insertions. The landscape is the starting and ending point and, therefore, may struggle to obtain data. It is because, in landscape effectiveness, the research guide is the search for the landscape in the past, understanding it as the result of current processes. The necessity of the historical process, whose elements are not passive, modifies the landscape in its movements and, consequently, its forms.

The second point to consider is the analysis of the relief as an integrating element of the landscape, independent of human influence degree. It is the vision of placing it as the only element considered natural, which is abstract, at least as matter. Based on this principle, relief is praised as an even supernatural element because it materializes in form and conditions or is conditioned by other natural features, influencing human decision-making concerning use and occupation.

Such reflection will be in an organization chart because it is a visual representation and will establish a hierarchical structure of the organization representing the different elements and their connections, being the relief the guide of this interlocution.

It is noteworthy to have in the relief the possibility of identifying connections among the elements, whether physical-natural or anthropic, in the constitution of the landscape, not only measuring or quantifying the flow of matter and energy but establishing relationships fruit of perception or materialization.

It is appropriate to resort to the understanding of the methodological proposal of geomorphological landscape proposed by Ab'Saber (1969), starting from the physiognomy, permeating the structure, and reaching the functionality, which is the physiognomy of the landscape.

On the initial proposal, the relief will be relevant as the stage for human activities, whatever they may be, from the constitution of dwellings or the use and exploration of the land, being the relief indispensable to constitute such activities.

Relief is a category of abstraction, but it materializes as the support of the activities of human actions. Its forms influence the decision-making of many activities, whether in engineering, rural planting, or urban planning. It materializes as something perceptible through mountains, dune features, or desert environments.

The truth is that relief is not a natural resource as soil, vegetation, rock, water, or even the climate in its aspects of wind and precipitation. The relief is simply a non-resource, not tangible but visible, not concrete as a material but more than concrete for decision-making at the planning level, whether it be for the construction of dams, reservoirs, port facilities, tourist architectural expressions, or other forms of structural fixtures.

In research, with the attention directed to relief, it is essential to have the notion of scale. The question of scale is fundamental since, on a small scale, it is unlikely to have a homogeneity of the landscape elements of a place. However, it becomes feasible to find similarities when on a large scale. Thus, the idea of place takes on a connotational focus. It usually relates to the area, facilitating a more analytical connection among the elements.

Therefore, the relief expresses incipience in the local dynamics with various expressions and, consequently, in the organization of the landscape. Whether windward or leeward, even in a small massif in the semi-arid region of the Brazilian Northeast, its characteristics are notorious through the visualization of the landscape, whose mobilization has natural and anthropic aspects.

Bertrand (1968), when presenting the geosystem, showed particular concern with relief concerning the overall landscape.

> "We, then, look for general landscape mechanisms, in particular geosystems and geo-facies. Chorlley's erosion system directly inspired this methodological order. Why not extend the concept of erosion system to the whole landscape? Thus, a narrowly geomorphological fact would turn into a broader and complete geographical notion of the general system of the landscape evolution".
>
> He also says that:
>
> "The geosystem is between the 4th and 5th spatial-temporal magnitudes. It is, therefore, a dimensional unit between a few square kilometers and a few hundred square kilometers. On this scale, most interference phenomena among the landscape elements and the dialectical combinations are interesting to the geographer evolve".

Bertrand's proposal is in topographic compartments, adopting magnitudes compatible with its empirical basis. Therefore, the relief should be the starting point for understanding the landscape, even before perceiving the landscape as a whole, contributing to the dimension of the landscape for analytical purposes and not only as an elementary unit.

In Fig. 1.2, the intention is to situate the relief as a guide for interactions of natural elements and anthropic actions.

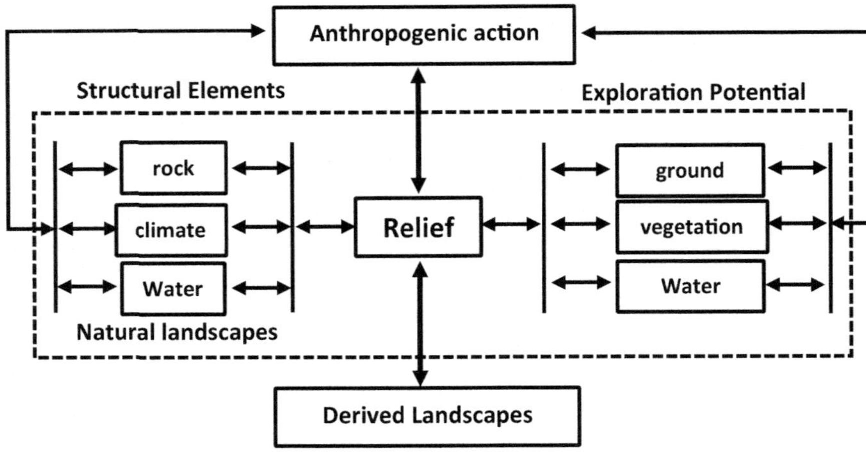

Fig. 1.2 Relief, integrated landscape stage. *Source* Falcão Sobrinho (2006)

The natural landscapes are, here, the physical elements. Their dimensions happen in two approaches: (a) structural elements and (b) exploration potential (Fig. 1.2).

The structural elements (rock, climate, and water) relate to the physical-natural elements that form the basis for other elements. It does not mean that they are or are not independent of others, nor are they conditioning factors. Initially, they are structural factors and, therefore, the target of exploration.

The geological structure is a prerequisite for understanding the analysis of the genesis and evolution of the landforms, that is, morpho-structure. However, the fact is that rock is the basis of relief support. In short, the structural base, whether of the crystalline, sedimentary, or metamorphic complex. The relief is according to the structure of the rock components, encompassing various aspects, sometimes covered by layers of soil or a vegetational mantle. It is worth highlighting that the rock, disposition, and lithological variation will determine the type of soil and vegetation.

Concerning climatic conditions, they determine the effects of weathering, associating mechanical/physical weathering and chemical/biochemical decomposition, which impacts the morpho-climate. This relationship relates to the morpho-sculpturing aspects of the relief. The elements and processes governed by the climate—temperature, humidity, precipitation, and atmospheric movements—maintain the dynamism of the relief modeling process.

The relief characteristics impact the surface water courses and fluvial environments, encompassing fluvial and interfluvial domains, the latter related to natural and occupational circumstances of the slopes, besides their morphological and morphometric conditions. In other words, the waters that flow, either on the slopes or the river waters, moving from the higher to the lower parts, promote the sculpting of the slopes and the intra-channel relief features.

The exploitation potential (soil, vegetation, and water) relates to essential elements for human subsistence.

The importance of the relationship between relief and soil gets clear in the studies of Tricart (1977), who explains the eco-dynamic proposal. The division into three units are (a) stable environments, where the pedogenesis supplants the mechanical processes in the model evolution; (b) intergrades, or transitional environments, where morphogenesis and pedogenesis act mutually in the landscape dynamics; (c) unstable environments, characterized by the predominance of morphogenetic processes in front of pedogenetic ones, either by natural or anthropic factors.

Concerning the vegetation and analyzing the geo-ecological research by Kling (1974), it establishes a detailed investigation of the area, which demonstrates a geo-complex that would have the following factors to be analyzed: relief, geological substratum, water balance, climate, vegetation, and fauna. Thus, one can perceive the functional vision of nature or put it in a sectored view. The relief supports the geo-complex elements, and the landscape expression in the vegetation accompanies the relief sectoring. Besides, every action, whether natural or human, when changes the natural vegetation cover, minimizing it or promoting a total eradication, causes soil erosion processes and, in many cases, changes the relief shapes, generating erosive features as a result of soil loss, such as ravines and gullies, and eventually even mass movements.

Human activity has significantly impacted the modulation of the landscape, being continuous actions and expressing itself slowly or intensely. It directly relates to the dimension over the place, the way of dealing with the land, having there its source of sustenance and occupation.

Thus, the cultural approach, which is also abstract and materializes as a form or result of an action, becomes decisive in constituting landscapes.

Culture is abstract, manifested through human activities—dances, folklore, planting, nature beliefs, animals, and places. It is the same as the one of Germanic origin, which treated the inhabitant with the land, producing his food. In semi-arid areas, the relief morphogenesis characteristics relate to the rain erosive power, soils with little thickness, and a sparse vegetation cover. The morphogenic processes are expressive through erosion, even if incipient. It falls to human action through inadequate practices in soil cultivation, which is common in the semi-arid region, an inheritance from a millenary culture.

In such a context, the work of Wilcock (2011)—Living Landscapes: ethnogeomorphology—discusses the fruit of human actions, consolidating new landscapes called ethnogeomorphology. Thus, the treatment of culture, specifically the rural area, has in agricultural production modes an inheritance of the landscape, which results from the relationship between humans and soil.

Considering the portrayal of the landscape through a cartographic base, which is a fundamental element of geographic science, the contours are scientific expressions and not artistic. The representations of the forms, when related to anthropic action, must be differentiated from the ones resulting naturally. It is not an additional taxon but a depiction of the taxon which Ross (1992) assigns to the 6th type.

Relief and culture are abstract and decisive in constituting landscapes. They are dynamic and integrative in the geographic space.

Thus, the landscapes succeed as the individual acts in the reality that the place presents in its nature of cultural and economic aspects and its nature as a physical and biological element.

References

Abreu AA (1993) A Teoria Geomorfológica e sua Definição: Análise Crítica. Instituto Geológico—SMA-SP n^{0}4 — São Paulo, pp 5–23

Ab'saber AN (1969) Um Conceito de Geomorfologia a Serviço das Pesquisas sobre o Quaternário. In: Geomorfologia. São Paulo. IGEOG-USP, n. 18.1969

Bertrand G (1968) Paysage et geographie globale: Esquisse Methodologique. Revue Geographique de Pyinées et du Sud-Quest

Cailleux, Tricart J (1956) Le problème de la classification des faits géomorphologiques. Annales de Géographie. n. 349, pp 162–186

Davis WM (1899) The geographical cycle. Geogr. Journ. London, pp 481–504

Demek J (1967) Generalization of Geomorphological Maps. In: Geomorfophological Mapping-Brno. Prague, p 3672

Falcão Sobrinho J (2006) O relevo, elemento e âncora, na dinâmica da paisagem do vale, verde e cinza, do Acaraú, no Estado do Ceará. Tese de doutorado. FFLCH/USP. São Paulo

Hack JT (1960) Interpretation of erosional topography. In: Humid-temperate regions. A New Haven. pp 80–97

Klink HJ (1974) Geocologia e Regional. In: Biogeografia. IGEO/USP. N. 17. São Paulo

Marques JS (1994) Ciência Geomorfológica. In: Geomorfologia uma Atualização de Bases e Conceitos. São Paulo, Bertrand.

Mecerjakov JP (1968) Les Concepts de Morphostruture et de Morphosculture: un nouvel instrumnt de l'analys geomorphologique. Paris. Annales de Geographie 77 année 539–532

Passarge S (1912) Physiologische Morphologie. Metteil. Geog. Gesellsch. Flamburg

Penk W (1953) Morphological analysis of land forms: a controbution of physical geology, Macmillian, London, 350p

Ross JLS (1992) O Registro Cartográfico dos Fatos Geomórficos e a Questão da Taxonomia do Relevo. In: Revista do Departamento de Geografia. São Paulo. USP/FFLCH. pp 17–29

Strahler AN (1950) Equilirium theory of erosional slopes Approached by frequency Distriuition analysis. Amer J Sci New Haven

Tricart J (1965) Principes et Methodes de 1a Geomorphologie. Masson et Cie, Paris

Tricart J (1977) Ecodinâmica. SUPREN, Rio de Janeiro

Troll C (1950) Die Geografische Landschsft und ihre schung. Berlin

Wilcock D (2011) Living landscapes: 'Ethno geomorphology' as an ethical frame of communication in environmental decision-making. PhD thesis. Co-tutelle PhD – York University, Canada, and Macquarie University, Australia

José Falcão Sobrinho He is a geographer, holding a post-doctorate in geography from the Federal University of Ceará (UFC) and a Ph.D. in physical geography from the University of São Paulo (USP). He is an associate professor of the geography graduate program at the Vale do Acaraú State University (UVA), the leader of the research and extension network of the semi-arid region/CNPq, a member of the editorial team of the Willian Moris Davis—Geomorphology Journal, and a research productivity fellow at CNPq (Level 2). He deals with geomorphology and erosion processes.

Part II
Western Northeast

Chapter 2
Geomorphology of Maranhão State

Helen Nébias Barreto, Marcelo Eduardo Dantas, Edilea Dutra Pereira, and Jorge Hamilton Souza dos Santos

Abstract The state of Maranhão is characterized with a diverse geological framework, from which remarkable geomorphological domains have developed. The coastal relief is characterized by flattened lowlands amidst extensive fluviomarine plains and the largest coastal dune fields in Brazil. The Coastal-continent transition presents a dissection sequence of extensive plateau surfaces in relatively modest elevations resulted from fluvial erosion processes on regional scale. Toward the Center-South of Maranhão, the landscape features a set of plateaus and mesas at various higher altitudes unfolded in a complex way, which reflects a differential tectonic uplift processes of the Parnaíba Sedimentary Basin as well as the action of different regional flattening events. The landscape of Maranhão reflects unique and peculiar landforms, which comprises a beautiful geomorphological heritage.

Keywords Geomorphological diversity · Heritage · Maranhão

H. N. Barreto (✉) · E. D. Pereira · J. H. S. dos Santos
Federal University of Maranhão (UFMA), São Luís, Brazil
e-mail: helen.barreto@ufma.br

E. D. Pereira
e-mail: edilea.dp@ufma.br

J. H. S. dos Santos
e-mail: jorge.hamilton@ufma.br

M. E. Dantas
Geological Survey of Brazil SGB/CPRM, Rio de Janeiro, Brazil
e-mail: marcelo.dantas@cprm.gov.br

V. Claudino-Sales and J. F. Sobrinho (eds.), *Geomorphology of the Northeast Region of Brazil*, The Latin American Studies Book Series,
https://doi.org/10.1007/978-3-031-58874-7_2

2.1 Introduction

The state of Maranhão presents a peculiar geodiversity (or geo-biodiversity) among the states of Northeastern Brazil. This singularity is because Maranhão is an area of remarkable phyto-climatic transition among lowlands covered by the humid Amazonian equatorial forest to the West, the tropical domain of semi-arid depressions occupied by the Northeast hinterland caatingas to the East, and the semi-humid tropical of the cerrado-covered plateaus to the South. Certainly, one of the most extraordinary ecological transitions in the country (Ab'saber 2003). Maranhão is located in an intricate crossroads of biomes with expressive climatic variability, especially concerning precipitation, with a sharp increase in rainfall from the Southeast to the Northwest. In summary, Maranhão and part of Piauí constitute the Mid-North, which stands out for its peculiar babassu and carnauba palm vegetation that make up a particular biome, the babassu forest.

Thus, in the South and Southeast of Maranhão, the predominant cerrado is under a semi-humid tropical climate, with a drought lasting five to six months in the winter and spring. It happens due to the less expressive influence of low-pressure atmospheric systems and an expressive action of the Atlantic Tropical mass (ATm), which gives climatic stability. On the other hand, in the West and Northwest, one can see the original domain of the Amazon Rainforest under an Equatorial Humid climate governed by the seasonal displacement of the Intertropical Convergence Zone (ITCZ) and the Continental Equatorial mass (CEm), both with marked action in summer and fall. Thus, a short dry period occurs in winter and part of spring, lasting two to three months (Nimer 1989). Therefore, there is a significant increase in the observed rainfall totals from the South and Southeast (1000–1200 mm/year) to the Northwest (2000–2500 mm/year) (Feitosa and Trovão 2006).

The Maranhão geological configuration is characterized by the generation of three sedimentary basins, partially overlapped, that occupy more than 90% of the state surface following, in general lines, the propositions of geotectonic compartmentation elaborated by Almeida (1967), Villas-Boas and Araújo (1999), and Góes and Rossetti (2001): (a) Parnaíba Gondwanic Basin, a vast intracratonic syneclise with prolonged sedimentation throughout the Paleozoic and Mesozoic, in the Silurian, with the deposition of the Serra Grande (Big Mountain) Formation, to the Jurassic, with the volcanosedimentary event demarcated by the basic spills of the Mosquito and Sardinha (Sardine) Formations, and the sandstones of the Corda and Pastos Bons (Great Grass) Formations, deposited in the early Cretaceous; (b) Post-Gondwanic Basin of Grajaú—São Luís, originated by the extensional tectonics of the Equatorial Atlantic during the Aptian (Cordani et al. 2000) with deposition along the Middle and Upper Cretaceous, constituting the Grajaú, Codó, Itapecuru, and Ipixuna Formations; and (c) Espigão Mestre (Master Ridge) Basin, the latter more recent and restricted to the extreme South of Maranhão, being deposited along the Upper Cretaceous with the sedimentation of the Urucuia Formation. Finally, the remaining 10% of the state area corresponds to the outcrop of crystalline rocks of the Precambrian age of the São Luís Craton (Bandeira 2013).

From this diversified geological framework, it occurs the development of a regional relief of tablelands unfolded in a complex way at various altitudes, which reflects a differential process of the post-Cretaceous tectonic uplift of the Parnaíba Sedimentary Basin and the action of different regional flattening events (Barbosa and Novaes Pinto 1973; Ross 1985). The continental uplift of the Parnaíba Basin during the Cenozoic originated a geomorphologic scenario represented by a discontinuous set of extensive plateaus in the Center-South of Maranhão, elevated on topographic levels that vary between 200 and 800 m of altitude, being progressively higher toward the South (Feitosa 2006). Near the coastline, these terrains consist of a set of terraces of the Barreiras (Barrier) Formation or flattened lowlands amidst extensive fluviomarine plains and dune fields (Dantas et al. 2013).

Despite the altitude variations, the plateaus in Maranhão—such as Mangabeiras, Alpercatas, and Itapecuru Plateaus—have a relation to an ancient planning surface of the Paleogene age correlated to the South American Surface of King (1956), later leveled off by differential uplift or basculation of blocks of the Parnaíba Basin during the Cenozoic (Barbosa et al. 1973). The record of detrital coverings or lateritic formations of the Paleogene and Neogene ages confirms the senility of the surface of the plateau tops. On the other hand, the differential erosion of the plateau surfaces delineated the development of wide-open valleys and inter-flattened areas, embedded tens to a few hundred meters below the plateau tops, generating a younger Neogenic plateau surface, correlated to King's (1956) Old Surface.

In the Paleogene—a period of remarkable morpho-dynamic stability and little tectonic activity—there was the development of thick alteration mantles of decametric thickness. The continuous weathering process and the remobilization of iron, manganese, and aluminum oxides and hydroxides originated, in vast extensions, mature bauxite-ferruginous lateritic profiles of significant mineral importance, correlated to the Eocene and Oligocene periods (Costa 1991; Kotschoubey et al. 2005). Such old and developed laterite profiles from the Lower Tertiary are present on the various flattened tops in plateaus corresponding to the South American Surface (Dantas et al. 2013).

For a better understanding of its geodiversity, the Maranhão territory comprises nineteen geomorphological domains (Dantas et al. 2013): Coastal Plain of Maranhão (subdivided into Reentrancy, Gulf and Lowland, Lençois Maranhenses, and Parnaíba Delta); Coastal Tablelands (subdivided into São Luís, Alcântara-Guimarães, and Little Plateau); Flattened Surfaces of Northwest Maranhão; Sub-Coastal Surface of Bacabal (Mesopotamia of Maranhão); Flattened Surfaces of the Parnaíba River Basin; Tablelands of the Itapecuru and Munim River Basins; Tablelands of the Parnaíba River Basin; Dissected Plateau Gurupi—Grajaú; Low Plateaus of Barra do Corda; Plateaus of the Upper Itapecuru River; Depression of the Middle Tocantins River; Plateaus and Mesas of Carolina Strait; Interplanal Depression of Balsas; Plateaus of the Upper Parnaíba River, and Mangabeiras Plateau (Fig. 2.1).

The chapter does not intend to elaborate on a detailed and systematic characterization of all the geodiversity of Maranhão, considering that it already exists due to the Geological Survey of Brazil (SGB/CPRM) since the past decade (Bandeira 2013). The intent of the chapter is, from the state-of-the-art established about ten years ago,

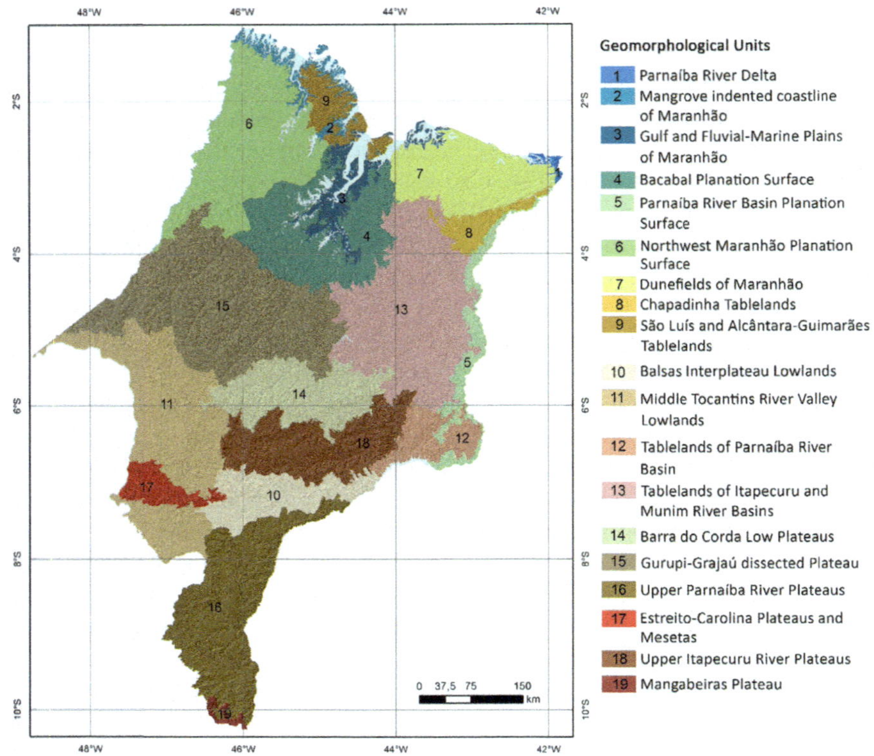

Fig. 2.1 Geomorphological domains proposed for the State of Maranhão (Dantas et al. 2013)

to highlight the new research and the evolution of scientific knowledge in Maranhão undertaken by researchers in universities and local research institutes, focusing on the geodiversity hotspots[1] that stand out in the landscape of the physical environment of Maranhão, namely: the Island and Gulf; the Reentrancies; the Lençois Maranhenses; the Dissected Plateau Gurupi—Grajaú; the Plateaus and Mesas of Carolina Strait (also known as Mesa Plateau); and the Plateaus of the Upper Itapecuru River.

[1] A geodiversity hotspot consists of an area of great variability of elements of the physical environment (lithological and geological structures; relief forms and morpho-dynamic activity; soil types; surface and groundwater) in peculiar intertwining, which gives a given area a remarkable geo-scientific interest (Gray 2008; Bétard and Peulvast 2019; Carvalho-Neta et al. 2020, among others) and geo-tourism potential and as symbiotic support to biodiversity (GRAY 2013; SILVA et al. 2018; Santos et al. 2019; among others). The recognition and identification of geodiversity hotspots of a state or region provides the geo-scientific community with a targeted focus on priority areas for research and for geoconservation in response to the growing threats of disorderly occupation derived from the expansion of multiple economic activities and

2.2 Maranhão Island

Maranhão Island comprises an area of 902.52 km^2 and an estimated population of 1,442,927 people (IBGE 2020) with a demographic density of 1599 hab/km^2 in its territory, divided between the municipalities of São Luís (capital of Maranhão), São José de Ribamar, Paço do Lumiar, and Raposa. It is in the center of the Maranhão Gulf, where it develops an estuarine environment with semi-diurnal macro-tides (two high and two low tides per lunar day)—a maximum amplitude of 7 m, an average of 6.6 m, but usually less than 5.5 m (Fundação Sousândrade 2009).

The Maranhão Island is in the geological context of the intra-continental rift basins of São Luís and New Island—during the Atlantic Ocean opening in the Eocretaceous—associated with an extended transcurrent process that intensified in the Albian. It resulted in the NW–SE and NE-SW faults, which relates to reactivations of ancient structures of the São Luís Craton (Aranha et al. 1990; Góes and Rossetti 2001; Vasquez et al. 2012). These graben systems extend for more than 500 km along the Equatorial Atlantic margin and comprise the Bragança-Vizeu, São Luís, and New Island Basins (Soares et al. 2008).

Based on sequence stratigraphy studies, the Itapecuru Group comprises the Undifferentiated Unit (Eo-Mesoalbian), Alcantara Formation (Neoalbian-Cenomanian), and the Cujupe Formation (Neocretaceous-Lower Tertiary) (Rossetti and Truckenbrodt 1997; Rossetti 2001). The sedimentary sequences were deposited in continental to transitional environments with marine influence (Rodrigues et al. 1994; Rosetti 2001). The Alcântaras Formation contains sandstones, argillites, and limestones of the sequence (S3) from the medium and upper shoreface profile, besides tidal channel, and lagoon environments. The Cujupe Formation comprises the inter-digitated arkose sandstones and argillites relating to the most superficial unit (S3f) formed in tidal environments and complex estuarine valley systems (Rossetti 1997, 1998, 2001).

The highest altitudes (60–40 m) are in the Northeast and Southwest of Maranhão Island and represent the recharge areas of the semi-confined Itapecuru Aquifers and the Free Barrier Aquifer, generally used for human consumption. Studies of the relief units through the geological-geotechnical model indicated the presence of intensely dissected plateaus (Pereira 2006) with flat to gently undulated tops supported by layers of lateritic crusts. In elevations lower than 40 m, hills develop from an intense denudational process (Fig. 2.2). These reliefs comprise rocks of the Itapecuru Group (Fig. 2.3) and the Cenozoic surface coverings of the Quaternary formed by Mangrove Deposits, Fluviolagunar Deposits, Coastal Deposits, Aeolian Coastal Deposits, and Alluvial Deposits (Barros and Bandeira 2020).

The tablelands from the Cujupe Formation appear in the Northwest and South of Maranhão Island along the drainage basin of the Anil, Bacanga, Cachorros, Tibiri, and Geniparana Rivers and in the coastal zone along the cliffs, which sometimes outcrop at the base the rocks of the Alcântara Formation at altitudes lower than 5 m, consisting of layers of yellowish sandstones and whitish carbonate argillites. Exposures of the Cujupe Formation and Post-Barrier sediments are present in the cliffs along the coastal fringe that comprise the Eye Water, São Marcos, One-Eyed,

Fig. 2.2 Intensely eroded mesa top—Maranhão Island—São Luís, MA. *Source* Personal file Pereira (2020)

Calhau, Middle, Araçagi, Guide, Little, Love, and Panaquatira Beaches. They present extensive strips of quartz sands, being possible, during low tide, to see the fractures directed to the Mid-Atlantic Ridge at São Marcos Beach (Figs. 2.2, 2.3, and 2.4). Fossil records of dinosaur footprints are also present at Guide Beach (Carvalho 2020).

The cross-shore profiles of the beaches in northern Maranhão Island have a slope of less than 5°, a medium degree of wave exposure, a wide intertidal band of about 50–100 m at Curupu Island, 200–500 m at Fox Beach, 300–400 m at Araçagi Beach, 220–400 m at Middle Beach, and 300–400 m at Calhau, São Marcos and Sandpoint Beaches. The sand strip of the latter reaches more than 2000 m. Guide Beach has a 1000 m width between Good End and Guide Sandpoint. The sediments along the beach prism are composed essentially of fine (0.177–0.128 mm) to quite fine (0.88–0.062 mm) quartz sands, finely selected and lacking in biogenic fragments, indicating fine sandy dissipative beaches (Santos et al. 2004).

The Northeastern Maranhão Island presents 60–40 m dissected terraces covered by Post-Barrier sediments along the watersheds of the Patience and Santo Antonio Rivers, constituting a ramp with smooth slopes toward São José Bay. Silva (2012) observed that the denudational forms—tablelands, sparse hills, cliffs, and hillsides—are in significant proportions at 64.9%, and the aggradational ones are at 31.8% with various dissection indexes. The coast is sub-rectilinear, observed principally to the

Fig. 2.3 Exposure of ferruginous rocks on the sandstone of the Alcantara Formation of the Itapecuru Group—Sao Marcos Lighthouse—Sao Luis, MA. *Source* Personal file Pereira (2019)

North and Northeast and less usually to the East and Northwest. The shape is due to the exposure of these areas to the action of waves and coastal currents.

The dunes and paleo-dunes in the coastal fringe are generally contiguous to the high-tide beach line in the post-shore zone and may have sandbank vegetation. The best example is São Marcos Beach, which has around 2 km long and height ranging from 7 to 39 m, predominating the barchan forms because of oceanographic and wind agents, besides Northeast trade winds. It stands out the presence of sandbanks along São Marcos Beach until the sandy point of the Anil River mouth and outcroppings of ferruginous sandstones on the Calhau, São Marcos, Panaquatira, One-Eyed, and Middle Beaches.

The abrasion platforms and terrace of irregular surface are sieved of holes by chemical dissolution and marmite, with the presence of blocks and clasts in Good End and Guide Points, besides a strip between Guide Point and Love Beach (thallus deposit), according to Santos et al. (2004), and also present in São Marcos and Cashew Tree Beaches.

Barros and Bandeira (2020), in the function of the geological particularities and relief forms of the island, divided the geomorphological domains into seven geological-environmental units: recent alluvial plains; mixed (marine/continental) environment; coastal marine environment; mobile and fixed dunes; alternation between layers of sediments of diverse composition; and intercalations of sandy

Fig. 2.4 Fractures toward the Mid-Atlantic Ridge at São Marcos Beach—São Luís, MA. *Source* Personal file Pereira (2020)

silty-clayey sediments and shales. These geological units result from an intense dynamic of morpho-genetic and morpho-dynamic processes at various levels on the island.

The physical-environmental processes acting in the relief modeling present in Maranhão Island also promote problems of environmental dimensions considering the use and occupation without prior knowledge of the vulnerabilities and potentialities such as coastal and urban erosions, mass landslides in the areas Sacavém Saline and Hope Summit, floods along the Anil (Silva 2012), Patience and Crowned Rivers in the Bacanga Drainage Basin. The salt wedge entrance, principally in the coastal fringe (Castro 2019)—lower Bacanga River (Martins 2019) and Patience River mouth—indicated moderate vulnerability (5.1–5.6) to salt intrusion, involving partially Raposa City, the tubular wells of the Serviço Autônomo de Água e Esgoto (Autonomous Service of Water and Sewage, SAAE) in Paço do Lumiar (Pereira 2017), and occupations in Areas of Permanent Preservation (APP), mainly along rivers and mangroves.

It is noteworthy that the rivers of the Maranhão Island have been suffering high environmental impact, principally by the release of domestic sewage without treatment, which reflects on the quality of seafood and fish and beaches usually unsuitable for bathing.

Maranhão Island presents a strong interaction between the continental and oceanic crusts, constituting a fragile estuarine ecosystem under high-tide amplitudes and a strong dynamic with erosion and accumulation of sedimentary deposits along the constantly changing coast, resulting in beautiful landscapes and potential for geotourism.

2.3 Maranhão Gulf

The Maranhão Gulf is an inlet situated between the São José and São Marcos Bays in the São Luís Drainage Basin. It was born from the Cretaceous fissure rift sediments implanted in the northern portion of the Paleozoic syneclise of Parnaíba as a result of the Wealdenian reactivation of the Brazilian platform (Almeida 1967). A series of horsts and grabens that make up the substrate of the continental shelf covered by the Cenozoic surface deposits stand out. Ab'Saber (1960) observed that the Maranhão Gulf presents geological and physiographic characteristics extraordinarily different from the Amazonian Gulf and Parnaíba Delta.

The Maranhão Gulf relief represents a complex estuarine system positioned at right angles to the coast, with hydrodynamics driven by the semi-diurnal tidal regime. The tidal wave is progressive in the outer areas (shelf), while in the inner ones, such as São Marcos and São José Bays, there is a tendency to a stationary behavior. The tides, waves, and coastal currents associated with the dominant northeast trade winds are the principal agents responsible for coastal morphodynamics.

In this context, it is worth highlighting that São José and São Marcos Bays consist, respectively, of the estuaries of the Itapecuru and Mearim Rivers, which extend for kilometers into the continent, as well as the whole drainage network that flows into these immense rivers, subjected to conditions of ebb tides from the Gulf (Ab'saber 1960; Barbosa and Novaes pinto 1973).

The Gulf is open to the North over the continental shelf, communicating directly with the South Atlantic Ocean through the opening between Cumã and Shark Bays. To the West of the coastline, the coastal plains with mangrove-rich indentations develop; to the East, the sub-retinal dune formations, sandy beaches, and the bay (El-Robrini et al. 2018); and to the Center, Maranhão Island.

Due to the coast dynamics, the Maranhão Gulf has four sectors that reveal the relief features typical of this environment (Teixeira and Sousa Filho 2009) (Fig. 2.5): wraps lowlands, Sao Marcos Bay, and Crab Island, which has salt and freshwater marshes, intermittent lakes, and an estuarine channel (Fig. 2.6); has a coastal tableland, a muddy tidal plain, an alluvial plain, and a sandy tidal plain, macro-marine beaches, built-up area, and artificial lagoons; has mangroves, paleo-dunes, and mixed tidal plais, and with mobile dunes.

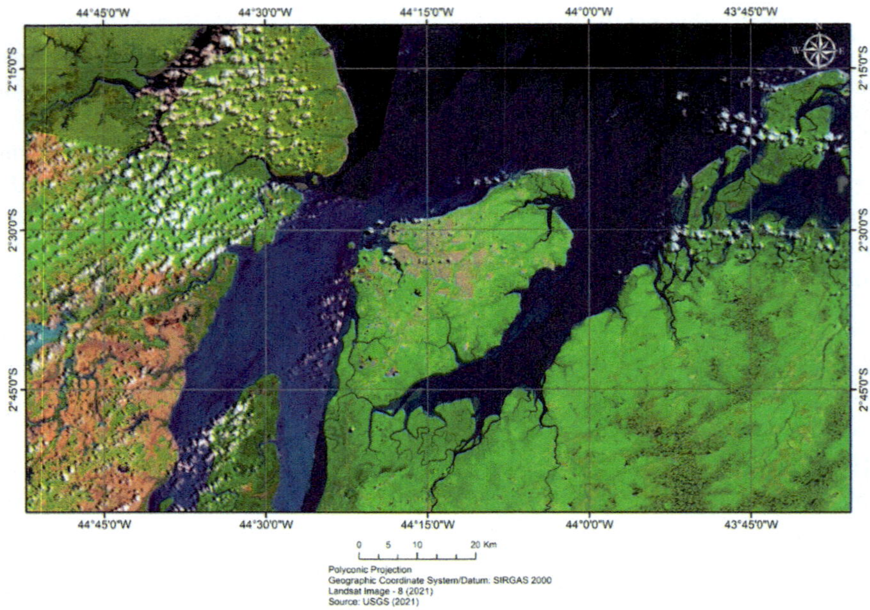

Fig. 2.5 View of the Maranhão Gulf highlighting São Marcos and São José Bays—Maranhão. *Source* LANDSAT-7 (2000). EMBRAPA Satellite monitoring. (www.zee.ma.gov.br)

Fig. 2.6 Aerial view downstream of the Bacanga River, São Luís—MA. *Source* Martins (2019)

São Marcos Bay, due to its genesis and configuration, presents navigability conditions with an average depth of 27 m in the bed of the access channel, characterizing the Itaqui, Ponta da Madeira, and VALE Harbors as a national and international reference in cargo movement. However, it is constantly subjected to the dominance of depositional processes of clastic sediments from the Mearim, Pindaré, Grajaú, and Dogs Rivers, forming sandbars.

The dynamics of the estuarine system promote a rich biodiversity with the richness of fish and shellfish, associated with a unique scenic beauty, where one can find potential areas for tourism and ecotourism.

In the Maranhão Gulf retro area, an extensive river and marine plain with extremely flat topography practically adjusted to the general base level is developed, which comprises the lowlands (Dantas et al. 2013). Finally, in the embayed terrain of the Gulf caves, the human occupation process is based on traditional fishing and subsistence livestock (cattle and buffaloes) without promoting significant environmental impacts in the flat and poorly drained lowlands (Souza and Feitosa 2009).

2.4 Maranhão Reentrancies

The coast of Maranhão is around 640 km long, in the West–East direction from the mouth of the Gurupi River, the border with Pará, to the Parnaíba Delta, the border with Piauí, the second most extensive of Brazil and Northeastern region, surpassed only by Bahia (Feitosa 2006). The Maranhão Reentrancies are between the Gurupi River mouths to the West until the West shore of the Cumã Bay to the East, limited by the Guajuru Edge in the municipality of Cedral (Feitosa 2006). The coastal segment goes from Carutapera to Cedral (Fig. 2.8), comprising cities close to the coastline or at the bottom of the creeks or bays.

The Maranhão Northwest coast and the Pará Northeast coast received from Souza Filho (2005) the name of Amazon Macro-tidal Mangrove Coast. The area has floodable lowlands, altitudes between 0 and 5 m, Tertiary, Pleistocene, and Holocene sediments, and morphology typical of a depositional coast.

Geomorphologically, it is a coast jagged in strips and false strips formed by tectonic and eustatic processes, causing the formation of extensive plains, muddy/sandy sites, and hundreds of islands (Mello et al. 1999). The West coast (Fig. 2.7) presents large recesses represented by several bays—Gurupi, Irimirim, Iriaçu, Tromaí, Pericumã, Piracaua, Carará, Maracaçumé, Mutuoca, Turiaçu, Lençois Maranhenses, Capim, Elderly Hair, and Cumã.

Besides the referred bays and countless islands, the reentrancies shelter in its jagged coastline the most complex range of depositional environments of the Brazilian coast, with estuaries, flood and ebb tide deltas, extensive muddy tide plains with or without mangrove vegetation, meandering channel systems, chenier plains, sandy spurs, coastal ridges, beaches, and sandy terraces, mobile and fixed dunes, extensive sandbars, tidal flats, lagoons, and mangroves (El Robrini et al. 2018;

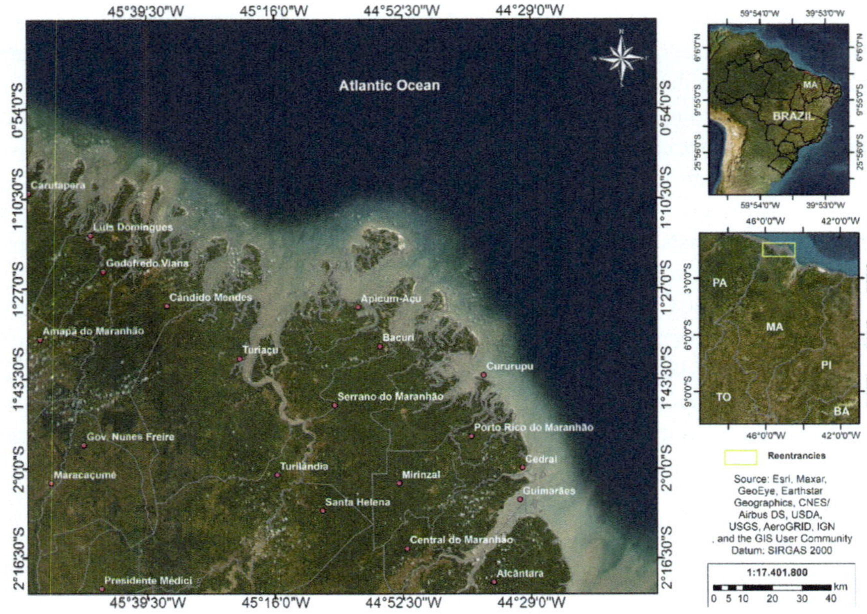

Fig. 2.7 Location of the Reentrancies from Carutapera to Cedral

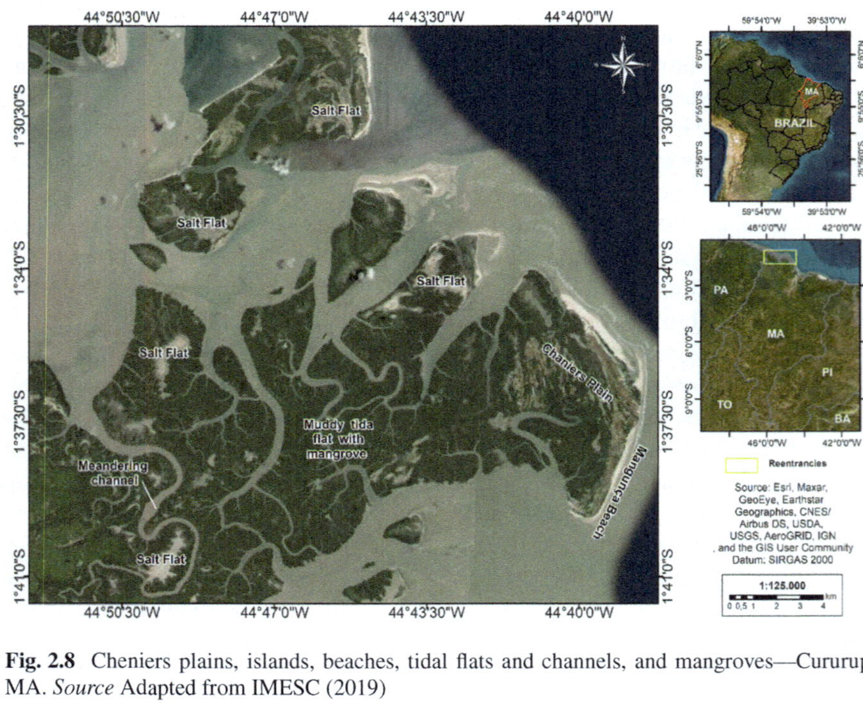

Fig. 2.8 Cheniers plains, islands, beaches, tidal flats and channels, and mangroves—Cururupu, MA. *Source* Adapted from IMESC (2019)

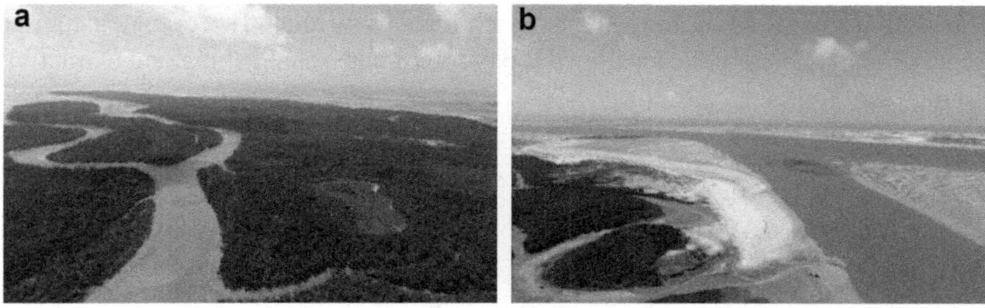

Fig. 2.9 Tidal plain with mangrove vegetation in Porto Rico do Maranhão (**a**) and sandy plain, beach, and sandy banks at the end of the indentations in Porto Rico do Maranhão (**b**). *Source* IMESC institutional archive (2019)

IMESC 2019). The cheniers plains (Fig. 2.8) are gradational coastal features represented by bundles of sandy coastal ridges, with heights between 3 and 6 m, which overlap and separate muddy tidal terraces with mangrove vegetation or saltmarshes.

To Ab'saber (2004), from the Guajará-Mirim Bay entrance until before São Marcos Bay, there is an estuarine coast (incised river mouths at the tableland edges drowned by the rising Flandrian sea waters) partially masked by the recent advent of tidal flats, comprising mangroves frontal and lateral to the successive estuaries originating from the marine invasion in the early Holocene. Furthermore, unlike the tidal plains in Brazil, generally located behind sandbank lagoons at the bottom of estuaries or in intra-lagunar deltas, the ones with mangroves (Fig. 2.9) in the coast grow perpendicular to the coastline widely and irregularly.

The tidal plains (mangrove deposits) have a dark greenish color and clayey and silty sediments and are muddy, rich in organic matter, and permanently waterlogged, subject to tidal dynamics. The sedimentary samples from the tidal plain presented an average particle size between 4.3 and 6.5 phi, classified as fine to coarse silt, with a selection parameter of 1.32–2.4. The samples were considered very poorly or not selected, with gravel contents between 0 and 0.21%, sand contents in 0.5–48.1%, silt contents between 48 and 92.3%, and clay contents between 2.2 and 21.6% (Maranhão 2018).

In the extremities of the reentrancies, extensive sandy depositional features develop, also known as Coastal Barriers. Part of the Maranhão Reentrancies can be sub-environments—beaches, frontal, transgressive, and parabolic dunes, deflation plains, tidal troughs, overflow fans, and lagoons (Maranhâo 2018). In the interdunal areas, there is the formation of temporary freshwater lagoons. In parts of the coast, the tidal terraces are covered by mangrove vegetation, except for specific sites in the mouths of tidal channels, where the positive balance of sediments prevails due to the influence of flood and ebb tidal deltas.

The reentrancies stand out for the diversity of geomorphological features associated with significant biodiversity represented by various plant and animal species (birds, reptiles, fish, mollusks, and crustaceans). Thus, this part of the coastline became an Environmental Protection Area in the 1990s by the government. Despite the beauty, the area has not consolidated itself for activity related to ecological tourism because the main activities developed today refer to artisanal fishing and extractivist and incipient tourism.

2.5 Lençois Maranhenses

The eastern coast of Maranhão is home to shallow lagoons, the largest coastal dune field in Brazil, developed during the Quaternary, presenting fossil and present-day dunes (Fig. 2.10). It is bordered to the North by the Atlantic Ocean, to the South by dissected plateaus of the Barriers Group, to the East by the Parnaíba/PI River, and to the West by the São José Bay/MA and the Itapecuru River.

In the Lençois Maranhenses region, the annual rainfall cycle is characterized by rates between 1600 and 1800 mm, predominantly NE winds, with speeds ranging from regular to moderate, and occasionally gusts exceeding 45 m/s in the dry season (Santos et al. 2005). The consistency of the migration direction of the current dunes and the ones fixed by the vegetation indicates the constancy of the NE winds. The wind direction at Lençois Maranhenses does not seem to have undergone significant changes during the Holocene. Corroborating this assertion, the paleo-dunes distant up to about 160 km from the coastline maintained the preferential NE/SW direction

Fig. 2.10 Location of the present and past dune fields of Lençóis Maranhenses. *Source* Personal files Santos and Silva (2023)

of displacement because they are usually in front of abandoned migration tracks arranged in the mentioned direction.

Geologically, Lençois Maranhenses is in the Cretaceous Little Barrier Basin, to the Northeast between São José Bay and Parnaíba Delta. According to Pamplona (1969), the emersed area is 15,000 km^2–250 km long and 60 km wide. Also, according to the author, the basin is of Cretaceous age (Albian-Champanian), with a maximum thickness of around 7000 m. The predominant sediments are of eolic, marine, and fluvial origin. They have a fine/medium sandy texture, low moisture retention capacity, and very low natural fertility.

The mobile dune field genesis, according to MMA (1996), Muehe (1998), MMA/IBAMA (2003), and Santos (2008), relates to the sediment selection due to the retrograde of the Barrier Formation sedimentary deposits and the widening of the continental caused by marine transgressions that occurred in the Pleistocene. Besides, there is a contribution of fluvial sediments from the main rivers, such as the Parnaíba and Preguiças. Due to the ecosystem exuberance and diversity, on June 2, 1981, at the suggestion of the RADAM Project, part of its area was transformed into the Lençois Maranhenses National Park (LMNP), with 155,000 hectares.

To Gonçalves et al. (2005), the gathering of climate, littoral drift, tide regime, and river currents constitutes the current coastal dynamics. According to Gonçalves et al. (2005) and Santos et al. (2005), the coastal plain under analysis receives sediments from the littoral drift, carried along the northeastern coast with a substantial increase at the Parnaíba River mouth, from the internal platform (covered with sands, with the occurrence of underwater dunes) by the waves, and from the sandy load reworked by the local drainages. The beaches facing the PNLM, according to Santos (2008), are dissipative, presenting a low slope with multiple wave breaks and a width between 200 and 400 m. The sediments are predominantly medium (0.354–0.250 mm) to fine (0.177–0.125 mm) sands, well selected.

The LMNP coastal plain and its surrounding area are characterized by a mild to moderately wavy relief with mobile dune fields of various configurations, with an average height of around 30 m, consisting of very fine or not sand (Santos 2015). The fixed dunes consist of vegetated wind forms and are older than the free ones, occupying most of the coastal plain and constituting approximately 70% of the existing wind system. They are to the South of the mobile dune fields, which gradually advance over them, covering them in part or totally.

The ages obtained by thermoluminescence of fourteen samples from the fixed dune fields (Santos 2008; Santos et al. 2009) suggest the existence of at least three alternating phases of intense Aeolian activity followed by high humidity, resulting in the stabilization of the dune forms. The sandy deposition is principally from the Holocene age (Fig. 2.11).

In the Lençois Maranhenses geomorphological domain, the morphological features that stand out are beaches, sandbanks, sandpits, sandy ridges and terraces, deflation, flood, and tidal plains, mangrove, besides the barchan, parabolic, and oblique dunes, and the barchan, parabolic, and transverse chains (mobile and fixed).

The local hydrography contains bays, rivers, streams, and lagoons, being São José, Tubarão, and Tutóia Bays quite relevant. The principal rivers on the Maranhão

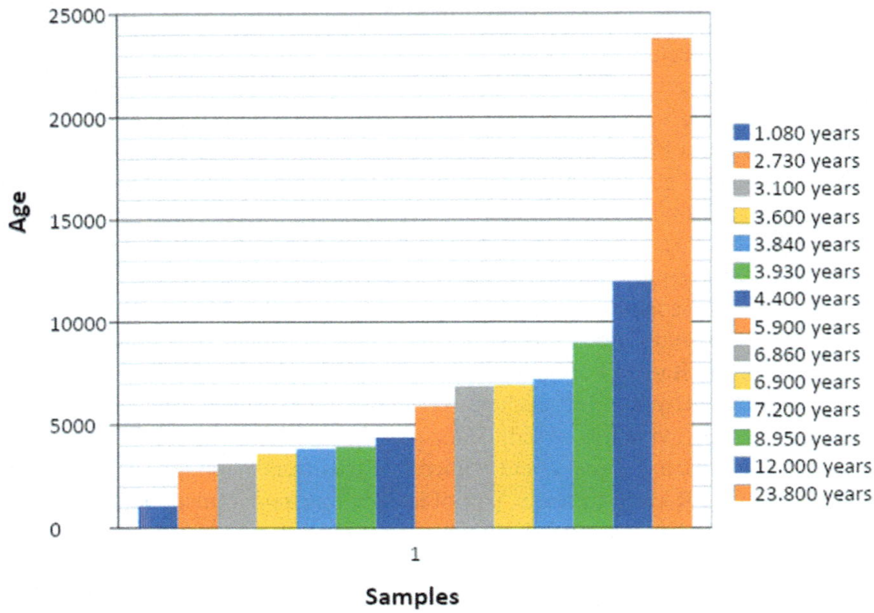

Fig. 2.11 Graphical representation of the samples of fixed dunes from the Lençois Maranhenses dated by thermoluminescence (TL). *Source* Personal file Santos (2008)

coastline are Itapecuru, Munim, Periá, Big, Black, Preguiças, Cangatá, Hard Clay, and the Parnaíba Rivers. In such a system, there are various temporary and permanent lagoons among the dunes, besides drainage channels of first and second orders. The rivers and the lagoons in this dune field are essential for controlling the wind movement (Santos 2008) since they hinder the migration over the paleo-dunes and the sub-tidal wind deflation plains, situated contiguously in front of the mobile dunes (Fig. 2.12).

2.6 Dissected Plateau Gurupi—Grajaú

The Gurupi–Grajaú Dissected Plateau covers a vast area of the central-western region, comprising a set of lowland surfaces elevated at modest altitudes and carved by some of the principal rivers of Maranhão, such as the Gurupi, Buriticupu, Pindaré, Grajaú, and Mearim Rivers.

Such a geomorphological unit explains the Maranhão geomorphology development with flat surfaces slightly elevated by Cenozoic orogenies. Such terrains show a gradual and progressive increase in altitude toward the South, where relief of higher plateaus prevails. In contrast, to the North, there is an abrupt contact (Fig. 2.13) with

Fig. 2.12 **a** Stream between the mobile and vegetation-fixed dune fields slowing the advance of the free ones over the fixed ones; **b** Dune field advancing over the Hope Lagoon in LMNP. *Source* Personal file Santos (2019)

the flat and low surfaces of the lowlands, Maranhão Gulf, and other geomorphological units in the North.

The regional landscape presents lowlands, between 200 and 450 m, delimited by erosive ridges through which incised and deepened valleys are fitted (Fig. 2.14) with sub-dendritic drainage patterns to the lattice, with structural control. The local unevenness sometimes exceeds 100 m (Dantas et al. 2013).

The tops of the low dissected plateaus present very deep soils (predominantly Red-Yellow Latosol) generated on developed regoliths that can reach tens of meters in thickness. Such thick regoliths are the result of a slow and progressive weathering process of the immature sandstones of the Itapecuru and Ipixuna Formations, which attest to the remarkable antiquity and tectonic stability of these surfaces supposedly from the Paleogene (Barbosa et al. 1973), which would correspond to the South

Fig. 2.13 **a** Disturbed relief of dissected hills in the contact between the flat and low terrains of the Maranhão Lowland and the Dissected Plateau Gurupi–Grajaú, Lago do Junco municipality. **b** Embedded valley of the Buriticupu River, a tributary of the Pindaré River in the middle of the plateau surface, registering local relief amplitude around 80 m. Highway BR-222 near Buriticupu City. *Source* Personal file: Dantas (2013)

Fig. 2.14 Thick alteration profile of the Itapecuru sandstone and the development of laterite formation. Flat relief in vast extensions of the plateau tops. Highway BR-222, between the cities of Buriticupu and Bom Jesus das Selvas. *Source* Personal file Dantas (2013)

American King Surface (1956). Despite the observed tectonic stability, it is undeniable that the uplifting process during the Neogene promoted the plateau dissection, resulting in an energetic fluvial erosion process on a regional scale through the vertical incision of the trunk channels and the readjustment of the drainage system to the lowered local base level (Dantas et al. 2013). The valley bottoms of the Pindaré and Buriticupu Rivers present broad fluvial plains.

The intensity of chemical weathering on ancient weathering mantles promoted the generation of thick lateritic, alumino-ferruginous profiles (Fig. 2.16), similar to those described in northeastern Pará (Kotschoubey et al. 2005). These ferruginous shell form tough cornices that retard the dismantling and destruction process of the plateaus by erosive action.

The relief evolution model for the region rests on a progressive dismantling of a low plateau sustained by detrital-lateritic crusts overlain by strongly weathered friable sandstones and can be perfectly explained by etchplain processes in humid and semi-humid tropical regions, as postulated by Büdel (1982). The removal of the laterite, which delays weathering and denudation, causes accelerated erosion and regional readjustment of the underlying regolith (Dantas et al. 2013).

A more detailed analysis of the Dissected Plateau Gurupi–Grajaú reveals a regional relief moved, characterized by frank dissection of extensive plateau surfaces

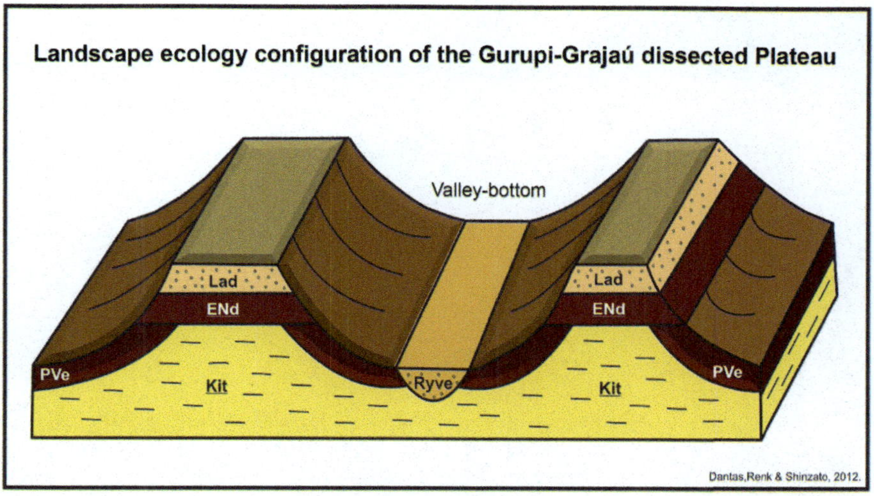

Fig. 2.15 Schematic model of the geo-ecological configuration of the Dissected Plateau Gurupi–Grajaú, underlain by Cretaceous-age Archean sandstones of the Itapecuru Formation (Kit): Tops of plateaus supported by lateritic crusts (ENd) and the development of deep and well-drained soils—dystrophic Yellow Latosol (dYL). Intense dissection in notched valleys, exhibiting steep slopes and erosion processes, in a place of less thick soils—Eutrophic Red Acrisol (eRA). Valley bottoms show expressive alluvial sedimentation and soils good in natural fertility, prevailing the eutrophic and vertex Fluid Neosol. *Source* Dantas et al. (2013)

raised in relatively modest elevations (Dantas et al. 2013). In this scenario, the following geomorphological compartmentalization is (a) the flat tops of the plateaus, covered by thick and well-drained soils, such as Latosol, and the highest ones, positioned at elevations between 250 and 400 m of altitude; (b) the strongly carved slopes that, through the progressive retreat of the declivous erosive edges, have been destroying the low plateau surfaces; and (c) the undulating surfaces, flattened or reshaped into collines forms, which spread through the valley bottoms (Fig. 2.15).

The environmental fragility observed on the rugged terrain of the Dissected Plateau Gurupi–Grajaú is striking, and the occupation process for the most diverse forms of human intervention requires careful attention to the occurrence of environmental degradation phenomena.

The thick regolithic mantle generated from the weathering of sandstone rocks is very friable and highly erodable. In this sense, the linear erosion processes analyzed by Santos and Guerra (2003) stand out, which registered large-sized gullies amid the relief and dissected hills, consisting of the most deforested area and susceptible to erosive processes on a regional scale. Castro (2019) also records the erosion processes induced by the BR-222 highway in Açailândia (Fig. 2.16). The widening work of this highway, which connects São Luís and Açailândia, exposes several stretches susceptible to laminar erosion processes, ravines, and landslides along the newly implemented slopes (Fig. 2.17).

Fig. 2.16 Generalized landslide along the BR-222 highway. Sharp linear erosion in areas of hills and very dissected hills. Soils and weathering layers of the Itapecuru sandstone present high vulnerability to weathering. Highway BR-222 between Santa Luzia and Buriticupu. *Source* Personal file Dantas (2013)

The terrain in the Dissected Plateau Gurupi–Grajaú is, to the West, in the forested area of the Eastern Amazon, where some of the principal forest remnants of Maranhão subsist and where the Gurupi Biological Reserve and indigenous lands—Awá, Krikati, and Arariboia territories—are also located.

2.7 Plateaus and Mesas of Carolina Strait (Mesa Plateau)

Maranhão plateaus have various flat relief forms, known regionally as mesas, and are in the transition from the semi-humid tropical climate of the cerrado to the humid Amazonian climate. Dantas et al. (2013) divided the plateau compartments into four geomorphological domains, being one of them the Estreito-Carolina Plateau and Mesetas to the South (Fig. 2.18) and the Itapecuru Upper River Plateau (tablelands) to the Center-South.

The Estreito-Carolina Plateaus and Mesetas (tablelands) correspond to one of the most breathtaking landscapes in the Mesas Plateau National Park, Maranhão (Fig. 2.19). The relief forms of flat tops with rocky and abrupt slopes and extensive planed surfaces stand out. Most residual reliefs, known regionally as mesa, are supported by the sandstone coverings from the Paleozoic–Mesozoic transition,

Fig. 2.17 Steep slopes and road embankments over alteration mantle with high susceptibility to erosion of stratified sandstones of the Itapecuru Formation in relief of dissected hills frankly carved by a network of channels of high drainage density. Highway BR-222 between Santa Luzia and Buriticupu. *Source* Personal file Dantas (2013)

such as the Mosquito and Sambaíba Formations (Balsas Group) and by the Neogene detrital-lateritic coverings on the top and more resistant to erosion. The tops present a regionally differentiated topographic level with altitudes from 300 to 400 m.

Flat surfaces (200 m) sculpted by erosive processes of sandstone disaggregation stand out on the top of the mesas and surrounding lower areas (Fig. 2.20) (Barreto et al. 2015). These thick surface formations correspond to the Pleistocene colluvium and alluvium that border the domain of the Middle Tocantins Valley Depression and develop as ramps from the edges and foothills of the hills toward the depression. The colluvium becomes thicker at the base of the escarpments, where it takes the form of cones and fans.

The regional morphogenesis is associated with karst processes (Figs. 2.19 and 2.21) in sandstones, indicating a more complex long-term evolution of the plateaus. Martins et al. (2017) propose that the morphogenesis of the landscape of regional plateaus is due not only to surface processes with the incision of the river network widely recognized in the literature but also to the total or partial rupture of cave ceilings, which suddenly exhumed river courses that were once underground. Besides the flat features, the arches on hilltops, canyons, caves, ducts, and cavities of different sizes stand out, which are examples of forms associated with a paleo-karst in areas of the non-carbonate substrate. The waterfalls also resemble processes resulting from

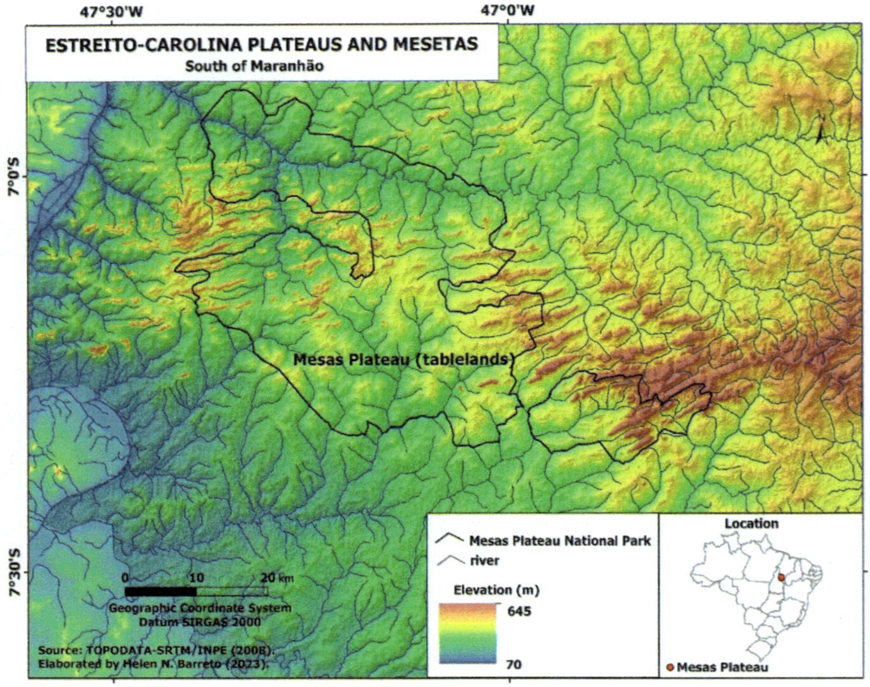

Fig. 2.18 Location of Mesa Plateau National Park, MA

the collapse of sinkholes and cave roofs, such as the Santa Bárbara, Caverna, and Santuário waterfalls.

The mesas assume similar elongated shapes and differentiated sizes, with preferential direction NE-SW. This positioning geologically indicates the profusion of fractures, faults, and fault zones, which show the influence of structural lineaments in the spatial arrangement of these features. The drainage network is configured similarly with a parallel drainage pattern controlled by the geological structures but with diversified directions. The relief presents itself in an advanced dissection process since, besides mesas, there are features such as buttes, towers, and pinnacles. The low regional drainage density (Martins et al. 2017) also reflects the characteristics of the sandstone substrate, which favors infiltration at the expense of surface runoff.

2.8 Upper Itapecuru River Plateau

This set of plateaus and mesas emerges in the middle of the extensive flattened surfaces of the Parnaíba Sedimentary Province in the Center-South of Maranhão, integrating the Mirador State Park (Fig. 2.22). The Upper Itapecuru River Plateau comprises the watersheds of four large basins: Itapecuru and Mearim Rivers, the

Fig. 2.19 **a** Two Siblings Hills; **b** Paleo-conduit: evidence of a karst drainage paleo-system; **c** Rock wall with cuneiform cross-stratification (Sambaíba Formation); **d** Hat Hill surrounded by an extensive planed surface. *Source* Personal files Barreto (2015) and Dantas (2013)

upper course of the Parnaíba River, and the middle course of the Tocantins River. The regional landscape presents relief forms of flat tops (Fig. 2.23)—elongated or rounded—carved in predominantly sandstone formations with weathered and abrupt rocky slopes and extensive flat surfaces with the presence of structural control (Barreto et al. 2015). The significant remnants of these surfaces are the Itapecuru Plateau (tableland), with a maximum altitude of 587 m and about 900 km^2 in extent; the Wild Plateau; the Croeira Plateau; the Red Plateau; and the Alpercatas Plateau. The elevations are around 180 and 580 m, sheltering the upper Itapecuru River and its tributary, Alpercatas River.

In the higher elevations (300–580 m), the relief presents flat tops, maintained by a thick layer of Neogene laterite coverings (23 Ma), whose origin relates to the alternation of dry and humid climates. In the middle slope portions, in the openings of the plateaus, the reddish sandstones, and siltstones of the Itapecuru Formation (~99 Ma) outcrop, where the erosive processes act promoting the vertical incision of the valleys and the retreat of the upstream headwaters. The lower relief areas (180–220 m) present colluvium ramps, broad flat valleys covered by thick laterite layers, and sandy-siltstone sediments (Pleistocene colluvium of ~ 5 Ma).

The relief configuration of the Upper Itapecuru River Plateau presents a diversity of shapes carved in the landscape and active erosive processes that show the regional

Fig. 2.20 Testimonial hill amidst vast pediplain. *Source* Dantas (2013)

dynamics. The structural framework has a strong influence on the fluvial dynamics and spatial configuration of the drainage network. Most rivers draining the Itapecuru Mountain fit into systems of structural lineaments of preferential direction NE-SW, in the North slope, the fluvial valley of the Alpercatas river; and NW–SE direction, in the South slope, the fluvial valley of the Itapecuru river (Góes and Rossetti 2001).

From the fluvial incision of the main channels under structural control, the lateral retreat of the scarped slopes of the plateaus stands out, generating detrital sedimentary features at the foot (debris-slopes), and opening of flattened valleys adjusted to the regional base level. It is also noteworthy episodic occurrences of isolated testimony hills, consisting of remnant features of the flattening in these valleys.

The water courses act directly in the configuration of the regional relief. In the regional context, the presence of fluvial captures in the drainage network stands out, such as the capture of the Itapecuru and Alpercatas rivers. These processes show the importance of the mechanical erosion rates present in the area that promote the evolution and reorganization of the drainage network. Another aspect related to the drainage rearrangement processes is the segmentation of the extensive regional plateaus into smaller blocks, characterizing the evolutionary stages of the tabular relief, as highlighted by Migòn et al. (2018, 2020).

Jesus et al. (2019) investigated the weathering behavior and evolution of slopes in Itapecuru Plateau by quantifying the geochemical denudation and rates of landform lowering in fourteen sub-basins. Annual chemical denudation was similar regionally

Fig. 2.21 Geomorphological features of the Mesa Plateau: **a** Presence of deep cavities in the sandstone walls, forming ducts; **b** Ruiniform relief, highlighting the pinnacles; **c** Santa Bárbara Waterfall: a typical feature of a sinkhole. *Source* Personal file Barreto (2014)

(Alpercatas river sub-basins with 1.35 tons/km^2 and Itapecuru river sub-basins with 1.36 tons/km^2). The chemical lowering on the South slope (0.593 m/Ma in Itapecuru sub-basins) was higher than those on the North (0.575 m/Ma in Alpercatas river sub-basins).

The results indicated low levels of dissolved ions in the river water, showing an intense leaching process of cementing minerals, with a predominance of quartz sands resistant to weathering. The rates showed that currently, the regional morphogenesis receives influence from mechanical erosive processes, where laminar erosion on the top of the plateaus and the presence of furrows and small gullies that form on the scarps stand out. The chemical lowering rates and regional morphometric parameters indicate that the headwaters of the Itapecuru river basin behave as catchment channels, intercepting the headwaters of the Alpercatas river basin (Jesus et al. 2019).

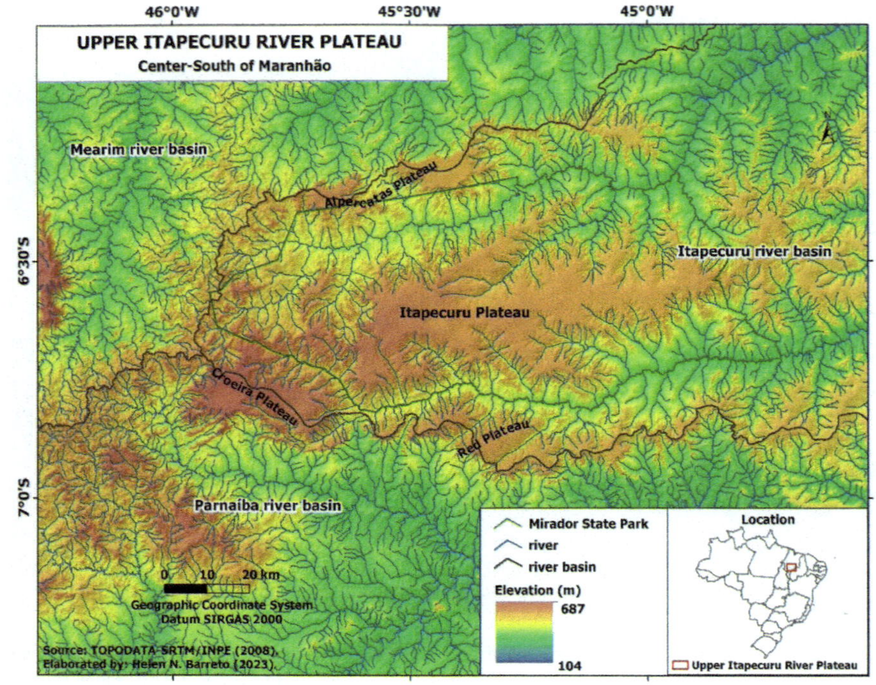

Fig. 2.22 Location of the Upper Itapecuru River Plateau. Elaborated by Barreto (2023)

Currently, the principal environmental impacts of this domain are the burning processes that extend during the dry season (August to November) and the productive arrangements that make up the agricultural frontier of MATOPIBA: the sugarcane and soy planting and forestry on top of the Maranhão plateaus in the South and Center-South (Fig. 2.24). The improper management of the productive areas has caused the acceleration of silting up and clogging processes in the river valleys and the reduction of water flow in the headwaters of the Maranhão basins (Itapecuru and Mearim) and federal basins (Parnaíba and Tocantins).

Fig. 2.23 **a** Elongated tabular forms high in the Itapecuru Plateau; **b** River valley of the Alpercatas—municipality of Mirador; **c** Typical regional relief; **d** Colluvial valleys. *Source* Personal files Dantas (2013) and Barreto (2013)

Fig. 2.24 **a** Soybean planting in the upper course of the Itapecuru River Basin; **b** Irrigation by a central pivot in the Croeira Mountain; **c** Burned Cerrado area on top of Itapecuru Plateau. *Source* Personal file Barreto (2013)

References

Ab'saber AN (2004) A Amazônia: do Discurso à Práxis. 2.ed. São Paulo: Editora da Universidade de São Paulo. 319p

Ab'saber AN (1960) Contribuição a geomorfologia do Estado do Maranhão. Notícias Geomorfológicas 3(5):35–44

Ab'saber AN (2003) Os Domínios de Natureza no Brasil: potencialidades paisagísticas. São Paulo, Ateliê Editorial, 160p

Aranha LGF, Lima HP, Souza JMP, Makino RK (1990) Origem e Evolução das Bacias de Bragança-Viseu, São Luís e Ilha Nova. In: Raja Gabaglia GP, Milani EJ (eds) Origem e Evolução de Bacias Sedimentares. Petrobrás, Rio de Janeio, pp 221–233

Bandeira ICN (2013) Geodiversidade do Estado do Maranhão. CPRM—Serviço Geológico do Brasil, Teresina/PI, 294p

Barbosa GV, Novaes Pinto M (1973) Geomorfologia das Folhas SA-23 (São Luís) e parte da Folha SA-24 (Folha Fortaleza). In: Projeto RADAMBRASIL, 3, DNPM, Levantamento dos Recursos Naturais, Rio de Janeiro.

Barbosa GV, Boaventura RS, Novaes Pinto M (1973) Geomorfologia das Folhas SB-23 (Teresina) e parte da Folha SB-24 (Folha Jaguaribe). In: Projeto RADAMBRASIL, 2, DNPM, Levantamento dos Recursos Naturais, Rio de Janeiro

Barreto HN, Silva JP, Pereira ED, Santos JHS (2015) Chapada das Mesas: unknown geomorphological heritage. In: Vieira BC, Salgado AAR, Santos LJC (eds) Landscapes and Landforms of Brazil. Springer, Amsterdam, pp 201–210

Barros S, Bandeira ÍCN (2020) Geodiversidade da Ilha do Maranhão Teresina: CPRM, 149p

Bétard F, Peulvast JP (2019) Geodiversity hotspots: concept, method and cartographic application for geoconservation purposes at a Regional Scale. Environ Manage 63(6):822–834

Büdel J (1982) Climatic geomorphology. Princeton Univ. Press, Princepton, 443p

Carvalho-neta ML, Correa ACB, Bétard F 2020. O Geoparque Mundial Unesco Araripe e seus Hotspots de Geodiversidade. In: Claudino-Sales V (Org.) Geodiversidade do Semiárido, Cap.1, Ed. Sertão Cult, Sobral/CE, pp 19–37

Castro RMS (2019) Avaliação da Vulnerabilidade à Intrusão marinha na Franja Costeira de São Luís utilizando o método GALDIT. Universidade Estadual do Maranhão.UEMA-PPGeo, São Luís, Dissertação, p 105p

Cordani UG, Sato K, Teixeira W, Tassinari CCG, BASEI, M.A.S, (2000) Crustal evolution of the South America Platform. In: Cordani UG et al (eds) Tectonic Evolution of South America, 31st Intern. Geol. Congr, Rio de Janeiro, pp 19–40

Costa ML (1991) Aspectos geológicos dos lateritos da Amazônia. Revista Brasileira de Geociências, 21(2), SBG, p 146–160

Dantas ME, Shinzato E, Bandeira ICN, Vargas L, Renk JFC (2013) Compartimentação geomorfológica do Estado do Maranhão, cap. 3. In: Bandeira ICN (ed) Geodiversidade do Estado do Maranhão, CPRM – Serviço Geológico do Brasil, Teresina/PI, pp 31–62

de Almeida FFM (1967) Origem e evolução da Plataforma Brasileira. DNPM/DGM, Rio de Janeiro, 38p (Boletim, 241)

dos Santos JHS (2008) Lençóis Maranhenses atuais e pretéritos: um tratamento espacial. Doutorado. UFRJ/PPGG, Rio de Janeiro, p 248

dos Santos JHS, Castro JWA, Gonçalves RA, dos Santos NFB 2005. Modelo evolutivo do sistema de dunas eólicas do setor leste do Parque Nacional dos Lençóis Maranhenses – MA/Brasil. In: Congresso da Abequa, 10., 2005, Guarapari. Anais Eletrônicos... Guarapari: Abequa, CD. 5p

El robrini M, Santos JHS, Lima LG, Santos ALS, Santos MCFV, Souza UDV (2018) Maranhão In: Brasil. *Panorama da erosão Costeira no Brasil.* (Org.) Dieter Muehe. Brasília, DF: MMA, pp 166–240

Feitosa AC, Trovão JR (2006) Atlas Escolar do Maranhão: Espaço Geo-Histórico e Cultural. Editora Grafset, João Pessoa

Feitosa AC (2009) Relevo do Estado do Maranhão: uma nova proposta de classificação topomorfológica. In: Simpósio Nacional de Geomorfologia, 6, Anais – CD/ROM, UGB, Goiânia/GO, 11p

Fundação Sousândrade de Apoio ao Desenvolvimento da Universidade Federal do Maranhão (2009) *Estudo de Impacto Ambiental Refinaria Premium I.* Bacabeira: UFMA. Convênio UFMA/ Petrobrás, Vol. II, p 440

Góes AM, Rossetti DF (2001) Gênese da Bacia de São Luís–Grajaú, Meio-Norte do Brasil. In: Rossetti DF, Góes AM, Truckenbrodt W (eds) O Cretáceo na Bacia de São Luís–Grajaú. Belém, Museu Paraense Emílio Goeldi, pp 15–29

Gonçalves RA, Santos JHS, de. Castro JWC (2005) O papel da dinâmica costeira no controle dos campos de dunas eólicas do setor leste da planície costeira do Maranhão-BR-Lençóis Maranhenses. In: Congresso da Abequa, 10., 2005, Guarapari. *Anais Eletrônicos...*, Guarapari: Abequa, CD. 5p

Gray M (2008) Geodiversity: developing the paradigm. Proc Geol Assoc 119:287–298

Gray M, Gordon JE, Brown EJ (2013) Geodiversity and ecosystem approach: the contribution of geosciences in delivering integrated environmental management. Proc Geol Assoc 124:659–673

Instituto Brasileiro de Geografia e Estatística – IBGE (2020) *Censo Demográfico.* Brasília: IBGE, *onine.* Disponível em https://cidades.ibge.gov.br/brasil/ma/sao-luis/panorama. Acesso em julho. 2021.

Jesus VC, Barreto HN, Pereira ED (2019) Processos denudacionais em domínios de chapadas areníticas no centro sul do Maranhão. Revista Equador. 8(2):309–331

King LC (1956) A Geomorfologia do Brasil Oriental. Revista Brasileira De Geografia, IBGE, Rio De Janeiro 18(2):147–266

Kotschoubey B, Truckenbrodt W, Calaf JMC (2005) Evolução geológica da Província Bauxitífera de Paragominas durante o Neógeno/ Pleistoceno (Noroeste da Bacia do Grajaú, Nordeste do Pará e Oeste do Maranhão). Revista Brasileira de Geociências 35(2), SBG, pp 263–272

Maranhão. Instituto Maranhense de Estudos Socioeconômicos e Cartográficos—IMESC (2019) Sumário Executivo do Zoneamento Ecológico Econômico do Estado do Maranhão –ZEE: etapa Bioma Amazônico. São Luís: IMESC, 578p

Martins FP, Salgado AAR, Barreto HN (2017) Morfogênese da Chapada das Mesas (Maranhão-Tocantins): paisagem cárstica e poligenética. Revista Brasileira de Geomorfologia 18(3)

Martins BN (2019) Avaliação da Vulnerabilidade à Intrusão Salina no Curso Inferior do Rio Bacanga Através do Método Galdit. 103 p. Dissertação (Mestrado em Geografia, Natureza e Dinâmica do Espaço) - Departamento de História e Geografia, Universidade Estadual do Maranhão, São Luís, 2019, p 60

Mello De CF, Mochel FR, Silveira OFM, Prost MT, Mendes A, Bastos MN (1999) Estuários, manguezais e lagoas costeiras/Norte. In: BRASIL/MMA. *Relatório Técnico*. Subprojeto de Avaliação e Ações Prioritárias para a Conservação da Biodiversidade das Zonas Costeira e Marinha. Rio de Janeiro: Ed: Fundação BIO RIO, pp 18–29

Migòn P, Rózycka M, Jancewicz K, Duszynski F (2018) Evolution of sandstones mesas – following landform decay until death. Prog Phys Geogr 42:588–606

Migòn P, Duszynski F, Jancewicz K, Kotwicka W (2020) Late evolutionary stages of residual hills in tablelands (Elbsandsteingebirge, Germany). Geomorphology 367(Suppl.1):107308

MMA (1996) Macrodiagnóstico da Zona Costeira do Brasil na Escala da União Brasília: UFRJ/ FUJB / LAGET: Programa Nacional do Meio Ambiente. 280p

MMA/IBAMA (2003) Plano de Manejo do Parque Nacional dos Lençóis Maranhenses. Ministério do Meio Ambiente, Instituto Brasileiro do Meio Ambiente e Recursos Naturais Renováveis. São Luís, MA. 499p

Muehe D (1998) O litoral brasileiro e sua compartimentação. In: GUERRA, A.J.T.; CUNHA, S.B. (Orgs.). *Geomorfologia do Brasil*. Rio de Janeiro: Bertrand Brasil. Pp 273–349

Nimer E (1989) Climatologia do Brasil. IBGE, Rio de Janeiro, p 448p

Pereira ED (2017) (Org.) Avaliação da vulnerabilidade à intrusão marinha do aquífero costeiro da bacia hidrográfica do Rio Paciência utilizando o método GALDIT. Relatório Final de Pesquisa, convênio UFMA/FAPEMA

Pamplona HRP (1969) Litoestatigrafia da Bacia Cretácea de Barreirinhas. Rio de Janeiro. Bol Tec Petrobras 12(3)

Pereira ED (2006) Avaliação da vulnerabilidade natural à contaminação do solo e aquífero do reservatório Batatã - São Luís (MA). Instituto de Geociências e Ciências Exatas, Universidade Estadual Paulista, Rio Claro, Doutorado - Programa de Pós-Graduação em Geociências e Meio Ambiente

Rodrigues TLN et al (Org) (1994) Programa de Levantamentos Geológicos Básicos do Brasil*:* São Luís, Folha SA-23-2-A, Cururupu Folha SA-23-X-C, escala 1: 250.000, Brasília: CPRM, 185p

Ross JLS (1985) Relevo Brasileiro: uma nova proposta de classificação. Revista do Departamento de Geografia, 4, São Paulo, pp 25–39

Rossetti DF (1998) Facies architecture and sequential Evolution of an incised-valley estuarine fills: the Upper Itapecuru Formation (São Luís Basin), northern Brazil. J Sediment Res 68(2):299–310

Rossetti DF, Truckenbrodt W (1997) Classificação estratigráfica para o Albiano-Terciário Inferior (?) na Bacia de São Luís, MA. Boletim Do Museu Paraense Emílio Goeldi, Série Ciências Da Terra 9:31–43

Rossetti DF (1997) Fácies analysis of the Lower Sucession of the Upper Itapecuru Formation, São Luís Basin, northern Brazil. In: Costa ML, Angélica RS (ed) Contribuições à Geologia da Amazônia. Falângola, Belém, pp 241–284

Rossetti DF (2001) Arquitetura deposicional da Bacia de São Luís-Grajaú. In: ROSSETTI, Dilce de Fátima; GÓES, Ana Maria; TRUCKENBRODT, Werner (ed.). O Cretáceo na Bacia de São Luís - Grajaú. Belém: Museu Paraense Emílio Goeldi, pp 31–46. (Coleção Friedrich Katzer)

Santos MM, Guerra AJT (2003) Indicadores ambientais relevantes para a análise da suscetibilidade à erosão dos solos em Açailândia (MA). Revista Brasileira de Geomorfologia 4(2)

Santos JHS, Santos NFB (2015) The Lençóis Maranhenses: a paradise of dunes and ponds. In: Landscapes and Landforms of Brazil, vol 1, 1st edn. Springer, New York, pp 79–90

Santos JHS, Silva JX (2009) Datação e evolução dos campos de dunas eólicas inativas dos Lençóis Maranhenses. In: Simp. Bras. Geogr. Fis. Aplic., 13, Viçosa, 2009. Anais... Rio de Janeiro, UFV. URL: http://www.geomorfologia.ufv.br/simposio/simposio/trabalhos/trabalhos_completos/eixo12/021.pdf. Acesso 07.07.2021

Santos JHS, Tarouco JEF, Souza BBQ, Mochel FR (2004) Características geológicas e geomorfológicas da Baia de São Marcos, Golfão Maranhense - MA. In: IV Simpósio Nacional de Geomorfologia, São Luís - MA. Geomorfologia: interfaces, aplicações e perspectivas. v 2. p 1–7

Santos DS, Mansur KL, Ribeiro E, de Arruda JR, Dantas ER, Shinzato ME (2019) Geodiversity mapping and relationship with vegetation: a regional-scale application in SE Brazil. Geoheritage 11:399–415

Silva MLN, Mansur KL, Nascimento MAL (2018) Serviços ecossistêmicos da natureza e sua aplicação nos estudos da geodiversidade: uma revisão. Anuário do Instituto de Geociências – UFRJ 41(2):699–709

Silva QD (2012) Mapeamento Geomorfológico da Ilha Do Maranhão. Tese (Doutorado em Geografia), FCT – UNESP, Presidente Prudente

Soares júnior AV, Costa JBS; Hasuy Y (2008) Evolução da margem Atlântica Equatorial do Brasil: três fases distensivas. Geociências, São Paulo, 27(4):427–437

Souza Filho PWM (2005) Costa de manguezais de macromaré da Amazônia: cenários morfológicos, mapeamento e quantificação de áreas usando dados de sensores remotos. Rio de janeiro. Revista Brasileira de Geofísica 23(4):427–435

Teixeira SG, Souza Filho PW (2009) Mapeamento de ambientes costeiros tropicais (Golfão Maranhense, Brasil) utilizando imagens de sensores remotos orbitais. *Revista Brasileira de Geofísica* 2009) 27(Supl. 1):69–82. Disponível em https://www.sbgf.org.br/revista/index.php/rbgf/article/view/1846, acesso em 05.01.2020

Vasquez ML, Klein EL, Lopes ECS (2012) Compartimentação tectônica. In: Klein EL, Sousa CS (Org.) Geologia e recursos minerais do Estado do Maranhão: texto explicativo do mapa geológico e de recursos minerais do Estado do Maranhão. Belém: CPRM – Serviço Geológico do Brasil. 150p. Escala 1:750.000. Sistema de Informações Geográficas - SIG; Mapas Geológicos Estaduais; Programa Geologia do Brasil - PGB.

Villas-Boas JM, Araújo CC (Orgs.) (1999) Açailândia. Folha SB.23-V-A, escala 1:250.000. Estados do Pará e Maranhão. Programa Levantamentos Geológicos Básicos do Brasil. CPRM – Serviço Geológico do Brasil, Brasília

Helen Nébias Barreto She holds a bachelor's degree (1999) and a master's degree (2003) in geography from the Federal University of Minas Gerais (UFMG) with an emphasis on environmental analysis, a Ph.D. in crustal evolution and natural resources (2012) from the Federal University of Ouro Preto (UFOP), a Ph.D. in geoscience (2012) from Université Aix-Marseille—France, and a post-doctorate (2015–2016) at Centre Européen de Recherche et d'Enseignement des Géosciences de L'Environnement (CEREGE), Aix-en-Provence (France). She is an associate professor of the geography course, a permanent professor of the graduate program in geography at the Federal University of Maranhão (UFMA), and part of the Union of Brazilian Geomorphology (UGB) as a councilor (management 2018–2020; 2020–2022). She has experience in geoscience, working mainly on geomorphology and relief evolution, geochronology (cosmogenic isotopes), fluvial geomorphology, water resources and geoprocessing.

Marcelo Eduardo Dantas He holds a degree (1991), a bachelor's degree (1992), and a master's degree (1995) in geography from the Federal University of Rio de Janeiro (UFRJ). He is a natural resources analyst at the Geological Survey of Brazil SGB/CPRM and an assistant professor at Augusto Motta University Center. He has experience in geoscience with an emphasis on physical geography, working mainly on geomorphology, Paraíba do Sul River Valley, local base levels, sediment storage, environmental changes, and the coffee cycle.

Edilea Dutra Pereira She holds a bachelor's degree in geology (1987) and a master's degree in science (1992) from the Federal University of Pará (UFPA), a doctorate in geoscience and environment (2006) from the Júlio de Mesquita Filho Paulista State University, and a post-doctoral degree in geology (2010) from the Wageningen University. She is an associate professor at the Federal University of Maranhão (UFMA) and has experience in geology, hydrogeology, and geotechnics, working mainly on watershed management, socio-environmental impact, geological-geotechnical mapping, and vulnerability to soil, water, and aquifer contamination.

Jorge Hamilton Souza dos Santos He holds a bachelor's degree in geography (1989) and a specialization in applied geography for environmental planning (1993) from the Federal University of Maranhão (UFM), a specialization in environmental expertise and licensing (2017) from the Federal University of São Carlos (UFSCar), and a master's degree (1996) and a Ph.D. (2008) in geography from the Federal University of Rio de Janeiro (UFRJ). He is an associate professor at the Federal University of Maranhão, having experience in geoscience with an emphasis on physical geography, working mainly on geomorphology, water resources, environmental expertise, planning, and environmental management.

Chapter 3
Geomorphology of Piauí State

Iracilde Maria de Moura Fé Lima

Abstract The present proposal for the geomorphological study of Piauí aimed to update the classification carried out in the 1980s (LIMA, 1987). The IBGE (2009) method was used, with the methodological procedures being the reading and analysis of theoretical texts related to the area studied, thematic mappings and fieldwork reports, satellite images, and digital cartography. The main results obtained made it possible to classify the relief of the State of Piauí and its respective mapping presented here.

Keywords Classificação do relevo do Piauí · Método IBGE · Mapeamento geomorfológico

3.1 Introduction

Geomorphology studies the relief of the Earth's surface, investigating its genesis, evolution in time and its spatial relationships. Thus, it results from the interaction among the atmosphere, hydrosphere, geosphere, and biosphere, from the exchange processes of energy and matter on the surface, ranging from the geological to the human scales (FLORENZANO, 2008).

Thus, the study of the relief has great importance for the sciences that study geographic space because, in addition to the relationships among the components of the surface, it allows the identification of the susceptibility of the environment to its use, expressing the possibilities and limitations in the process of occupation and use of land by humans (COOKE; DOORNKAMP, 1974; SANTOS, 2004; RAMOS et al., 2012; BRASIL/IBGE, 2019).

The study of litho-structural elements becomes indispensable because they compose the substrate of the surface as essential indicators for understanding the genesis and evolution of the processes that have acted on the materials and the relief

I. M. de M. F. Lima (✉)
Federal University of Piauí, Teresina, Brazil
e-mail: iracildefelima@ufpi.edu.br

V. Claudino-Sales and J. F. Sobrinho (eds.), *Geomorphology of the Northeast Region of Brazil*, The Latin American Studies Book Series,
https://doi.org/10.1007/978-3-031-58874-7_3

forms (BRITO NEVES, 1991; BIZZI et al., 2003). Based on that, Bigarella et al. (2007) consider the importance of observing the geological structure through the arrangement of mineral groups of rocks, faults, folds, and levees in stratigraphic columns and geological sections to understand the relationships between rocks and other components of the geological evolution at local, regional, and global levels.

Regarding relief sculpturing, the clime performs direct and indirect actions. The first action occurs with the intensity and distribution of the climate and the second one with the soil and vegetation cover (PENTEADO, 1983). Thus, mechanisms qualify morphoclimatic systems for the specificity of a climatic zone and the influence of regional and local factors resulting from the variations in the quantity and concentration of these elements in time and space. In tropical zones, these processes relate to the intensity and distribution of temperature and precipitation since the chemical action of the water increases with the increase in temperature and humidity. Besides, the infiltration and river regimes result directly from precipitation systems in continental areas, while the wind constitutes the outstanding element that influences the relief on the coast, where it performs deflation and sediment transport (FERREIRA; MELO, 2005; CARVALHO; CLAUDINO-SALES, 2016).

Concerning the essential indicators that explain the genesis and evolution of processes that have acted on the materials and the relief forms and considering the context of atmospheric circulation and dynamics, the studies of Pereira (2013), Souza et al. (2015), and Wanderley et al. (2019) analyze the influence of the climatic action centers in northeastern Brazil along with local geographic factors, especially the relief, as responsible for the dynamics and spatial distribution of landscapes where there are microclimatic types ranging from humid to semi-arid. On the other hand, it can also occur that the relief influences the change in climate dynamics, identified in large cartographic scales, from factors such as hillside orientation and shape, slope, and altitude. Such factors will increase or reduce the environmental water availability, implying physiographic changes related to different relief levels (OLIVEIRA; GALVANI, 2015).

Ab'Saber's morphoclimatic domain classification (1969) in Brazil is a valuable contribution based on how the climate sculpture the relief. In this classification, the domains are Amazon, Cerrado (tropical savanna), Sea of Hills, Caatinga, Araucaria, and Prairie, interspersed by transition areas (ecotones) in the contact between them.

Regarding the importance of studies and mapping, Bocco et al. (2001) emphasize the perspective of applied geomorphology, especially regarding its use in the monitoring of cultivated areas, because the data on vegetation cover and land cultivation is variable in time, while the relief compartments remain as the primary analytical spatial unit. These authors report that the Mexican National Institute of Ecology uses regional ecological mapping based on geomorphology for land use planning at national and local scales. They mention, as an example, that in Michoacán, regional geomorphological mapping is the basis of new planning adopted in the Cuitzeo Drainage Basin, the second largest lake in Mexico, whose basin is intensely degraded, with the effects of soil erosion reflected in the water body. In Brazil, applied or environmental geomorphology has supported the urban and regional planning and the feasibility of undertakings that impact the environment, such as housing projects, bridges, highways, mining, commercial plantations, and those that require environmental licensing or studies and reports on environmental impacts (AUGUSTIN; SAADI, 1985; CUNHA; GUERRA, 1996; ROSS, 1996; 2006).

Considering the theoretical discussions pointed out so far about the importance of relief knowledge for the development of life in its various aspects, the work presents a relief classification and mapping of the State of Piauí as a way to update the previous one from the 1980s (LIMA, 1987). Among the advantages, the proposal bases—research, data update, and cartographic technology—decrease mapmaker subjectivity. Using the IBGE (2009) method, it initially situated the State concerning the geological provinces and the morphoclimatic domains of northeastern Brazil to identify the influences of the geological substratum in the formation and configuration of the relief, as well as the climatic action in the sculpture of the relief units and the definition of the type of local modeling. The methodological procedures involved the reading and analysis of theoretical texts, thematic maps and satellite images, organization of climatic data, and previous fieldwork experiences of the author in Piauí at various times. Thus, the use of graphics, maps, and digital elevation models (DEM) from satellite images assisted the performance of cartographic representation of the area, a crucial tool for the analysis of the relief (TRICART, 1965; KOHLER, 2002).

The geomorphological mapping counted on geoprocessing techniques on a 1:250,000 scale, revealing four morpho-structural units, two morphoclimatic domains, six geomorphological regions, and twenty-one relief units, highlighting two types of modeling: denudation and accumulation.

3.2 Location of the State of Piauí

The State of Piauí is in northeastern Brazil, covering an area of 251,755.485 km^2 and is located between the geographic coordinates: at North: $-2°44'49''$ and $-41°48'18''$; to the South: $-10°55'41''$ and $-44°55'48''$; at East: $-6°48'01''$ and $-40°22'12''$; and at the West: $-8°55'39''$ and $-45°59'42''$ (PIAUI/CEPRO, 2012).

Its geological history comprises practically all Eons/Eras and Periods, that is, in portions of geological structures that were submitted to evolutionary processes of amalgamation and configuration of the continents, from long term to short term, from the Archeozoic to the Cenozoic.

The territory of Piauí lies in portions of four structural provinces, three dating from the Archean to the Mesozoic and one from the Cenozoic. They are: (a) Borborema, (b) São Francisco, (c) Parnaíba, and (d) Coastal. The structural partitioning keep elements from the genesis and evolution processes concerning tectonic, sedimentological, and flattening events (ALMEIDA et al., 1977).

Concerning climatic aspects, Piauí is in a transition area between the Amazon humid climates to the West and the semi-arid to the East. Thus, due to its geographical position, Piauí has part of its area in the Cerrado and Caatinga morphoclimatic domains, which are in transition areas (AB'SABER, 1969). In these zones (contact between domains or ecotones), several ecosystems reflect the environmental diversity, such as mixed deciduous broadleaf forest, babassu forest, major field complex, Cerrado, Caatinga, and others (CASTRO, 2007).

The current climates of Piauí are the tropical (Center and West), the variant subhumid tropical (North), and the semi-arid (East and Southeast), based on the Köppen classification (MEDEIROS et al., 2020). According to Costa Filho (2021), the space is under different systems. These climates result from the incidence of various atmospheric circulation systems responsible for the rainfall regimes, which are visible in the distribution of precipitation rates in space and time.

Concerning the coverage area of the climates, it is noticeable that, when adopting the Thornthwaite (1948; 1955) classifications for Piauí (ANDRADE JR. et al., 2004), spatial variations occur, that is, the areas under tropical and semi-arid climates sometimes expand and sometimes shrink over the year. It happens because they consider the classification data of potential evapotranspiration, resulting in four annual scenarios: rainy, dry, average, and regular.

3.3 Method

This work adopted the IBGE method (2009), having as initial procedures the analysis of theoretical texts, thematic maps, satellite images, and data of Piauí. It also considered previous fieldwork experiences the author had in Piauí. Besides, it used geoprocessing techniques for mapping the relief compartmentation.

The compartmentation map results from the crossing between hypsometry and geology. It is a method—by Puebla, Chávez, and Damián (2010); Puebla, Chávez, and Noa (2009)—for the semi-automatic delimitation of landscape units in a Geographic Information System (GIS) environment. The maps and images were worked on the scale of 1:250,000 in the program ArcGIS Pro 10.8, following the steps:

a) Acquisition of the matrix and vector data for the mapping—These data corresponded to the Hydrographic Base of the State of Piauí (ANA, 2021), to the Digital Elevation Model (MDE) AsterGDEM v2 using a 30 m spatial resolution (acquired in the Earth Explorer database from the United States Geological Survey (USGS)) and to the SA23, SA24, SB23, SB24, SC23, and SC24 sheets at the scale of 1:250.000 from the Brazilian RADAM Project updated by IBGE in 2018;
b) Pre-processing of the cartographic base—This step consisted in creating a mosaic of 42 MDE scenes and georeferencing, according to the local coordinate system and horizontal Datum adopted in Brazil, the SIRGAS 2000, and clipping the limits of Piauí in the RADAM/IBGE geology sheets;
c) The calculation of the morphometric variables of the relief and the geology map generated data on the slope, shaded relief—which assists the delimitation and classification of the units—and hypsometry. It classified them into eight classes with altitude intervals of 100 m for the first seven classes and 175 m for the eighth

one. The altimetry varied from 0 on the coast to 875 m in plateaus. Concerning geology, it regrouped the seventy-nine units into ten large units according to the ages and the litho-structural nature of the terrains;

d) Crossing between hypsometry and geology—After the geological classes regrouping and the hypsometrical reclassification, converting the hypsometric value intervals into corresponding numerical groups, the crossing of the two maps was performed: first joining the two vector files, then the attribute tables of both, from the union of the raster files, and, finally, crossing the eight hypsometry classes, individually with the ten geology classes. The result was a new map, with the union of the attribute tables of the previous maps, plus a new column containing 54 subclasses of features;
e) Regrouping, analysis, editing, and classification of the relief units—They occurred from the compartmentation of morpho-structural domains by recognizing the relief textural and structural elements in the images—organizing themselves in homologous zones or relief patterns—since the texture allows the identification of tops, slopes, valleys, and drainage (MOREIRA, 20,003). Besides, the procedure overlay was performed in the drainage environment and shaded relief for interpretation of features by redefining the polygons delimited in the processing and limits for geomorphologic regions and relief units.

It is noteworthy that the use of geoprocessing played a significant role in supporting the analysis and definition of the relief compartmentalization in this work because.

> Digital image processing and multivariate statistical techniques to reduce the subjectivity of textural analysis and correspondence among homologous zones will favor the discrimination of imaged elements and find a model to explain more satisfactorily the compartmentation and the equivalence between zones with similar textures (MOREIRA, 2003, p. 15).

The process culminated in the organization of a table with three columns named morpho-structural domains, geomorphological regions, and relief units, with two more columns identifying current morphoclimatic domains and predominant local modeling types.

The geomorphological units, which correspond to the third taxonomic level, are an arrangement of altimetric and physiognomically similar forms in their diverse types of modeling. Their explanation lies in paleoclimatic factors and lithological and structural conditioning. "These units show their original processes, surface formations, being the drainage behavior, their patterns and anomalies taken as a reference as they reveal the relationships between current or past climatic environments and lithological or tectonic conditioning" (IBGE, 2009, s/p).

3.4 Result and Discussion

The geomorphologic mapping in this work resulted in the Piauí relief compartmentation. It allowed the identification of four morpho-structural units, six geomorphological regions, and twenty-one relief units. Besides, it identified two morphoclimatic domains associated with the compartmentation and the local models (Table 3.1) and (Fig. 3.1).

3.4.1 The Morpho-Structural Domains of Piauí

The morpho-structural domains of Piauí correspond to portions of four structural provinces of the Brazilian space, namely (a) Borborema Structural Province, (b) São Francisco Structural Province, (c) Parnaíba Structural Province, and (d) Coastal Structural Province.

The brief characterization of the province is a way to contextualize the portions of the areas of Piauí and relief compartmentations, which correspond to the morpho-structural domains of Piauí.

3.4.1.1 Archean/Neoproterozoic Mobile Belts and Craton

The morpho-structural domain mobile belts and cratons correspond to two portions of crystalline terrains, the utmost extension in a strip of the Borborema Mobile Belts and the shortest extension in the northern strip of the Craton and Rio Preto Folds of the São Francisco Province, both in southeastern Piauí. The geomorphological region in this domain corresponds to the hinterland depression peripheral to the Parnaíba Sedimentary Basin under current semi-arid climatic conditions (Fig. 3.2).

The morpho-structure of the mobile belts of the Borborema Province comprises a complex mosaic of mobile belts and lithological associations submitted to several tectonic events of predominantly transcurrent character, resulting in a regional framework of blocks and bands of different metamorphic degrees (Lima and Brandao, 2010; Correa, 2015). Although the basement shape is predominantly from the Paleoproterozoic, it keeps some Archean cores, continuing the evolution in stages dated from the Mesoproterozoic and Neoproterozoic, having culminated in the association of all these events to the Brasilian Orogenesis, when the formation of the Gondwana Supercontinent occurred (BRITO NEVES et al., 1995; CORRÊA et al., 2010; CLAUDINO SALES, 2016). The Gondwana fragmentation process and the evolution of the regional morpho-structures in this Borborema Province during the Mesozoic will be commented on in the item morphostructural domain Mesozoic Sedimentary Basins because among them occurred the formation of the Araripe Sedimentary Basin, which has part of its area in Piauí.

Table 3.1 Relief compartmentation of the state of piauí

Morpho-structural domain	Geomorphologic region	Relief unit	Current morphoclimatic domain	Prevailing modeling type
Cenozoic sediments	Piauí coast	Delta, coastal, and fluviomarine plain	Sub-humid	Accumulation
		coastal tablel and		Denudation
Mesozoic sedimentary basins	Tableland and pediment	Grajaú plain	Tropical	
		Grajaú pediment		
		mangabeiras plateau		
		Araripe plateau	Semi-arid	
Paleozoic parnaíba sedimentary basin	North-central piauí	Lower north-central highlands	Tropical	
		Dissected sedimentary surface with residual hills		
		North-central valleys and plains		
	South of piauí	Southern plateaus		
		southern lowlands		
		southern interplanar depression		Denudation/ accumulation
		Southern valleys and plains		
	East of piauí	Secondary cuesta of ibiapaba	Tropical to semi-arid transition	Denudation
		Orthoclinal depression of ibiapaba		
		Pediments with ruiniform features	Semi-arid	

(continued)

Table 3.1 (continued)

Morpho-structural domain	Geomorphologic region	Relief unit	Current morphoclimatic domain	Prevailing modeling type
		Eastern high plateaus		
		Eastern low plateaus		
Archean/ Neoproterozoic mobile belts and craton	Peripheral hinterland depression to the parnaíba sedimentary basin	Crystalline mountains		
		Dissected hinterland surface with residual hills		
		Southeast river valleys and floodplains		Denudation/ accumulation

Source The author (2021)

The strip of the Borborema Province which comprises Piauí corresponds to a western portion, represented by flattened relief, on the periphery of the Parnaíba Sedimentary Basin, where Piauí is bordered east/southeast/south by the States of Ceará, Pernambuco, and Bahia.

About the flattened region of Borborema, which comprises part of the intermontane hinterland depression (AB'SABER, 1969), it is worth mentioning that.

> the stages of landform rejuvenation were sequenced by long phases of a tectonic lull – on the order of several hundred million years, which potentially allowed the erosion of uplifted landforms to produce plateau surfaces. [...] Along with differential erosion, this fact may explain the representativeness of planform surfaces in the current relief (Maia et al., 2010) since the relief forms tend to create a geomorphological resistance, not being easily dismantled by erosion (Peulvast and Claudino-Sales, 2004) (CLAUDINO-SALES, 2016, p.18).

Thus, the Piauí portion of the mobile belt morpho-structure of Provincia Borborema forms a topographically depressed strip on the periphery of the Parnaíba Basin in southeastern Piauí. Small outcrops of these crystalline terrains also occur in the coastal strip of Piauí with granite blocks to the West at Atalaia Beach (Delta Island) and in the lower course of northeastern rivers close to the sea, besides other stretches at the contact of the Barreiras Group with the Parnaíba Sedimentary Basin in the upper coastal rivers. These outcropping rocks belong to intrusive suites from the Neoproterozoic (CPRM, 2006).

The second portion of this morpho-structural domain corresponds to the northwestern belt of the Craton and the Rio Preto Folds that make up the São Francisco Structural Province. The province contains the agglutination of Archean and Paleoproterozoic rock cores and, therefore, part of the oldest rocks in South America (HASUI, 2012). Thus, the Craton.

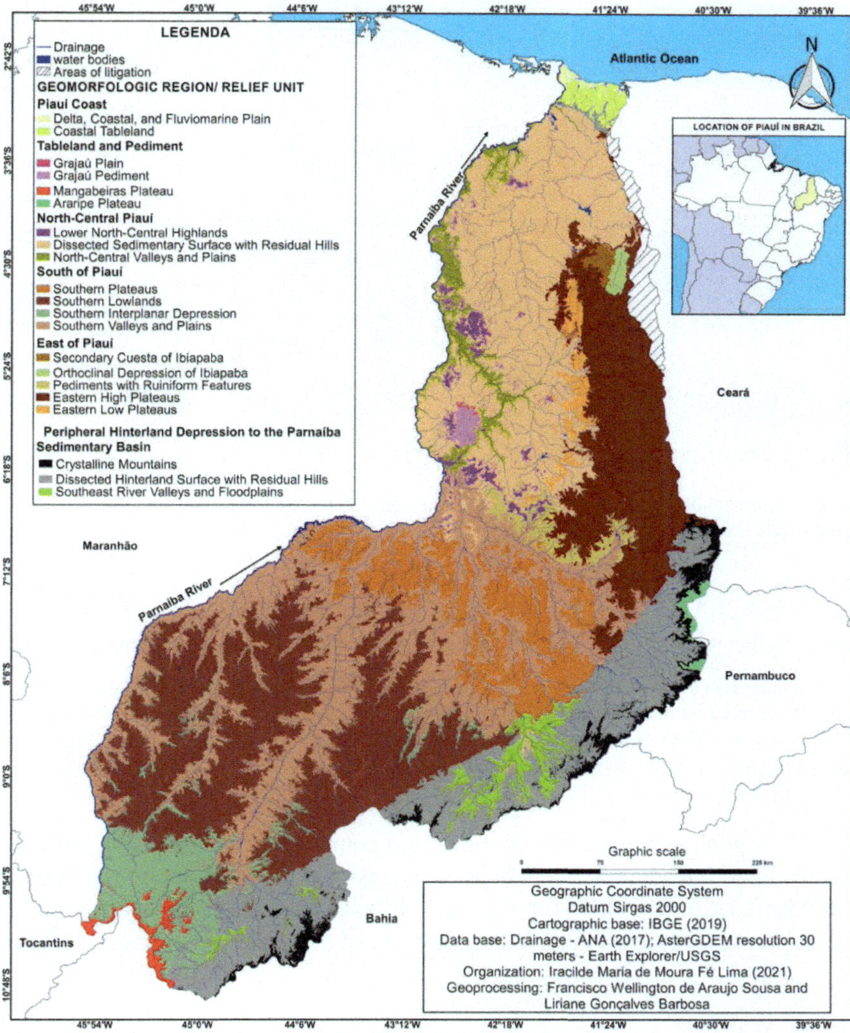

Fig. 3.1 Relief compartmentation map of Piauí. *Source* prepared by the author

connects the Borborema and Tocantins Systems and is exposed in a window in the middle of the Phanerozoic cover [...] being approached by including it sometimes in the former (Brito Neves et al., 2000), sometimes in the latter. It involves a pre-Brazilian basement, exposed in the region of Cristalândia do Piauí and Neoproterozoic metasedimentary packages (HASUI, 2012, p.200).

The Rio Preto Belt limits itself with this outcropping portion of the Craton, located in the north-northwest of the State of Bahia and south of the State of Piauí, corresponding to a metasedimentary sequence composed of phyllites, schists, and gneisses

Fig. 3.2 Relief and lithology of the Precambrian area between Paulistana and Acauã (Piauí). *Photographs* Iracilde Lima (2010)

(BARBOSA; BATISTA, 1971). About the tectonic-structural units of the belt, more recent studies cited by Barros (2019, p.22–23) identify that they correspond to the.

> Cristalândia do Piauí and Júlio Borges Complexes, both represented by 23 TTG-greenstone belts type associations, intruded by small poly-deformed and migmatized plutons; and by the Mansidão and Volta Grande granitoid bodies composed of Paleoproterozoic age Tonalites Granodiorites and Monzogranites, also deformed (Arcanjo & Braz Filho, 2001; Aquino & Batista, 2011; Camacho et al., 2017).

The two geological units in São Francisco compose the continuity of the intermontane hinterland depression belt (as an extension of these depressions that occur in northeastern Brazil) that is in the western belt of the Borborema Province and southeastern and southern Piauí in the periphery of the Parnaíba Sedimentary Basin.

3.4.1.2 Paleozoic Parnaíba Sedimentary Basin

The morpho-structural domain Paleozoic Sedimentary Basin of Parnaíba in the Parnaíba Province occupies a significant area in Piauí and presents three geomorphological regions, namely Center-North of Piauí; East of Piauí; and South of Piauí.

Considering the genesis and evolution characteristics, the three units of the Parnaíba Province are the Parnaíba Sedimentary, Alpercatas, and Grajaú Basins (GÓES, 1995). Such a domain corresponds to the first litho-structural unit formed in the Parnaíba Province, and 80% of Piauí is in this unit—the Parnaíba Sedimentary Basin.

In total, the Parnaíba Province occupies about 600,000 km^2 in sedimentary rock packages that reach a thickness of about 3500 m in its central portion. It is between the

Amazon Shield to the West and a positive flexural feature to the North (concerning the opening of the Equatorial Atlantic), bordering to the Northeast, East, and part of the Southeast with the Borborema Province, being the other southeastern and southern portion with the São Francisco Province (CUNHA, 1986; OLIVEIRA, 2008).

The Parnaíba Province's history started with the separation of Pangea in the Paleozoic, the event responsible for forming the continental margin and the marginal and interior basins. Similar to the Paleozoic Sedimentary Basins of the Amazon (including Solimões and Acre) and Paraná, its origin happened in cycles with differing genesis, tectonic styles, sedimentary filling, and ages of the basins (BIZZI et al., 2003, p. 67). After the geological-structural classification initially adopted by Almeida et al. (1977), these large sedimentary basins appeared as three of the ten structural provinces in such a classification.

The following chart represents, in general terms, the sequence of sedimentation cycles from the Paleozoic to the beginning of the Mesozoic, the uplift of the basin's eastern edge, the dip of the formations, the outcropping area, and the topography of the present relief units shaped by the erosion, differential, denudation, and resistance of the rocks (Fig. 3.3).

The Serra Grande Group geologically (Big Highland) represents the first sedimentation sequence (the oldest) dated to the Silurian, corresponding to a complete transgressive–regressive cycle deposited on the Precambrian structure, forming the eastern border of the Parnaíba Province. Concerning rock composition, it stands out conglomerates and coarse and medium sandstones in the base, fine sandstones in the surface layer, called Itaim, and a sedimentation gap (GÓES; FEIJÓ, 1994).

Fig. 3.3 East edge of the Parnaíba Sedimentary Basin, dating from the Paleozoic: Serra da Capivara seen from the Precambrian Depression, in São Raimundo Nonato and Coronel José Dias (Piauí). *Photography* Iracilde Lima (2015)

The second sequence, the Canindé Group, dates from the Mesodevonian to the Carboniferous and comprises the Pimenteiras, Cabeças, Longá, and Poti Formations, all presenting alternating layers of sandstones, siltstones, shales, and sometimes limestone. The deposition cycle ended in a deltaic and littoral environment with tidal and storm influences. From this period on, there has been an upward epeirogenic movement and a global regression (Eohercinian Orogeny), which installed erosion in the middle Carboniferous, characterizing another intermediate period with a sedimentation hiatus (VAZ et al., 2007).

The third sequence, called the Balsas Group, has its first sedimentary layers dated to the upper Carboniferous and Permian, corresponding to the Piauí and Pedra de Fogo Formations, respectively, with the former composed mainly of sandstones, and the latter of sandstones, shales, silex, and limestone. The sedimentary settling occurred in a fluvial environment with eolic contribution and brief marine incursions in a semi-arid to desert climate (CPRM, 2006).

The fourth sedimentary sequence occurred during the Mesozoic, already in the second unit of the Parnaíba Province (Alpercatas Sedimentary Basin), covering the Parnaíba Basin in small stretches, whose discussion appears in the Mesozoic Sedimentary Basins Microstructural Domain.

3.4.1.3 Mesozoic Sedimentary Basins

The morpho-structural domain called Mesozoic Sedimentary Basins, which occurs in Piaui, corresponds to small portions of the Borborema and São Francisco Provinces and the Alpercatas Sedimentary Basin (Parnaíba Province). Thus, it is discontinuous in space since it corresponds to a group of basins of different structural provinces considered geomorphological region tabular plateaus and pediments.

The formation of these Mesozoic basins began in the Triassic when occurred the closing of a supercontinental cycle (Pangea), with the establishment of another that led to the unleashing of taphrogenesis and fission, still in progress, responsible for the formation of the Atlantic (and other Meso-Cenozoic oceans) (BIZZI et al., 2003, p.45). These Mesozoic tectonic, sedimentary, and magmatic events are reflexes of actions and interactions—accretion, subduction, transformations, and micro-collisions—at the margins and interior of the continent, which continue nowadays under attenuated conditions. In such a period, the pre-Ordovician Brazilian structures were strongly affected, both in continental margins and interior domains, activating lineaments generated in the extrusion phase. It restructured areas of Paleozoic and Mesozoic sedimentation along fault lines, including the Transbrasilian Lineament that started in northwestern Ceará, crossed Piauí, and extended until Tocantins (BIZZZI et al., 2003). It is worth noticing that the fault systems in Piauí associated with this one and the Picos-Santa Inês lineament (orthogonal to each other) directed the regional fitting of the Poti River drainage, for instance, one of the tributaries of the Parnaíba River, of regional extension.

In the evolutionary process of regional morpho-structures during the Mesozoic, the Gondwana fragmented, in stages that lasted about 100 million years, when South

America started a rotation by distending efforts toward the West in continental drift. It culminated in the abortion of the intra-continental rifts and generated interior Cretaceous sedimentary basins, including the Araripe Basin in the Borborema Province (CLAUDINO SALES, 2016). The Araripe Sedimentary Basin, like others, was uplifted in the Tertiary (PEULVAST et al., 2008), having a small western portion of the top and the craggy edges facing a depression strip still in Borborema but already in Piauí.

Concerning the evolution of the Parnaíba Province with the formation of the Alpercatas Basin, the sedimentation of the Sambaíba Formation overlaid the Pedra de Fogo Formation, which outcrops in a small portion of southwestern Piauí, at the border with Maranhão. Thus, the sequence of the Balsas Group ended in the Triassic, with the predominance of sandstones and the intrusion of dikes of rocks during a volcanic event (Mosquito Formation) (LIMA; LEITE, 1978). A fourth sequence happened in this Sedimentary Basin after another hiatus of sedimentation, followed by the settling of the Pastos Bons and Corda Formations in the Jurassic Period. They are composed of fine to medium sandstones, showing cross-stratification and rare lenses of limestone, siltstones, and shales, and on the surface, thin red to pink sandstones, harrowing to siltstones with levels of shale (Fig. 3.4).

Fig. 3.4 Mesozoic plateau, known as Serra do Grajaú (graben with inverted relief), in Agricolândia (Piauí). *Photographs* Iracilde Lima (2012)

These formations extrapolated the Alpercatas Basin and covered part of the Parnaíba Sedimentary Basin in the central region of Piauí, overlying the Pedra de Fogo Formation. The deposition occurred in an alluvial-lacustrine environment, under desert climate conditions, ending with a new volcanic event characterized as the intrusion of rocks (Sardinha Formation) (VAZ et al., 2007).

The Mosquito and Sardinha Formations, called Orozimbo Formation by RADAM (1973), formed diabase/basalt dikes and sills with ages varying between 215 and 110 million years, according to Góes and Feijó (1994). According to this source, the Ar–Ar method identified the age of 134 Ma for these rocks in the central part of Piauí, in the municipality of Agricolândia and surroundings, placing them in the Cretaceous Period. Diabase/basalt dikes appear throughout Piauí, one of the largest outcrops in the municipalities of Amarante and surrounding areas in the center of Piauí (LIMA, 2013).

Concerning the Mesozoic portion of the São Francisco Province, which occurs in the southern boundary of the Parnaíba Province, it outcrops in contact with the Paleozoic formations of the Parnaíba Sedimentary Basin in Piauí and the Alpercatas Basin toward Maranhão. The sedimentary portion consists of a small portion of the Sambaíba Formation in the Parnaíba Province and a sedimentary portion represented by the Areado and Urucuia Formations of the São Francisco Basin, whose sedimentation started in the Triassic and ended in the Cretaceous, in an environment essentially dominated by eolic systems (LIMA; LEITE, (1978); BIZZI et al., (2003); CPRM, (2006)). It is noteworthy that this zone is relevant for Piauí because it corresponds to the area of springs and underground feeding of the largest river of the Parnaíba Sedimentary Basin, which also forms the regional base level of Piauí, the Parnaíba River (LIMA et al., 2017).

3.4.1.4 Cenozoic Sediments

This Morphostructural Domain is located in the Coastal Province, where the Coastal Geomorphological Region of Piauí is located.

The coastal province corresponds to a strip along the Brazilian coast with a length of about 8000 km from the North equatorial region to the South sub-tropical latitudes, bordering the Atlantic Ocean (Muehe, 2005). The province is dated from the Cenozoic, subdivided into three geological periods: Paleogene and Neogene (that substitute the Tertiary) and Quaternary (BIZZI et al., 2003).

The Coastal Tableland is formed by sediments of Barreiras Group (Paleogene and Neogene) that rest discordantly on the crystalline terrains in contact with the Parnaíba Sedimentary Basin, forming a strip limited by the variable thickness of sandy-clay sediments, not or little lithified, with granulation that varies from fine to medium, conglomeratic horizons, and lateritic levels associated with the groundwater percolation (LIMA; BRANDÃO, 2010, p.22) (Fig. 3.5).

The depositional character, continue Lima and Brandão (2010), is considered predominantly continental, in which the sediments were deposited under semi-arid climate conditions when the sea level was lower than the current one.

Fig. 3.5 Commercial sediment exploration area of the Barreiras Group, next to the access to the Portinho lagoon, in Parnaíba (Piauí). *Photography* Iracilde Lima (2016)

In Brazil in the Paleogene and Neogene Periods, the Barriers Group sedimentation occurred in a coastal strip of variable width that goes from State of Amapá to northern the State of Espírito Santo, bordered by active and inactive cliffs, with sandy, muddy, and mangrove deposits of the coastal or fluvial-marine plains dating from the Quaternary (IBGE, 2019). Regarding the formation of the Barreiras Group, Nunes et al. (2011, p.19) affirm that the isostatic adjustment that occurred in the Neogene uplifted the South Atlantic edges, with a tilting of the plates toward the continents, exposing to erosion a strip 100 km to 300 km wide along the coast of the continents. Thus, it became a source of sediments for the inland and coastal basins.

The most recent period, the Quaternary, is a reference for the studies of global and local climate variations, by the record of sea level fluctuations, through traces of ancient beach lines found above and below the current sea level (PAULA, 2013). The author points out that sea levels were much lower than today, which relates to the Quaternary cold climate phases when glaciers and ice sheets became extensive. On the other hand, the opposite also happened when the Ice Age climate gave way to milder conditions, enabling the rise in sea level. Thus, "resulting coastal forms have been largely shaped over the past 6000 years when the sea stood at or near its current level, and the climate approached its current characteristics" (PAULA, 2013, p.29).

In Piauí, Quaternary sediments correspond to narrow bands of river sedimentation, following state rivers. In the coastal region, they correspond to the sediments that overlie the Barreiras Group, in varying thicknesses, following the coast of the Atlantic Ocean. They form a narrow west–east strip 66 km long that begins at the contact with the Canary Bay (Delta do Rio Parnaíba), where it borders the State of Maranhão, and extends to a bay where the Ubatuba River flows, which forms the coastal limit of the States of Piauí and Ceará.

3.4.2 Recent Weather Conditions in Piauí

The climatic data representation highlights that State of Piauí is currently in the transition range from higher to lower rainfall rates as it moves away from the bed of the Parnaíba River and the coast toward the East. From the sub-humid Western Northeast to the semi-arid Eastern Northeast, high annual temperatures persist with minor variations even in different climates.

In relation to the annual rainfall rates, current climatic conditions of Piauí, data from the 30-year climatological normal for Brazilian municipalities (INMET, 1981 to 2010), indicate large variations mainly in the average annual rainfall rates. The data from the following municipalities are examples of this variation: in Parnaíba and Esperantina, located in areas with a sub-humid tropical climate in the North, approximately 1,400 mm occurs; in Floriano and Cristino Castro, located in areas with a tropical climate in the Center and South, approximately 1000 mm occurs; and in Picos and Paulistanana, located in the semi-arid climate zone in the East and Southeast, 700–600 mm occur.

It should be noted that the temperature data (°C) corresponding to the reference for the period from 1981 to 2010 indicated the highest values achieved and not the average of the maximum values, always above 30°C throughout the State.

It is noticeable that, following the trend of variations in precipitation rates, the values of relative humidity of the air also present a curve of significant variation during the year (with rates around 87–67%), corresponding to the sub-humid tropical climate, a subtype of the northern region. In the Center and South, despite the increased rainfall levels, the relative humidity of the air is almost semi-arid, whose values vary from 82 to 75% in the rainiest quarter to 44–38% in September and October, the driest months of the year. Thus, it is necessary to study the climatic conditions, not only the current but also the previous variations that manifest themselves in the relief forms as paleoclimatic inheritance and that are reworked by the current climates.

It is relevant to understand the relief because it will reflect in the flow regime and the dynamics of the rivers, significant agents of relief sculpture. Rivers are predominantly perennial in the sub-humid climate zone, also associated with underground feeding, and are temporary rivers in the semi-arid zone.

The different levels of humidity and relief reflect the types of weathering and erosion processes, with saprolites or deeper soils occurring in tropical and sub-humid climates, such as Oxisols on plateaus and on alluvial terraces. In the semi-arid region, except for alluvial valleys and other localized areas, mechanical weathering and shallower, litholic soils occur, in addition to stretches of varying extents of undecomposed or partially decomposed rocky outcrops. Therefore, variations in climatic conditions associated with litho-structural resistance will be reflected in the dynamics of morphogenetic processes and in the evolution of the shape of Piauí's relief units (LIMA, 2013).

3.4.3 The Geomorphological Regions and the Relief Units of Piauí

In the four morpho-structural domains, the six geomorphological regions correspond to the second taxonomic level of geomorphological compartmentation—the hinterland depression peripheral to the Parnaíba Sedimentary Basin, East, South, Center, North, tablelands, pediments, and the coast of Piauí. The characterization of such regions and their respective groupings of relief units follow below.

3.4.3.1 Hinterland Depression Peripheral to the Parnaíba Sedimentary Basin

The geomorphological region is in the range of the intermontane hinterland depression of the crystalline basement. It corresponds to the area of the morpho-structural domain mobile belts and Archean/Neoproterozoic craton in the periphery of the Parnaíba Sedimentary Basin under the domain of the semi-arid climate. The three local relief units are crystalline mountains, dissected hinterland surface with residual hills, and southeastern valleys and alluvial plains.

The crystalline mountains are on the border of Piauí with the States of Ceará, Pernambuco, and Bahia, subjected to long processes of differential erosion and maintaining altitudes that vary from 400 to 700 m, with the highest ones distributed in small strips in a discontinuous way. The dissected hinterland surface with residual hills—flattened by long erosive processes—is topographically a depression with inselbergs between the mountains and the Paleozoic Sedimentary Basin of Parnaíba, whose altitudes vary from 301 and 600 m. The southeastern valleys and river plains are accumulation models associated mainly with the two large temporary rivers whose sources and high course are located in this area, the Piauí River and Canindé River.

3.4.3.2 Eastern Region of Piauí

The geomorphologic region comprises five relief units named Secondary Cuesta of Ibiapaba, Orthoclinal Depression of Ibiapaba, Pediments with Ruiniform Features, Eastern High Plateaus, and Eastern Low Plateaus. The entire region has a semi-arid climate.

The Secondary Cuesta of Ibiapaba, known as the Matões Mountains, was shaped in the reverse of the regional cuesta that forms the eastern edge of the Parnaíba Sedimentary Basin. The front makes a klint at contact with the terrains of the crystalline basement in Ceará (MOURA-FÉ, 2017), having the other side facing West in Piauí. The Poti River sections the front, forming a canyon in an epigenesis process that comes from Ceará, where it has its high course in the hinterland depression, and enters Piauí, where it has its middle and low ones (LIMA; ALBUQUERQUE, 2020).

The Secondary Cuesta in this reverse remained elevated, even with regional denudation, highlighting the resistance of the Cabeças Formation in contact with the depression formed for removing a voluminous rock package of the Pimenteiras Formation and the Itaim Member of the Big Mountain Group. In this reverse, there is the formation of the orthoclinal depression of Ibiapaba between the primary and secondary cuestas (Fig. 3.6). The Secondary Cuesta presents altitudes between 600 and 800 m under intense erosive processes, while the second one is around 300 m high under predominantly accumulation processes (LIMA; GUERRA, 2020).

In the contact between the geomorphological and the North-Central regions, the unit of pediments with ruiniform features appears in the Seven Cities National Park, Valença, and other areas further south in this region. Such features stand out on the pediments forming pinnacles and towers of rare beauty, still in the Cabeças Formation, under intense processes of differential erosion.

Fig. 3.6 Litho-structural and relief aspects of the Secondary Cuesta and the Orthoclinal Depression of Ibiapaba, in Pedro II (Piauí). *Photography* Iracilde Lima (2014)

The Eastern High Plateaus continue to form the reverse portion of the regional cuesta on the eastern edge of the Parnaíba Sedimentary Basin, with tabular tops, steep slopes, and westward slopes reaching altitudes of 301–875 m. The plateaus are modeled on the same litho-structure and aligned in a North–South direction, in the continuation of the regional cuesta reverse following the plunge of geological layers. They are isolated by deep valleys from the drainage of the drainage basin of the Longá and Poti Rivers, predominantly, and from some affluents of the right margin of the Canindé River, being under intense erosive processes.

The low plateaus correspond to varied extensions of tabular tops, as a continuation of the high ones, forming the topographic divisors of the affluents of the rivers mentioned above, whose altitudes decrease toward the Parnaíba River channel, being between 200 and 300 m. They are under intense erosive processes, observed mainly in the surroundings of the fluvial valleys of small rivers with a temporary regime exhibiting rocky beds, predominantly.

3.4.3.3 Southern Region of Piauí

In this geomorphological region, the four relief units are Southern Tableland, Southern Lowlands, Southern Interplanar Depression, and Southern Valleys and Plains. They are all under the influence of the tropical climate.

The Southern Tableland corresponds to plateaus of great extension, presenting tabular tops under laminar erosion, limited by steep hillsides subjected to intense erosive processes. The altitudes vary from 301 to 700 m, being isolated by deep valleys of large tributary rivers of the upper course of the Parnaíba River, the largest of which is the Gurguéia River. The river bed corresponds to the limit of the perennial fluvial regime (to the West) of the temporary one, as it is noticeable that from the affluents of the right margin to the extreme East/Southeast of Piauí, they are temporaries. The Southern Lowlands correspond to plateaus that decrease in altitude between the tablelands and Parnaíba River, extending eastward and isolated by valleys of rivers that drain the area, especially the lower Gurguéia, middle-bottom Piauí and Itaueiras, and other small affluents of the left margin of the middle Canindé. The Southern Interplanar Depression extends from the South Tableland base to the Mangabeiras Plateau Depression, where the large rivers have their sources and are fed in a trail environment with topography between 350 to 500 m. The Southern Valleys and Plains correspond to small and large tributaries valleys in the upper Parnaíba River, especially the Uruçuí-Preto alluvial valleys and the medium and low courses of Piauí and Guerguéia Rivers (Fig. 3.7).

The latter, mainly, presents flat-bottomed valleys for long stretches—resulting from the retreat of the hillsides by pediment processes—and low amplitudes, that is, altitudes between 350 m in the spring to 150 m at the mouth, characterized predominantly by the accumulation modeling.

Fig. 3.7 Uruçui plateau and descent to the Gurguéia valley, Palmeira do Piauí (Piauí). *Photography* Iracilde Lima (2000)

3.4.3.4 North-Central Piauí Region

The three relief units in this geomorphologic region are the Low North-Central Plateaus, Dissected Sedimentary Surface with Residual Hills, and North-Central and Valleys River Plains. They are all under sub-humid and tropical climates.

The North-Central Low Plateaus correspond to the set of low plateaus of the middle course of the Parnaíba River, which are isolated and vary between 151 and 300 m. They present tabular (as the Santo Antônio Plateau in Campo Maior) and convex tops (Palmares National Forest Plateau in the limit between Teresina and Altos). The mass movements in Monsenhor Gil are rock falls and landslides of disaggregated material on the top and high slopes, resulting in residual hills. It frequently happens on hills sectioned between the plateaus since the vegetation removal intensified the erosive processes (Fig. 3.8).

The orthogonal alignment of two plateaus in the municipality of Amarante-Piauí, known as Santo Antonio and Muquila Mountains, associated with the faults of the Transbrasilian and Picos-Santa Inês regional lineaments, reactivated in the Mesozoic, stands out (Fig. 3.9).

The faults are related to the formation of plateaus and the strong arch curvature of the Parnaíba Riverbed, between the strips where the two lineaments cross, between Amarante and Teresina. Besides, the diabase and basalt outcrop in Amarante and Angical is a market product in some places (Fig. 3.10).

The Dissected Sedimentary Surface with Residual Hills comprises a large area in the geomorphological region that suffered erosion, which brought to surface parts of

Fig. 3.8 Residual feature of low plateaus in Monsenhor Gil (Piauí). *Photography* Iracilde Lima (2012)

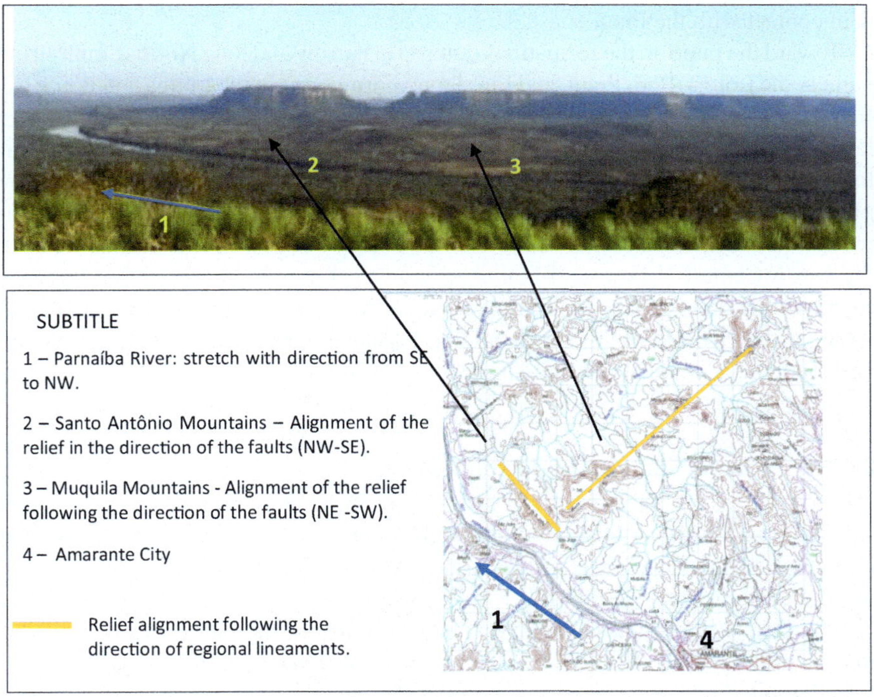

Fig. 3.9 Location of the Santo Antonio and Muquila Mountains, topographic divider of small tributaries of the middle Parnaíba, and a stretch of the arch formed by the bed of this river. Org. Lima (2013). Data bases: RADAM (1973); Carta Amarante (DSG, 1973). *Photo* Lira Filho (2011).

Fig. 3.10 Details of diabase/basalt exploration on top of a plateau in the municipality of Angical (Piauí). *Photographs* Iracilde Lima (2012)

the oldest formations. It begins with a narrow strip of the Big Mountain Group at the northern limit of this relief unit in the East–West direction, where the Pirangi River is in contact with the Piauí coast.

Toward the interior, the formations outcrop in narrow and long North–South strips, namely the Longá, Poti, Piauí, and Fire Stone Formations in the Longa and Poti River Basins and the whole area of diffuse basins of the lower Parnaíba. The local base level is around the Pirangi River mouth, with altitudes between 15 and 150 m in the residual hills and tabular and convex hilltops. It is noteworthy that, in the low plateaus and residual hills, there are superficial laterite layers that, despite undergoing mechanical disintegration, contribute to the support of these relief forms. Besides, the North-Central Valleys and Plains present great expression following the great rivers of this unit, such as the lower course of the Parnaíba, Longá, Poti, and small rivers in the diffused basins of the middle Parnaíba, with altitudes between 80 and 15 m and the accumulation modeling.

3.4.3.5 Tabular Plateaus and Pediments

The four relief units in this geomorphological region are the Grajaú Plateau, Grajaú Pediment, Mangabeiras Plateau, and Araripe Plateau. The first three units are tropical weather, and the last is semi-arid.

The Grajaú Plateau has small rivers that form one of the sub-affluents of the left margin of the Poti River, the São Francisco River (a Berlengas River tributary), with tops delimited by steep slopes and altitudes that reach 420 m. Weathering processes predominate on the tops, while on the North-facing slopes, from where the Fundo 1 and Natal Creeks are born, chemical weathering processes occur in greater intensity than on the south-facing slopes, where we have also observed blocks rolled away by the mass movement. The lithological composition corresponds to the Mesozoic Strip and Good Grass geological formations, where the stratigraphic and topographic positions reveal the limiting faults that generated a graben by the subsidence of a large block of the Mesozoic formations, which fitted into the Fire Stone and Piauí

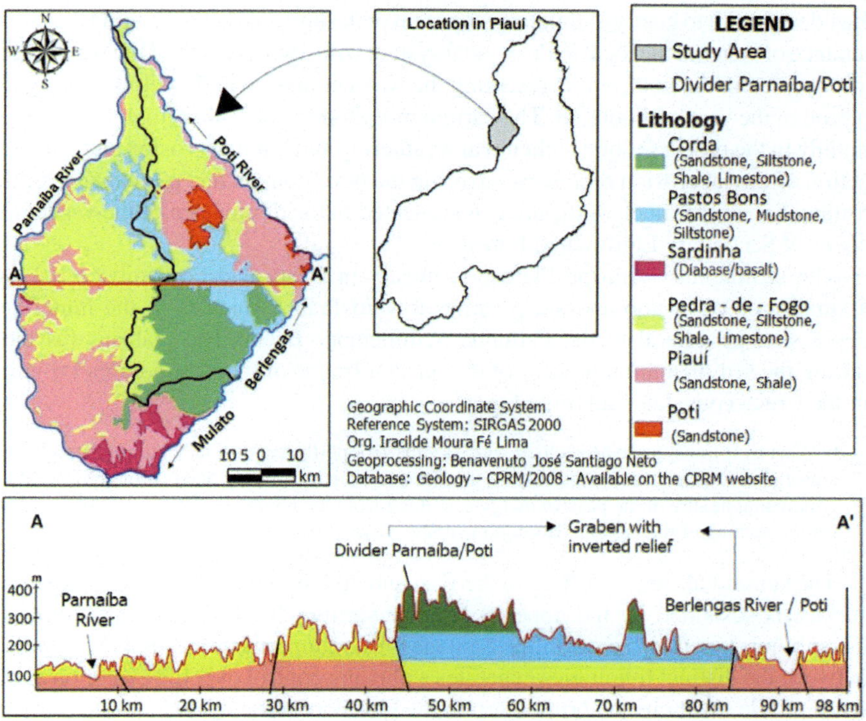

Fig. 3.11 Location of the Grajaú Plateau and Grajaú Pediment on the surface of the inverted relief graben in the middle Parnaíba area (Piauí). *Source* Lima (2013)

Paleozoic formations. The graben results from late Mesozoic tectonism, and from the lower or middle Cenozoic, it began to have its blocks lowered by denudational processes, which include paleo-drainage (BARTORELLI, 2012) corresponding to a graben with inverted relief, as observed in Profile A-A' (Fig. 3.11).

The Grajaú Pediment, contiguous to this plateau, formed in the geological formation called Pastos Bons in the Mesozoic, with altitudes between 160 and 380 m and restricted to the São Francisco Creek basin, a tributary of the Berlengas River on the Poti left bank.

This pediment corresponds to the continuity of the already lowered surface graben where the drainage installed from the Pleistocene to the Holocene formed a suspended basin in topographic terms, considering that it corresponds to the drainage basin confined in this graben and the whole area is higher than the surrounding basins. Such a condition prevented the São Francisco River basin from the erosion directly controlled by the Parnaíba River, which must have slowed down its evolution. Although the present climate of this area does not present a significant apparent variation in the humidity and temperature conditions concerning the other sub-basins, the cited conditions of the geologic basement present themselves with a higher permeability than those that compose the Fire Stone formation (AGUIAR; GOMES, 2004).

Besides, the flat to gently undulating slopes favor the infiltration process and the maintenance of water in the local hydrological system with a small depth of the water table, as observed in the field, even presenting the water courses with low flows during the period of the year without rain. The current morphodynamics, therefore, is expressed mainly in the processes of geochemical weathering and laminar erosion and with low action of fluvial erosion processes, favoring the development and permanence of the Yellow Latosol in much of the area and reflected in the dominance of the vegetation cover of Secondary Latifoliated Forest.

The sedimentary Araripe Plateau is mostly in Ceará and Pernambuco but has a small area of its top and steep slopes toward Piauí, which is in the hinterland depression peripheral to the Parnaíba Sedimentary Basin. The plateau was born during the sedimentation process of the interior basins of the Boroborema Province in the Cretaceous, later submitted to.

> the Cenozoic transpressive tectonics, which seems to be partly responsible for the topographic elevation of the plateau, especially in the East, to which certainly adds to the effects of continental flexure of the passive margins, coming from the North and East [...] responsible for the arching of the regional Borborema dome (SANTOS et al., 2017? p.5).

The summit altitude is 900 m to the East and 800 m to the West, whose structural surface is sustained by the upper permeable stratum (Exú Formation), following the sub-horizontal terrace with the Santana Group impermeable strata. The contact between the distinct lithologies generates a phreatic control over the evolution of the scarp profile, forming an abrupt scarp in the Exú Formation upstream of the springs line, while downstream is smooth, sometimes dissected by drainage and rainfall but other times covered by colluvial sediments and residual soil. The springs' line retreat favors the undermining of the permeable layer base, being responsible for the retraction and formation of talus deposits in the middle and lower slopes. On the other hand, the structural surface of the plateau summit remains practically immune to the effects of rainfall erosion (SANTOS et al., 2017?).

The third relief unit of this geomorphological region is the Mangabeiras Plateau in southern Piauí. It has a tabular top, steep slopes, 780 m altitudes in Piauí, and higher ones in Bahia, Tocantins, and Maranhão, forming the topographic divider of parts of the regional river basins—Parnaíba, Tocantins, and São Francisco Rivers (Fig. 3.12).

It is composed of the Urucuia Group and the Areado Formation, corresponding to a portion of the São Francisco Sedimentary Basin of Cretaceous age, consisting essentially of fine sandstones, with intercalations of shales in the superficial layer, having weak potential in terms of groundwater (AGUIAR; GOMES, 2004). At the slope base overlying this group, the Sambaíba Formation already outcrops in the Parnaíba Sedimentary Basin dating from the Triassic (CPRM, 2006), where it forms a structural step with paths from which the springs of a dense network of small rivers originate from the States of Piauí and Maranhão (LIMA, 2013).

Fig. 3.12 Overview of the Mangabeiras Plateau and one of the Parnaíba River springs formed in the paths at the base of the steep slope at an altitude of about 530 m. *Photographs* Iracilde Lima (2002)

3.4.3.6 Piauí Coast Region

The two relief units in this geomorphological region are the Coastal Tablelands and Delta, Coastal Plains, and Fluviomarine Plains.

The Coastal Tablelands are composed of sediments from the Barreiras Group, corresponding to a flat to gently undulating area, falling toward the north and occupying a strip between the Parnaíba Sedimentary Basin and the Piauí coastal plain. But following the Parnaíba River, it forms a wider strip, elongated toward the south, until approximately the −4°30' parallel, with the highest altitudes around 100 m. In some places, it is observed that its depth exceeds 30 m, for example, where it is temporarily removed for sale, close to Portinho Lagoon. In other places, it outcrops until contact with the sea, covered with thin layers of dunes, as can be seen on a low cliff in Cajueiro da Praia, unique on this coast, which is located close to the eastern border of Piauí with the State of Ceará. This cliff is approximately 200 m long and covered by a thin layer of sand, with its abrasion terrace formed by loose materials, predominantly ferruginous conglomerates (SILVA; LIMA, 2019) (Fig. 3.13).

The morphodynamics of this cliff follows a cyclic and episodic model in its evolution, following a sequence of three stages: detachment of grains or blocks from the cliff face, transport and deposition at the base of the cliff, and the removal of debris by marine processes (SILVA; LIMA, 2019).

The second relief unit of this region corresponds to a portion of the Parnaíba Delta, the Coastal and Fluviomarine Plains. Concerning the Delta formation, Ab'Saber

Fig. 3.13 Aspects of composition and morphodynamics at the Cajueiro da Praia cliff. *Source* Silva and Lima (2019)

(1960) points out that the Parnaíba and Longá Rivers were able to form a wide range of deltaic sediments embedded in a space opened by the post-Pliocene erosion on the sedimentary terrains of the Barreiras Formation, initiating the formation of the Parnaíba Delta in Piauí and Maranhão. The Santa Isabel archipelago is in this large island of the Parnaíba Delta, composed of intertwined channels forming several islands and islets, as well as several sandy beaches, but also with high levels of silts and clays, reflecting the fluvial contribution of the Parnaíba River (SILVA; LIMA, 2020). The island coast has influences from the Salt Stone promontory (outcrop of large granite blocks) because it causes downstream diffraction and retreat of the coastline, as highlighted by Paula et al. (2016).

The Coastal Plain of Piauí covers a narrow strip of Quaternary sediments between the tablelands and the coastline, representing a complex set of depositional environments of wind, fluvial, marine, and lagoon origins. The geomorphological features are the deltaic islands, beaches, fields of barchan and transverse dunes (including fossil dunes), fluviomarine and lacustrine plains, extensive mangrove swamps, estuaries, sandy reef ridges (BAPTISTA, 2016), and expressive areas of apicum (or salty). These features are reliefs of pleasure in the zone of current or sub-current accumulation composed of sandy-quartz sediments well selected and finely grained, with homogeneous profile and rounded shape due to the selective action of winds, which

Fig. 3.14 Photograph of a mobile dune moving over the access road to the Portinho River Lagoon. *Photography* Iracilde Lima (2016)

are the predominant transport agent of these environments (FERREIRA; DANTAS, 2010, p.49).

In the context of this dynamic, we highlight the formation of voluminous ridges of mobile dunes that move from the beaches to the interior of the Coastal Tablelands crossing highways (Fig. 3.14) and damming the flows of small coastal rivers with the formation of fluvial lagoons in rivers of greater length and flow, the largest being the Portinho and Sobradinho Lagoons.

3.5 Final Considerations

The proposed classification of Piauí's relief and its respective mapping presented in this work correspond to a need to update that made in the 1980s (LIMA, 1987), seeking to incorporate the results of new studies and knowledge about Piauí's physical environment.

Advances in research in the conceptual and methodological scope, associated with the development of new cartographic technologies, the experience accumulated by the author in studies and direct observations of environmental conditions throughout Piauí, were also taken into consideration. Thus, geoprocessing techniques were essential for carrying out the geomorphological mapping of Piauí presented in this study.

Such classification of Piauí's relief will certainly contribute to new studies, discussions, and knowledge, as well as to support land occupation planning and the sustainable use of state natural resources.

References

Ab'saber AN (1960) Contribuição à Geomorfologia do Estado do Maranhão. Notícia Geomorfológica. São Paulo: Instituto de Geociências. Universidade de Campinas 3(5):35–45

Ab'saber AN (1969) Participação das Superfícies Aplainadas nas Paisagens do Nordeste Brasileiro. Geomorfologia. São Paulo: Inst. de Geociências/USP 19

Aguiar RB de, Gomes JR de C (Org.). (2004) Projeto cadastro de fontes de abastecimento por água subterrânea, Estado do Piauí: diagnóstico do município de Agricolândia. Fortaleza: CPRM, Mar, 2004. Disponível em: http://www.cprm.gov.br/rehi/atlas/piaui/relatorios/002.pdf. Acesso em 30 mai 2021

Almeida FFM, Hasui Y, Brito Neves BB, Fuck RA (1977) Províncias Estruturais Brasileiras. In: SBG, Simp. de Geol. do Nordeste, 8, Campina Grande, PB, Atas, 363–391

Aquino MR, Batista LH (2011) Geologia e recursos Minerais da Folha de Avelino Lopes, Piauí SC 23-Z-A-111, Escala 1/100.000. CPRM/ Serviço Geológico do Brasil, Residência de Terezina

Arcanjo JBA, Braz Filho PA (2001) O mapeamento geológicodas folhas Curimatá/Corrente – Uma abordagem sobre osprincipais aspectos estruturais, estratigráficos egeomorfológicos. In: SBG, Congresso Brasileiro de Geologia,38, Balneário Camboriú, Boletim de Resumos Expandidos, vol 2, pp 106–107

Augustin CHRR, Saadi A (1985) Avaliação preliminar qualitativa dos impactos antrópicos na bacia do córrego do Mergulhão, Pampulha, Belo Horizonte-MG. Anais do simpósio Situação ambiental e qualidade de vida na RMBH, ABGE-MG/IGC-UFMG, Belo Horizonte-MG pp 261–287

Barbosa O, Batista MB (1971) Projeto Leste do Tocantins/Oeste do rio São Francisco. Petrópolis. DNPM/CPRM/Prospec. (Rel. Fotoint. Prel., Fase II)

Barbosa JA, Neumann VH, Lima Filho M, Souza EM, Moraes MA (2007) Estratigrafia da faixa costeira Recife-Natal (Bacia da Paraíba e Plataforma de Natal), NE Brasil. Estudos Geológicos, Recife, vol 17, no 2, pp 3–30

Barros RA (2019) Evolução Geológica e Contextualização Tectônica do Bloco Cristalândia do Piauí, Faixa Rio Preto, Piauí/Bahia. 2019, 212f. Dissertação (Mestrado). Programa de Pós-Graduação em Geologia. Universidade Federal de Minas Gerais. Instituto de Geociências. Belo Horizonte

Baptista EMC (2016) Geomorfologia e Geodiversidade do litoral piauiense para fins de Geoconservação. Revista de Geociências do Nordeste 2:1250–1258

Bartorelli A (2012) Contexto Geológico da Rede Hidrográfica do Brasil. In: Hasuí Y, Carneiro CDR, Almeida FFM de, Bartorelli A (Org.). Geologia do Brasil. São Paulo: BECA/Patrocínio PETROBRÁS pp 574–610

Bigarella JJ, Becker RD, Santos GF (2007) Estrutura e origem das paisagens tropicais e subtropicais. 2. ed., Florianópolis: Ed. UFSC 3

Bizzi LA, Schobbenhaus C, Vidotti RM, Gonçalves JH (Org.) (2003) Geologia, Tectônica e Recursos Minerais do Brasil. Texto, Mapas & SIG. Brasília: CPRM

Bocco G, Mendoza M, Velázquez, A (2001) Remote sensing and GIS-based regional geomorphological mapping a tool for land use planning in developing countries. Geomorphology, Elsevier 39(3–4):211–219. Disponível em: https://doi.org/10.1016/S0169-555X(01)00027-7. Acesso em: 29 ago 2012

Brasil.CPRM/Governo do Estado do Piauí. Mapa Geológico do Estado do Piauí. 1:1.000.000. Teresina: CPRM, versão 2006

Brito Neves BBB (1991) Os Dois "Brasis" geotectônicos. In: Simpósio de Geologia do Nordeste, 15, Recife. Anais... Recife pp 6–8

Brito Neves BB (2021) Crátons e Faixas Móveis. In: Boletin IG-USP. Instituto de Geociências da Universidade de São Paulo, pp 1–188, 1995. Disponível em: https://www.revistas.usp.br/bigsd/article/view/45352. Acesso em: 02 June 2021

Camacho CR, Sousa FRFR. (2017) O arcabouço estrutural da Bacia Sedimentar do Araripe, Província Borborema, baseado em dados aeromagnetométricos. Revista do Instituto de Geociências-USP. Geol. USP, Sér. cient., São Paulo, vol 17, no 3, pp 114–161

Castro AAJF (2007) Unidades de planejamento: uma proposta para o estado do Piauí com base na dimensão diversidade de ecossistemas. Publicações avulsas Conservação de Ecossistemas. (Série: Publicações Prévias). Teresina: UFPI/Programa BIOTEN, n. 18, pp. 1–28, set.2007

Claudino-Sales V (2016) Megageomorfologia do Estado do Ceará: história da Paisagem geomorfológica. Novas Edições Acadêmicas

Cooke RU, Doornkamp JC (1974) Geomorphology in Environmental Management. An Introduction. 2 ed. Oxford: Claredon Press

Corrêa AC de B, Tavares BAC, Monteiro K de A, Cavalcanti LCS, Lira DR de (2010) Megageomorfologia e Morfoestruturas do Planalto da Borborema. Rev do Instituto Geológico. São Paulo 3(1–2):35–52

Costa Filho WF (2021) Entrevista sobre a dinâmica da circulação atmosférica e regimes pluviais. Teresina

CPRM (2006) Mapa geológico do Estado do Piauí. Governo do Estado do Piauí/CPRM (Companhia de Pesquisa e Recursos Minerais). Teresina: CPRM

Cunha FMB da (1986) Evolução Paleozoica da bacia do Parnaíba e seu arcabouço tectônico. 1986, 107f. Dissertação (Mestrado). Programa de Pós-Graduação em Geologia do Instituto de Geociências, Universidade Federal do Rio de Janeiro, Rio de janeiro

Cunha SB, Guerra AJT (Org.) Geomorfologia: exercícios, técnicas e aplicações. Rio de Janeiro: Bertrand Brasil

Fernandes RF (2011) Estudo da Evolução Termomecânica da Bacia do Parnaíba. 2011, 116 f. Dissertação (Mestrado)—Programa de Pós-Graduação em Engenharia Civil, COPPE, Universidade Federal do Rio de Janeiro, Rio de Janeiro

Ferreira RV, Dantas ME Relevo. In: Pfaltzgraff PA dos S, Torres FS de M, Brandão R de L (Org.). Geodiversidade do Estado do Piauí. Recife: CPRM, pp 45–64

Ferreira AG, Mello NGS (2005) Principais sistemas atmosféricos atuantes sobre a Região Nordeste do Brasil e a influência dos Oceanos Pacífico e Atlântico no clima da Região. Rev Brasileira de Climatologia 1(1):15–28

Florenzano TC (2008) Geomorfologia: conceitos e tecnologias atuais. São Paulo: Oficina de Textos

Fonseca BM, Carvalho GA, Moura ACM, Ramos VDV (2007) A difusão de novas tecnologias de mensuração e dados do relevo como apoio ao ensino da representação topográfica. In: Simpósio Brasileiro de Sensoriamento Remoto, 13. Florianópolis, Brasil, INPE, 2007, Anais... Florianópolis, pp 1473–1479. Disponível em: <http://urlib.net/dpi.inpe.br/sbsr@80/2006/11.14.22.22.41>. Acesso: 21 Mar 2011

Fortes FP (1978) Geologia Estrutural e Tectônica da Bacia Sedimentar do Meio Norte do Brasil (uma síntese com implicações metalogenéticas). In: Congresso Brasileiro de Geologia, 30, Recife, 1978, Anais..., Recife: SBG 1:321–336

Góes AMA Formação Poti (Carbonífero Inferior) da Bacia do Parnaíba (1995), 180f. Tese (Doutorado). Programa de Pós-Graduação em Geologia Sedimentar do Instituto de Geociências. Universidade de São Paulo

Góes AMO, Feijó FJ (1994) Bacia do Parnaíba. Boletim de Geociências. Rio de Janeiro: PETROBRÁS 8(1):57–67

Gomes ER, Cortez ATC (2014) Gênese das nascentes encontradas no entornoda Serra dos Matões, Pedro II, Piauí. 47 Congresso Brasileiro de Geologia. Anais ... Salvador, BA

Hasui Y (2012) Compartimentação Geológica do Brasil. In: Hasuí Y, Carneiro CDR, Almeida FFM, Bartorelli A (Org.). Geologia do Brasil. São Paulo: BECA/ Patrocínio PETROBRÁS, pp 112–130
Hasui Y (2012) Cráton São Francisco. In: Hasuí Y, Carneiro CDR, Almeida FFM de, Bartorelli A (Org.). Geologia do Brasil. São Paulo: BECA/ Patrocínio PETROBRÁS, pp 200–227
IBGE (Instituto Brasileiro de Geografia e Estatística). Manual técnico de Geomorfologia. Coordenação de Recursos Naturais e Estudos Ambientais. 2. ed. Rio de Janeiro: IBGE, 2009. (Manuais técnicos em Geociências, n. 5)
Kohler HCA Escala na Análise Geomorfológica. RevBrasileira de Geomorfologia. Ano 3(1):21–31
Lima EAM, Leite JF (1987) Projeto estudo Global dos Recursos Minerais da Bacia Sedimentar do Parnaíba: integração Geológico-Metalogenética. Etapa III. Relatório Final. Recife: DNPM/ CPRM
Lima IMMF (1987) Relevo do Piauí: uma proposta de classificação. CARTA Cepro. Teresina. 12(2):55–84, ago/dez, 1987. Disponível em: http://iracildefelima.webnone.com
Lima IMMF (2013) Morfodinâmica e meio ambiente na porção centro-norte do Piauí, Brasil. 2013, 309f. Tese (Doutorado). Universidade Federal de Minas Gerais, Instituto de Geociências. Belo Horizonte, MG
Lima IMMF, Lima AB, Augustin CHRR (2017) Nascentes do Rio Parnaíba: Usos e Conservação da Terra e da água. In: Archimedes Perez Filho e Raul Reis Amorim. (Org.). E-book. Os Desafios da Geografia Física na Fronteira do Conhecimento. Ied. Campinas: UNICAMP 1:594–605
Lima IMMF, Guerra AJT (2020) Ambiente montanhoso e turismo em Pedro II, Piauí. In: Geosul, Florianópolis 35(74):518–538
Lima IMMF, Augustin CHRR (2010) O relevo do espaço piauiense: aspectos de sua estruturação e evolução. Simpósio Nacional de Geomorfologia 8, Sensitividade de paisagens: geomorfologia e as mudanças ambientais, Recife, 2010, Anais..., Recife
Lima IMMF, Albuquerque ELS (Org.) (2020) Rio Poti: caminhos de suas águas. Teresina: EDUFPI
Lima EAM, Brandao RL (2010) Geologia. In: Pfaltzgraff PAS, Torres FSM, Brandão RL (Orgs.). Geodiversidade do Estado do Piauí. Recife (Brasil): CPRM
Maia RP, Bezerra FHR, Claudino-Sales V (2010) Geomorfologia do Nordeste: Concepções Clássicas e Atuais acerca das Superfícies de Aplainamento Nordestinas. Revista de Geografia. Recife: UFPE-DCG/NAPA, v. especial VIII SINAGEO 1:1–19
Medeiros RM, Cavalcanti EP, Duarte JFM (2020) Classificação climática de Köppen para o Estado do Piauí Brasil. Rev Equador (UFPI) 9(3):82–99
Moreira MR (2003) Avaliação dos aspectos texturais na imagem Landsat como subsídio a compartimentação fisiográfica dos municípios de Peruíbe e Itanhaém, São Paulo. 2003. 129f. Dissertação (Mestrado). Instituto de Geociências e Ciências Exatas, Universidade Estadual Paulista, Rio Claro
Moura-Fe MM (2017) Planalto, cuesta, glint: a Geomorfologia da Ibiapaba Setentrional (região noroeste do Ceará, Brasil). In: Revista Brasileira de Geografia Física 10(6):1846–1858
Muehe D (2005) Aspectos gerais da erosão costeira no Brasil. Mercator - Revista de Geografia da UFC, Fortaleza, vol 4, no 7, pp 97–110
Nunes FC, Silva EF, Vilas Boas GS (2011) Grupo Barreiras: características, gênese e evidências de Neotectonismo. Boletim de Pesquisa e Desenvolvimento, 194. Rio de Janeiro: Embrapa Solos
Oliveira MRP, Galvani E (2015) Avaliação do efeito orográfico no perfil longitudinal Paraty (RJ) e Campos do Jordão (SP). Entre Lugar (UFGD) 6(11):133–151
Oliveira RG (2008) Arcabouço Geofísico, Isostasia e Causas do Magnetismo Cenozoico da Província Borborema e de sua Margem Continental (Nordeste do Brasil). 411f 2008. Tese (Doutorado). Programa de Pós-Graduação em Geodinâmica e Geofísica. Universidade Federal do Rio Grande do Norte, Natal
Paula JEA (2013) Dinâmica morfológica da planície costeira do estado do Piauí: evolução, comportamento dos processos costeiros e a variação da linha de costa. 2013, 242f. Tese (Doutorado). Universidade Federal do Ceará. Instituto de Ciências do Mar-Labomar. Programa de Pós-Graduação em Ciências Marinhas Tropicais. Fortaleza, Ceará

Paula JE de A, Morais JO, Baptista EM de C, Gomes ML, Pinheiro L de S (2016) Piauí beach systems. In: Short AD, Klein AHF (Eds.) Brazilian Beach Systems. Austrália. Springer
Penteado MM (1983) Fundamentos de Geomorfologia. 3 ed., Rio de Janeiro: IBGE
Piauí. Cepro (Superintendência de Estudos Econômicos e Sociais do Piauí). Piauí em números. 9a.ed. Teresina, 2012. Disponível em: http://www.cepro.pi.gov.br/download/201306/CEPRO07_8a8208d146.pdf. Acesso em 10 July 2021
Puebla AMR, Chavez ES, Noa RR (2009) Diseño Metodológico para laelaboración de mapas de Paisajes con el uso de los SIG: aplicación a la Cuenca Alta del Río Cauto, Cuba. Geografía y Sistemas de Información Geográfica (GeoSIG). Luján, Argentina, ano 1(1):95–108, 2009. Disponível em: http://www.gesig-proeg.com.ar. Acesso: dez 2017
Puebla AMR, Chavez ES, Damián JLP (2010) Diseño metodológico para laelaboración de mapas de paisajes con el uso del SIG ArcGIS 9.3. Tonantzintla, Puebla, México. CRECTEALC
Radam P (1973) Folha SB.23 Teresina e parte da folha SB.24 Jaguaribe: geologia, geomorfologia, solos, vegetação, uso potencial da terra. Rio de Janeiro. (Levantamento de Recursos Naturais, 2)
Ramos AM, Cunha L, Cunha PP (2012) Cartografia geomorfológica aplicada ao ordenamento do território Área da Figueira da Foz—Nazaré (Portugal Central). In: Revista Geonorte. Edição especial 3(4):1433–1449
Ross JLS (1996) Geomorfologia Aplicada aos EIAs-RIMAs. In: Guerra AJT, CUNHA SB da (org.) Geomorfologia e Meio Ambiente. Rio de Janeiro, Bertrand do Brasil, pp 291–336
Ross JLS (2006) Ecogeografia do Brasil. Subsídios para planejamento ambiental. S Paulo, Oficina de Textos
Santos RF (2004) Planejamento Ambiental. Teoria e Prática. São Paulo. Oficina de Textos
Santos CA, Neumann VH, Corrêa ACB (2017) Análise da Compartimentação Geomorfológica da Sub-Bacia Leste do Araripe. Disponível em: http://lsie.unb.br/ugb/sinageo/7/0070.pdf. Acesso em 12 2021
Silva, CHS, Lima IMMF (2020) Litoral do Estado do Piauí: proposta de compartimentação. In: Revista Brasileira de Geomorfologia (Online). São Paulo 21(1):17–32
Silva CHS, Lima IMMF (2019) Falésia de Cajueiro da Praia: testemunho da dinâmica da costa piauiense. Rev de Geociênc do Nordeste. Sobral, Ceará 5:30–41
Souza JOP, Almeida JDM, Correa ACB (2015) Caracterização e espacialização em bacia hidrográfica com relevo complexo: Sertão Central pernambucano–bacia do Riacho do Saco. Rev De Geogr (UFPE) 32(2):106–126
Tricart J (1965) Principes et méthodes de la géomophologie. Paris, Masson e Cie
Vaz PT, Rezende Ngam, Wanderley Filho JR, Travassos Was (2007) Bacia do Parnaíba. Boletim de Geociências da Petrobrás 15(2):253–263
Wanderley LSA, Nóbrega RS, Moreira AB, Menezes RSC (2019) A relação entre o relevo e o clima como proposição de caracterização da fisiologia da paisagem em Pernambuco, Brasil. Geo UERJ, Rio de Janeiro 34:1–24

Iracilde Maria de Moura Fé Lima She holds a degree in geography from the Federal University of Cearà (UFC), a master's degree in geography from the Federal University of Rio de Janeiro (UFRJ), and a Ph.D. in geography from the Federal University of Minas Gerais (UFMG). She is a perpetual member of the Academy of Sciences of Piauí (chair 53) and of the Historical-Geographical Institute of Piauì, and an associate professor in the master's degree course in geography at the Federal University of Piauì (UFPI). She works in the area of geoscience with an emphasis on physical geography, specifically geomorphology, hydrography, environment, and education.

Part III
Northern Northeast

Chapter 4
Geomorphology of Ceará State

Vanda Claudino-Sales, Rubson Pinheiro Maia, and Frederico de Holanda Bastos

Abstract The morpho-structures on Ceará State result from the break-up of Pangea in the Middle Cretaceous.The relief that developed later has notable erosive surfaces resulting from the denudation that occurred throughout the Cenozoic. The denudation happened in terrains of different ages and lithologies with various degrees of deformation and metamorphism, producing residual forms resulting from differential erosion over different structures, which are the expression of processes associated to crustal deformation, metamorphism, plutonism, and neotectonics. These processes are responsible for the presence of denudation forms, aggradation forms, and circunsdenudation forms, distributed in crystalline rocks, sedimentary basins, and sedimentary deposits.

Keywords Planation surfaces · Crystalline massifs · Sedimentary plateaus · Coastal plain · Northeast Brazil

4.1 Introduction

The relief of Ceará has notable erosive surfaces resulting from the denudation that occurred throughout the Cenozoic (Peulvast and Claudino Sales 2004; Maia and Bezerra 2014). This continental process distinguished the relief forms, originating a complex geomorphological mosaic that expresses on the surface the structural constraints responsible for the current morphology of the Borborema Province (Maia and Bezerra 2014).

V. Claudino-Sales (✉) · R. P. Maia
Federal University of Ceará, Fortaleza, Brazil
e-mail: vcs@ufc.br

R. P. Maia
e-mail: rubsonpinheiro@yahoo.com.br

F. de H. Bastos
State University of Ceará, Fortaleza, Brazil
e-mail: fredholanda@gmail.com

V. Claudino-Sales and J. F. Sobrinho (eds.), *Geomorphology of the Northeast Region of Brazil*, The Latin American Studies Book Series,
https://doi.org/10.1007/978-3-031-58874-7_4

However, the denudation happened in terrains of different ages and lithologies with various degrees of deformation and metamorphism. Thus, the residual forms resulting from differential erosion over deformational structures related to tectonic events that occurred along the formation of the Borborema Province—which is northeastern Brazil according to geological features and structural controls from the Neoproterozoic (Brito Neves et al. 2000)—constitute the surface geomorphological expression of processes associated to crustal deformation, metamorphism, plutonism, and neotectonics. Such a characteristic keeps the geological structures that preserve their evolutionary history exposed on the surface as structural reliefs.

These structures bear the marks of the four episodes of continental agglutination and dispersion that geological history records (Brito Neves et al. 2014; Brito Neves 1999). For northeastern Brazil, three episodes influence the relief evolution, notably concerning Ceará and its surrounding areas—Piauí, Rio Grande do Norte, and Paraíba. Such continental agglutinations and dispersions tell the regional tectonic history, especially about Ceará, as detailed below.

4.2 Tectonic History of Ceará and Northeastern Brazil

The first continental agglutination relating to the Columbia Megacontinent happened around 1.8 Ga, resulting from the collision of the Ur, Arctic, and Atlantis continents, probably the first one that geological history records (Shobbenhaus and Brito Neves 2003). Atlantis (Teixeira et al. 2007) composed part of the Borborema Province. It resulted from the collision of several continental masses of the Archean age (between 3.6 and 2.5 Ga). Such a collision occurred around 2.2 and 2.0 Ga (Teixeira et al. 2007). During the Columbia collision, Atlantis was already going through a splitting process.

Indeed, around 1.8 and 1.6 Ga (Paleoproterozoic), fission processes split Atlantis into several individual continental blocks (Brito Neves et al. 2014; Brito Neves 1999). Within the Borborema Province, intracontinental rifts were formed but did not evolve (Nogueira Neto et al. 1997; Brito Neves 1999), generating structures that endure in the regional geological fabric—such as the Orós-Jaguaribe Domain in Ceará.

Around 1.45 and 0.97 Ga (Paleo and Meso-Proterozoic), the continental segments isolated with the Columbia split came together again creating a new mega-continent (Brito Neves et al. 2014; Brito Neves 1999). This agglutination occurred along sutures developed over an extension of the order of 20,000 km, represented in all the present continents, and its result was the formation of Rodinia Megacontinent (Murphy and Nance 2013, 1995).

The position of the Borborema Province inside Rodinia is not clear, but it seems that the northern Northeast did not suffer the consequences of this continental agglutination episode. On the other side, it does not happen to the southern Northeast, which would have passed through the denominated Old Cariris Orogeny (Brito Neves et al. 2000).

Around 1.0 Ga and 750 Ma (Neoproterozoic), a new Wilson Cycle dispersed the continental masses in Rodinia (Brito Neves et al. 2014; Brito Neves 1999). Rodinia divided itself into large continental blocks called East and West Gondwana and

Laurasia. In the northern Northeast, geochronological data (Fetter et al. 1997) indicate that, around 800 million years ago, the current northwestern and central Ceará split from adjacent continental masses—the so-called West African/São Luis Craton, which represents a continental block inherited from the split of Rodinia (Shobbenhaus and Brito Neves 2003)—producing a small oceanic basin between them (Fetter et al. 1997).

A third continental agglutination episode occurred at the end of the Proterozoic and beginning of the Paleozoic (approximately 550 Ma), giving rise to the Gondwana Supercontinent, represented by the terrains that today make up South America, Africa, Oceania, India, and Antarctica (Teixeira et al. 2007). The agglutination of Gondwana occurred between 650 and 540 Ma in the Neoproterozoic (Brito Neves et al. 2014; Brito Neves 1999; Shobbenhaus and Brito Neves 2003), bringing together the East Gondwana and the West Gondwana. In Brazil, such a collage received the denomination "Brasiliana Orogeny," which structured the Borborema Province.

This orogeny formed a Himalayan mountain range inside the Borborema Province, including the terrains associated with what today represents the state of Ceará (Caby et al., 1995). It was also responsible for the emergence of shear zones of SW-NE and E-W directions that control the subsequent evolution of the relief and drainage network (Maia 2012). In the sequence, erosion and tectonic collapses destroyed the mountain range. During the tectonic collapse stages, there was the generation of late granites—the late-brasiliano granites, whose surface occurrence resulting from later tectonic processes and erosion is nowadays quite representative in Ceará (CPRM 2020).

After the tectonic collapse of the Brasilian Chain, the terrains presented a new thermal imbalance in the sense of cooling temperatures. It produced crustal thickening, resulting in crustal subsidence (Boillot 1996). The process generated an accentuated intracratonic depression filled with sediments, forming a wide sedimentary basin—the Parnaíba Basin, which goes from the Ceará-Piauí boundary until Maranhão, also comprising parts of Pará, Bahia, Tocantins, and Goiás (Góes and Feijó 1994). It was also significant in the evolution of the Parnaíba Basin, the occurrence of not-evolved rifts (Castro et al. 2014).

The Paleozoic Parnaíba Basin has sedimentary layers dating from the Silurian (Middle Paleozoic: 435 Ma) to the Cenozoic (Tertiary: 30 Ma less) (Castro et al. 2014; Radambrasil 1981). It has a marine incursion from the westward penetration of Pacific waters (Oliveira and Mohriak 2003) in the early phase and probably with sediments from the erosion of the remnants of the Brasilian Chain.

Gondwana stayed individualized as a separate supercontinent (Brito Neves 1999) for about 400 Ma. Thus, the Borborema Province, situated in a central continental context within Gondwana, remained amalgamated all this time, forming a large flattening surface on the tectonically stable terrains covered by the Parnaíba Basin layers.

Between the end of the Paleozoic and the beginning of the Mesozoic (around 240 Ma), the mega-continent Pangea merged. It contained continental masses in the northern and southern hemispheres surrounded by a single ocean—Panthalassa. The Borborema Province, situated in a central continental position within Gondwana, did not undergo orogenesis during the Pangea agglutination process, which occurred in the marginal areas. Pangea remained amalgamated for about 50 Ma. The division took place in stages starting in the Jurassic (around 180 Ma), which individualized South America as a separate continent and, at the same time, defined the creation of the Northeast (Brito Neves 1999).

This fragmentation, responsible for the opening of the South Atlantic, was carried out in stages (Chang et al. 1988; POPOFF 1988). Initially, it formed the Equatorial Atlantic Ocean, separating North America and South America (Marinho and Mascle 1987). The division initially occurred only up to the limit of the Guiana Shield (Szatmari et al. 1987) and was interrupted. Meanwhile, fragmentation started in the South, around 145 Ma, through rifting processes (Conceição et al. 1988). Subsequently, between 145 and 124 million years ago, the opening spread axially northward (Conceição et al. 1988), such that around 120 Ma, a series of intracontinental rifts were opened in this territorial segment, forming a structural axis—the Cariri-Potiguar Axis—of predominantly SE-NE orientation (Chang et al. 1992; Matos 2000).

The reactivations of the Brasiliano shear zones and older ones during the initial Pangea splitting efforts may have put portions of the Brasiliano granites in outcropping condition. During this, the terrains to the West of Ceará and the East of Borborema—in Rio Grande do Norte, Paraíba, and Pernambuco—were uplifted in the form of rift shoulders (Peulvast and Claudino Sales 2004; Claudino-Sales and Peulvast 2007). The separation caused a reorientation in the continental division efforts, resulting in the abortion of the rifts and generating Cretaceous sedimentary basins—the Araripe and Apodi Basins (Matos 2000), invaded by oceanic waters, the newly formed Atlantic and the Tethys Sea (Carvalho, I.S., oral communication, 2018)—and other smaller ones (Matos 2000).

With the abortion of the intracontinental rifts and the change in orientation of the distending efforts, South America underwent a rotation toward the West, initiating the continental drift process. Around 100 Ma, the separation had a conclusion through transforming faults (Matos 2000). Therefore, the final division in Gondwana in the Brazilian Northeast did not happen through rifts, which explains—in the absence of rift shoulders—the inexistence of Sea Mountain-type reliefs in the coastal zone of this territorial segment.

Afterward happened the uplifting of the Araripe and Apodi Sedimentary Basins (Apodi Basin at the end of the Cretaceous, around 70 million years ago (Cremonini 1995) and Araripe Basin in the Tertiary (Peulvast et al. 2008)). The area also underwent the process of marginal flexure, with the lowering of the coastal zone and uplift of the southern terrains. This process, accompanied by differential erosion, removed sediments from the inner continental segments and deposed the material along the coastal zone, forming, since the Tertiary (around 30 Ma (Arai et al. 1988)), the so-called Barreiras Formation.

In synthesis, the main morpho-structural conditioning factors of the regional landscape are (1) the Atlantis orogeny (2.2 Ga), represented by the majority of the rocks of the crystalline basement; (2) the Brasiliano orogeny, which defined the shear lines of SW-NW and E-W directions that control the drainage network and the installation of the Brasiliano granites that sustain part of the reliefs; and (3) the cretaceous division of the Pangea that uplifted the crystalline terrains as well as the Parnaíba Basin in a solidary way, in the form of rift shoulders.

The present morpho-structural organization of the northern Northeast appears to have changed little since the Cretaceous episodes of split Gondwana, from which it has a direct inheritance. However, the morpho-structural devices have been relocated, often in a localized way, by magmatic (Tertiary volcanism), tectonic (marginal flexure), erosional (climate change, sea level variations, and differential erosion commanded by dry climates, which is a significant factor as it exploited the pre-existing structural mesh elements) and sedimentological (accumulation of sediments over the hinterland surface and, in particular, in the coastal zone) episodes.

Figure 4.1 depicts the dominant hypsometry and topographic compartmentalization of Ceará as a consequence of the action of tectonic stages. Moreover, Fig. 4.2 represents, in simple terms, the geological characteristics resulting from these processes. The following paragraphs synthesize the resulting relief characterization.

4.3 Morpho-Structural Domains and Morphodynamics of Ceará

The Cretaceous reorganization of the northern Northeast relief resulted, in the territory of Ceará, the structuring of three well-defined geomorphological and morpho-structural domains. These are (1) denudational forms (planation surfaces and crystalline massifs); (2) circum-denudational forms (sedimentary plateaus); and (3) depositional forms (coastal and pre-coastal front, which cuts across or incorporates the different domains, while presenting specific geomorphological characteristics). Figures 4.3 and 4.4 present an approximate spatialization of these relief compartments and morpho-structural domains.

4.4 Denudational Forms

4.4.1 Planation Surfaces

Planation surfaces of various altitudes truncate the crystalline rocks of the basement all over the territory of Ceará State (Souza 1988). Dated through discordant deposits that partially cover them, they have a meaning different from that in the theoretical schemes in the literature.

The Post-Brasiliano (and pre-rifting Cretaceous) surface is preserved between 40 and 500 m altitude underneath the Siluro-Devonian sedimentary deposits of the Parnaíba Basin (Ibiapaba Glint (Fig. 4.5)) to the West and in the Cariri Basin (Araripe Plateau) and its periphery to the South. It has practically no role in the dynamics of the present relief, or locally, as exhumed and rapidly reworked elements that generate

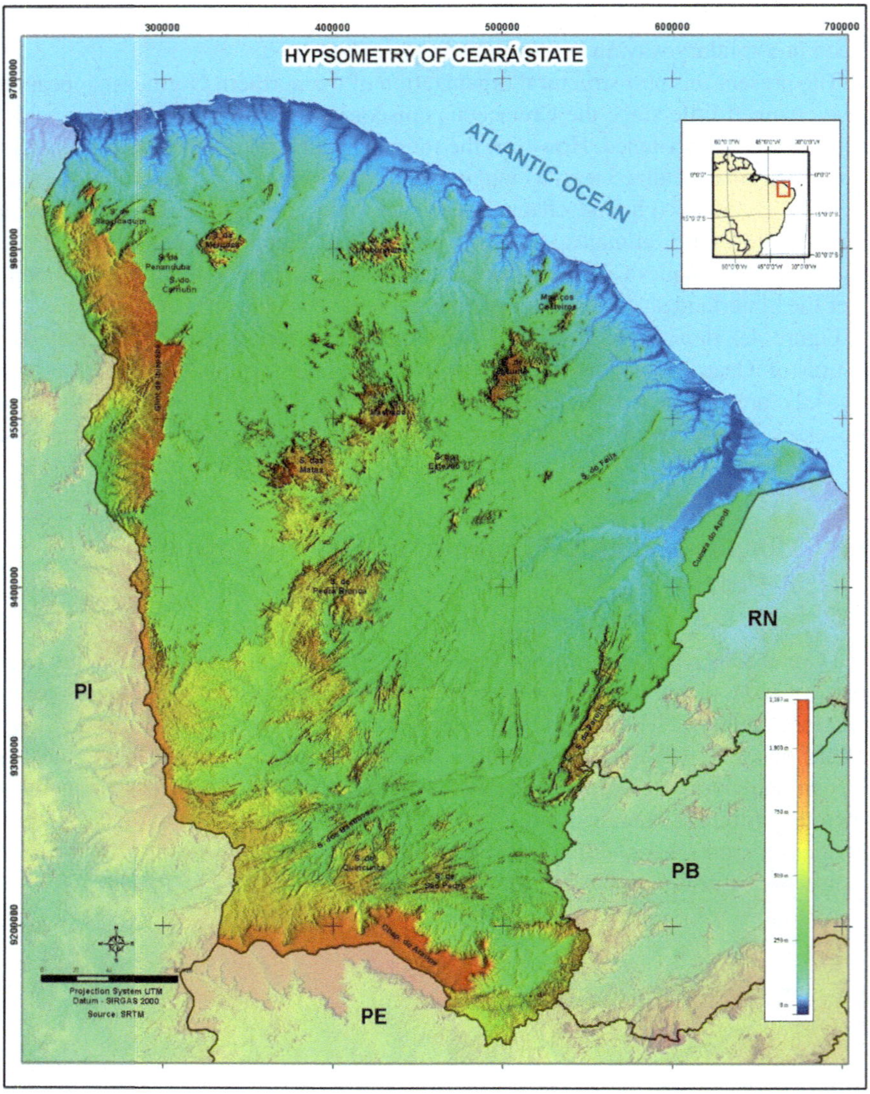

Fig. 4.1 Hypsometric map of Ceará. *Source* Elaborated from SRTM, NASA, 2020

other surfaces to the Southwest, Northwest, and South of Ceará. In the other sectors, during Cretaceous deformation and the action of Tertiary erosive processes, it was deformed and destroyed by dissection and sculpturing of depressions of differential erosion—for example, the depression of Crateús (identified as Meridional Peripheral Depression in previous studies: AGB 1988; Moreira and Gatto 1981), dissected by the Poti River—being substituted by new surfaces with residual reliefs (as the hinterland surface: see definition below).

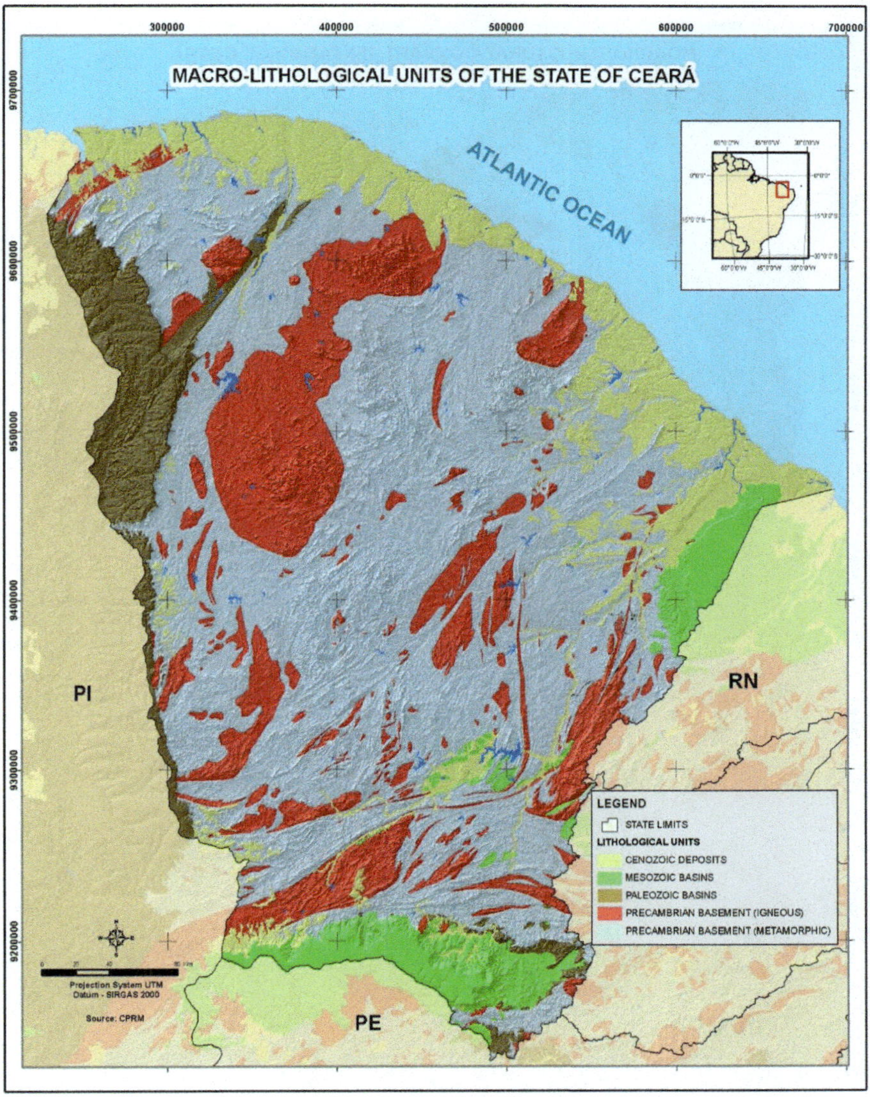

Fig. 4.2 Simplified geological map of Ceará. *Source* CPRM (2003, 2020)

The Post-Brasiliano Surface is under Paleozoic cover, witnessing the pediplanation processes that occurred in geological periods after the development of the Borborema Province (caused by the long stability of the Gondwana Supercontinent) and before the formation of the transforming passive continental margin, which resulted from the fission of the Pangea Megacontinent (Fig. 4.5).

The Jaguaribe Surface is in eastern Ceará. It is outdoors, 10 to 30 km wide, on the periphery of the Potiguar Basin (Apodi Cuesta (Fig. 4.6)), between 40 and 200 m

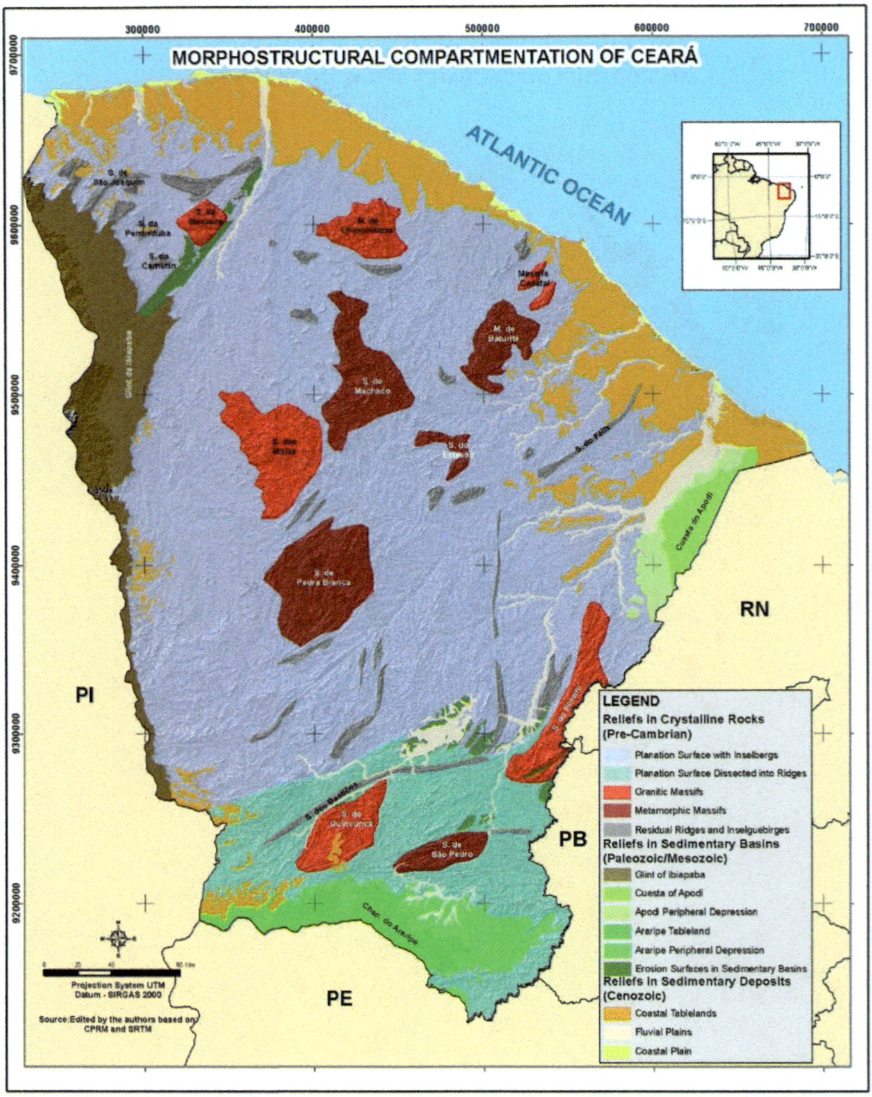

Fig. 4.3 Morpho-structural compartmentation of Ceará. *Source* the authors

altitude, containing exhumed inselbergs of Cretaceous age. Modeled in a post-rift Cretaceous context, before the Cenomanian (before 97 Ma), it suffered inhumation by post-rift deposits of the Upper Cretaceous, which overlie the aborted Potiguar Rift. Subsequently, it suffered partial exhumation during the Tertiary (Peulvast and Claudino Sales 2004). At the basin contact, the surface is still in the process of exhumation. On the other hand, it evidences that the classical ideas about the elaboration of inland surfaces—the oldest ones on the tops of the reliefs and the most

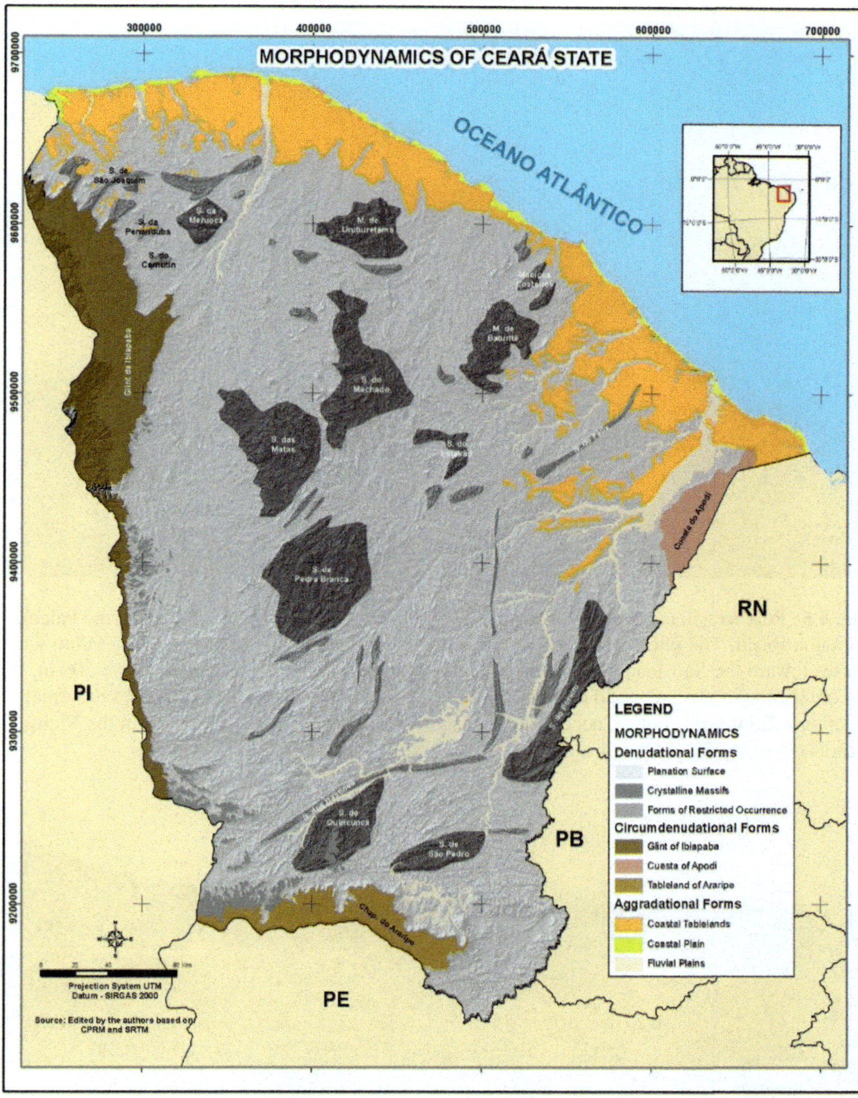

Fig. 4.4 Morphodynamics of Ceará. Prepared by the authors

recent ones on lower altimetry—require relativization since it is an ancient surface in modest altimetric levels.

The Jaguaribe Surface shows how fast the pediplanation may be, which, in this case, happened between the intracontinental rifting (once it borders the aborted Potiguar Rift) around 120 Ma (Matos 1992) and the deposition of the post-rift cover, represented by the Açu and Jandaíra Formations, whose termination (Souza 1997)

Fig. 4.5 Post-Brasiliano Surface inhumed over the Serra Grande Formation, part of the Paleozoic Parnaíba Basin. The photograph shows the dissected northern slope of the Ibiapaba Glint, which passes toward the São Joaquim Mountain, a crystalline massif with an altitude above 700 m. The top of this massif was exhumed from the Paleozoic cover, probably during Cretaceous deformation, creating a Glint-type contact, according to Claudino-Sales et al. (2020). View from the Viçosa do Ceará (*photo by* Vanda Claudino Sales)

Fig. 4.6 Jaguaribe Surface with the view from the foothills of the Apodi Cuesta, also exposing two Cretaceous inselbergs which were inhumed during the Pangea splitting processes and exhumed during the Tertiary. *Photo by* Jean-Pierre Peulvast

was 98 Ma. It means that around 20 Ma, the relief adjacent to the Potiguar Rift underwent complete flattening.

The Jaguaribe Surface extends in the same plane beyond the limits of the peripheral Jaguaribe Depression into the Sertaneja surface (Ab'saber 1964), which serves as a pedestal for most reliefs of Ceará. The hinterland surface contains numerous inselbergs (Figs. 4.7 and 4.8) and goes until the coastal region.

Corresponding to the broadest of the flattened surfaces in Ceará, going back to the Tertiary, the Sertaneja pediplain extends toward the interior until the foot of the Araripe Plateau to the South, presenting a gentle slope toward the coastal zone, masked by sedimentary deposits of Cenozoic age (the Barreiras sediments). To the Southeast, it cuts through several small basins of the Neocomian age (Icó, Malhada Nova, and Iguatu Basins) that also represent small intracontinental rifts aborted during the opening of the Atlantic Ocean. The Sertaneja surface also develops into the coastal zone, where it sometimes presents inselbergs inhumed by the Barreiras Formation, as in Cascavel in the East and Chaval in the West (Fig. 4.9).

The large planation surfaces and inselbergs in Ceará reveal a Cretaceous-Tertiary morphoclimatic evolution commanded by mechanical morphogenesis associated with dry climates (Souza et al. 1979). The situation continues today, making planation surfaces be dissected by intermittent rivers and colonized by caatinga vegetation, adapting itself to litholic Neosol and low rainfall levels from July to December.

The Sertaneja surface constitutes the largest unit in territorial expression in Ceará, represented by three topographic levels from 50 to 400 m. Using the classification of Costa et al. (2020), which distinguished the Sertaneja surface in two, according to altimetric criteria, the following text will present the general characteristics of these surfaces according to their elevation, topographic gradient, and degree of dissection.

Fig. 4.7 Sertaneja surface with the Uruburetama Massif in the background. *Photo* Rubson Pinheiro Maia

Fig. 4.8 Sertaneja surface with inselbergs in Quixadá. *Photo* Jean-Pierre Peulvast

Fig. 4.9 Chaval Inselbergs above the Sertaneja surface in the Western coastal segment of Ceará. *Photo* Frederico de Holanda Bastos

SS1 is typical of lower areas, constituting the areas where the flattening processes are more evident (Fig. 4.10). Concerning altimetry, it is located between elevations of 50 and 250 m approximately, with flat or slightly undulating topography. SS2, with altimetric levels above 250 m, reaches 400 m at the base of the residual massifs.

Generally, at this level, there is a more dissected morphology with the formation of more prominent and collinear interfluves (Souza 1988). The lithological complexity associated with the planation surfaces makes their geological characterization difficult. In this case, the metamorphism associated with these zones conditions the processes of differential erosion between the planed surfaces and the crystalline massifs.

The dissection processes are commanded mainly by the river currents that follow the tectonic directions marked by rupestrian and ductile structures of the basement (Maia and Bezerra 2014). The dendritic pattern presents irregular branches on all sides where the main channel joins the tributaries, impacting the slope of the area and the strength of crystalline and metamorphized rocks, which results in a framework difficult to permeabilize with dense and ill-defined drainage. The pattern exists on the Sertaneja surface and crystalline massifs (Fig. 4.11).

The structural control of fluvial dissection presented in Fig. 4.11 shows up spatially in the present proposal for morpho-structural compartmentalization in Fig. 4.3. The two large Sertaneja surfaces are (1) flattened in the northern part of Ceará and dotted with inselbergs (Fig. 4.11a) and (2) dissected in ridges to the South (Fig. 4.11b), being controlled by the Patos Lineament.

The lowermost Sertaneja surface altitude (200–300 m) presents a surface dissected by the 8th–4th-order channels. Surrounding this surface, the higher Sertaneja surface (300–450 m) stands out, limited by the higher reliefs (Silva and Maia 2017). The dissection is in all directions caused by lower-order channels, resulting in an extensive flattened area, starting from the base of the crystalline massifs and sedimentary plateaus and extending inland.

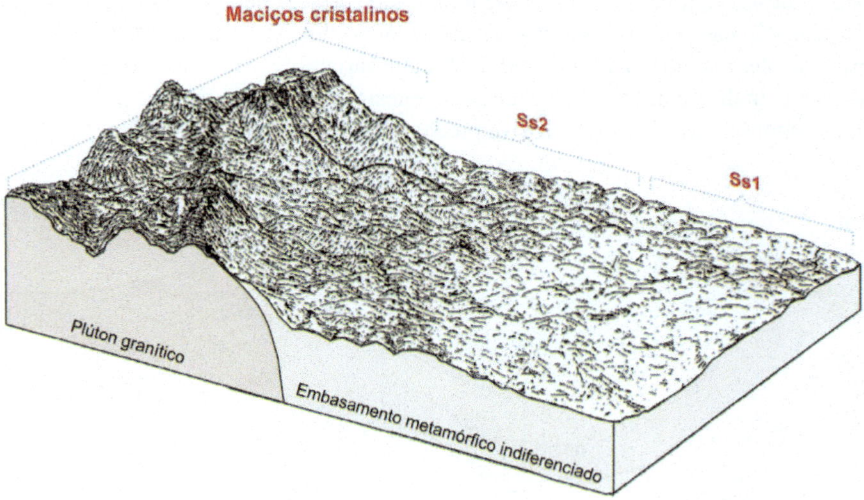

Fig. 4.10 Block diagram demonstrating aspects of the geomorphology of the Sertaneja surfaces and crystalline massifs. *Source* Maia et al. (2018)

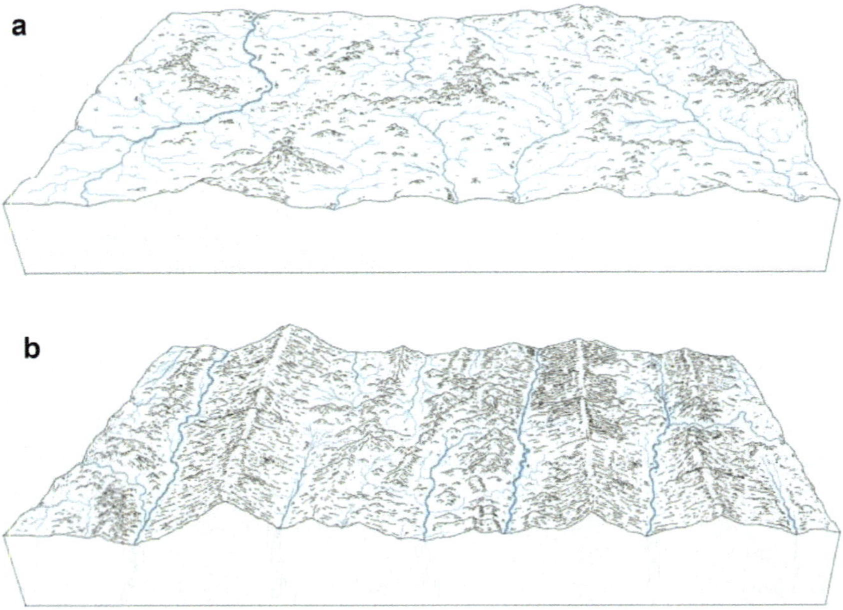

Fig. 4.11 Examples of drainage networks on the hinterland surface without structural control (**a**) and with structural control (**b**). *Source* Silva and Maia (2017)

The high erosion rates triggered mainly by the semi-arid climatic context limit pedogenesis, making the soils of the hinterland surface shallow. In this context, the parent rock (basement) commonly outcrops. According to Bigarella (2003), the high erosion rates result from the context of climatic variability, where wet and dry periods alternate. In the humid phases, when pedogenesis predominates over morphogenesis, thick regoliths rose, limiting erosive processes to the fluvial valleys (Fig. 4.12).

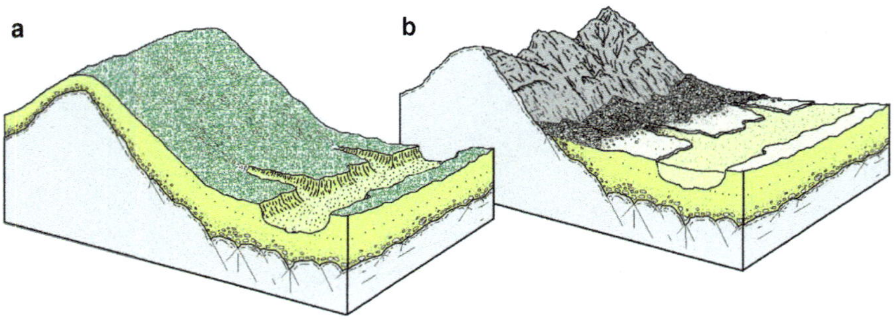

Fig. 4.12 Model of landform exhumation based on climate change. *Source* Maia et al. (2018)

In the cases of the predominance of morphogenesis over pedogenesis, intense erosive processes have remobilized the alteration mantles located on the scarps to the valley bottoms. The erosive processes associated with the climatic regime favor the removal of the thinner facies, leaving only the partially weathered blocks of the saprolite. It occurs from the diffuse surface runoff, whose capacity to remove sediments is directly proportional to the runoff volume.

These erosive processes are associated with the semi-arid climate, where long drought periods expose the alteration mantle to erosion that occurs by the sporadic rainfall regime. The joint basement exhumation reveals the partially weathered blocks that characterize the hinterland landscape.

4.4.2 Crystalline Massifs and Spatially Restricted Residual Forms

The Southwest and Northeast structural axis—where the shoulders of the aborted Cretaceous Apodi and Cariri Rifts are—and the Southeast and Northwest borders present the highest elevations, occupying about 1/3 of Ceará. With average altitudes between 500 and 900 m and dispersed among the flattening surfaces, these massifs relate to more resistant lithological groups, such as granitoid, quartzites, and orthogneisses (Bastos and Cordeiro 2021), that functioned as mechanical obstacles to erosion and elevated in the landscape as a result of differential erosion.

They are common in areas of granitic rocks related to plutons in the exhumation phase. Orogenic granites result from intrusion into extensional regions associated with local and regional tectonic structures (Neves 2012). In Borborema Province, these structures are evident through the shear zones of Northeast-Southwest and East–West directions (Vauches et al. 1995).

Shear zones are conduits through which a significant volume of fluids circulate (Trindade et al. 2008). Most granitic bodies observed on geological maps relate to shear zones (Almeida and Ulbrich 2003). In such areas, the occurrence of the massifs is substantial, thus composing the highest elevations Ceará.

Crystalline massifs are mountainous reliefs, usually isolated and with accentuated dissection caused by past and present weathering and erosive actions. The differential erosion action stands out, conditioned by lineaments, forming slopes, scarps, and topographic unevenness. Due to particularities, the massifs underwent individualization, receiving six large subunits: Central Ceará Massif, Upper Jaguaribe Massif, Pereiro Massif, Baturité Massif, Uruburetama Massif, Meruoca Massif, besides other elevations mapped as crystalline massifs.

The Central Ceará Massif has a preferential alignment in the South-Southwest and North-Northeast directions. It is a set of high residual reliefs with altimetry around 600 m and Precambrian lithology (Souza 1988). The Serra das Matas and Pedra Branca mountains are outstanding, with crests and hills (Fig. 4.13). Sometimes the tops present a flat topography, resulting mainly from little effectuation of linear processes due to hydro-climatological conditions homologous to the semi-arid.

The Upper Jaguaribe Massifs are composed of Precambrian crystalline materials of the Central Ceará and Jaguaribean sub-domains, being a group of mountain ranges with altimetry lower than 600 m and convex and sharp tops distributed separately in the region of the Jaguaribe River upper course. The area presents a significant tectonic influence demonstrated by folds, faults, parallelisms, and alignments of rectilinear and semicircular ridges (Radambrasil 1981; Souza 1988). The crystalline elevations in this unit are the Tataira and Quincuncá Mountains, with Southwest-Northeast slope orientation, suffering influence from the Senador Pompeu and Orós Shear Zones; and the Furtado/Santa Maria, Bastiões, and São Pedro ranges, whose Southwest-Northeast and East–West orientations also show influence from the Patos Shear Zone (Fig. 4.14).

Pereiro Massif is composed of crystalline rocks of the Doutor Severiano and São Miguel Suites, with convex and sharp shapes. The altimetry varies between 500 and 900 m with Southwest-Northeast slopes (Radambrasil 1981), bordered by the Jaguaribe and Porto Alegre Shear Zones. The latter, in turn, acts in the structuring of the Pereiro Massif, questioning the residual nature itself, to highlight aspects that show factors of tectonic influence in the area as the occurrence of Cretaceous and Quaternary basins on the massif (Gurgel et al. 2013) (Fig. 4.15).

Fig. 4.13 Serra das Matas Mountain view from the surrounding western hinterland surface (Nova Russas). *Photo* Frederico de Holanda Bastos

Fig. 4.14 Morpho-structural context of the sector between the Quincuncá Mountain (East) and the Araripe Plateau (West), controlled by the Patos Lineament. *Photo* Frederico de Holanda Bastos

Fig. 4.15 Pereiro Massif with the Sertaneja surface at the foot. *Photo* Rubson Pinheiro Maia

The Baturité Massif, besides the small Maranguape, Juá-Conceição, and Aratanha Massifs, stands out due to their proximity to the littoral region, being crystalline elevations closer to the coast of the northern Northeast. They are composed of quartzites, paragneiss, meta-calcites, mica schists, and granitoid of the Independência and Canindé Units of the Ceará Complex of Paleoprotezoic age (CPRM 2003). The composition of scarps, ridges, and summits influences the limitation of regressive erosion, sustaining higher areas (Bétard and Peulvast 2011). The Baturité Massif is North-Northeast—South-Southwest oriented, with altimetry ranging from 500 to 800 m, convex tops, humps, and ridges, V-shaped valleys, and alveolar plains in the more humid portion (Souza 1988) (Fig. 4.16), reaching elevations above 1000 m, such as High Peak in Guaramiranga (1115 m).

The Uruburetama Massif presents the Neoproterozoic porphyritic meta-granitoid of the Santa Quitéria Granitoid Unit—Northeast-Southwest aligned—which represents one of the most significant intrusive manifestations of the Brasiliano orogeny (Fetter et al. 2003). This highland relief presents weak deformation structures (faults or fractures) of predominant West-Northwest and Northwest directions that diverge from those of the main shear zones (Brasilian deformations of ductile/fragile character) of the Borborema Province, which direction is predominantly Northeast-Southest and East–West (Lima et al. 2019). The average altitudes revolve around 500–800 m, occurring in places with elevations higher than 1000 m. The relief suffers

Fig. 4.16 Morphological context of the transition sector between the plateau and the windward slope of the Baturité Massif. *Photo* Frederico de Holanda Bastos

dissection in elongated ridges and humps that alternate with V-shaped valleys and alveolar plains, with low slopes and composed of colluvial-alluvial sediments. On the plateau and the dry dissected slope, the highest density of geofeature is evidenced (Lima et al. 2019).

The Meruoca Massif contains monzonites, granodiorites, and syenites of the sub-alkaline to alkaline Meruoca intrusive suite from the Paleozoic (CPRM 2003). Its extension is limited to a rhombus shape due to faults and shear zones in the peripheries with a strong influence of the Trans-Brasiliano Lineament (Fig. 4.17).

According to Claudino-Sales and Peulvast (2007), the Trans-Brasiliano Lineament was reactivated in the Cretaceous and, after erosive processes, contributed to the sculpturing of the Meruoca Massif and Chaval Inselbergs in Ceará. The orientation of the Meruoca Massif is predominantly Southwest-Northeast, with maximum elevations varying from 500 to 1000 m and sharp and convex tops (Souza 1988; Radambrasil 1981). The northern sector has greater humidity, with V-shaped valleys and alveolar plains. In the southern sector, there is the occurrence of a smaller compartment, denominated Rosário Mountain, where the hydro-climatological conditions are similar to the Sertaneja surface (Souza 1988).

Thus, the crystalline massifs were modeled on late Proterozoic and Cambrian granitoid related to the Brasiliano orogeny and contain metamorphic complexes where gneisses and quartzites appear as resistant rock volumes. They stand as

Fig. 4.17 Contact between the northern and oriental slopes of the Meruoca Massif concerning the surrounding Sertaneja surface. The faults control the contact between the hillside and the lowlands. *Photo* Frederico de Holanda Bastos

vestiges, probably re-softened in more recent periods, of the shoulders of the intracontinental rifts (Cariri/Potiguar structural axis) that preceded the transforming opening of the Atlantic Ocean in the Upper Cretaceous (Matos 1992; Claudino Sales 2002; Peulvast and Claudino Sales 2004).

The crystalline massifs are very dissected, generally exposing sharp ridges. In some cases, however, there are also convex slopes of the hill seas, as in the higher sectors of the Baturité (Bétard et al. 2007) and Pipocas (Claudino Sales 2002) Mountains. Due to the high relief, the rivers, to reach the base erosion level represented by planation surfaces, incisively dissect the slopes, creating V-shaped valleys (Souza 1997). Frequently, the crystalline massif elevations create areas of climatic differentiation, the highest ones representing humid enclaves in the hinterland domain of the caatinga (Souza 1997), in which chemical morphogenesis dominates and vegetation of the Atlantic Forest type occurs.

The low-altitude massifs, however, and the leeward slopes of the coastal massifs have climatic behavior, that is, close to the semi-arid lowland patterns, with the dominance of the caatingas, and represent dry mountain ranges. In addition, smaller ridges and massifs dot the Ceará territory here and there (Fig. 4.18).

Fig. 4.18 Smaller residual reliefs in more resistant metamorphic rocks. These forms commonly stand out on the hinterland surface throughout the state. *Photo* Vanda Claudino-Sales

4.5 Circum-Denudational Forms (Sedimentary Plateaus)

The circum-denudation of uplifted sedimentary basins constitutes a post-cretaceous morphogenetic feature in Brazil (Ab'sáber 1975). The circum-denudational forms considered in the present analysis represent the sectors of sedimentary basins uplifted and sculpted in typical reliefs of this type of structure as cuesta, plateaus, peripheral depression, and glint.

The sedimentary plateaus correspond to three large sedimentary basins—the Paleozoic Parnaíba Basin and the Mesozoic Araripe and Apodi Basins. They correspond mainly to sedimentary layers of sandstones and limestones, which were either deformed (Parnaíba Basin) or produced during the deformations that accompanied the separation of South America from Africa. The Parnaíba Basin represents a Paleozoic-aged syneclise, formed during the existence and split of the supercontinent Gondwana (Brito Neves et al. 2014), which produced numerous rifts in the northern and eastern northeastern segment of Brazil.

The sedimentary plateaus and crystalline terrains were uplifted between the end of the Cretaceous and Tertiary (Peulvast and Betard 2015; Claudino-Sales 2002; Cremonini and Karner 1995) and are today in an elevated topography concerning the surrounding crystalline terrains. Besides these basins, Ceará also presents small Phanerozoic sedimentary basins (Jaibaras, Cococi, and Iguatu/Lima Campos) that are in the same altimetric quotas of the Sertaneja surfaces and, in this way, were related to the denudational forms.

The Apodi and Araripe Basins cover aborted Cretaceous rifts, witnessing the first phase of the separation process between South America and Africa. The rifting did not advance in these segments due to the thickness of the terrains of the Borborema Province (Popoff 1988), in a way to inhibit the continuation of the rifts of the Apodi/Araripe structural axis, which contain post-rift sedimentary layers (of Cenomanian to Turonian age, around 95–88 ma).

With reliefs above 700 m, the Parnaíba Basin escarpment, situated close to the border of Piauí and Ceará to the West, stands out as a modeled form of a spectacular glint concerning the low interior surfaces of Ceará (glints represent a relief formed by a sedimentary package deposited in discordance under the crystalline basement, indicating that there was an erosion of the basement due to the resistance of the sedimentary rocks concerning the crystalline). It presents a dissected front, maintained by rocks from the Serra Grande sedimentary formation, and slopes kept by crystalline basement stones.

4.5.1 Ibiapaba Glint

The Ibiapaba Glint constitutes one of the most expressive reliefs of the northern sector of the Brazilian Northeast, being situated at the limit between the Borborema and Parnaíba Provinces and covering western Ceará. The morphological behavior of this relief is related to the presence of Silurian sandstones of the Serra Grande Group discordantly overlying the Precambrian basement (Claudino Sales et al. 2020). It is a dissymmetric relief with an East-facing escarpment (Fig. 4.19) and a reverse gently sloping to the West, configuring a morphology similar to a cuesta.

On the reverse side of the glint, which has a gentle slope toward Piauí, the cuestiform reliefs unfold. It is, especially in the northern part, a segment with humid climatic conditions, differentiating itself from the rest of Ceará from this point of view. Toward the South, the altimetry decreases, giving way to lower scarps. The glint is dissected in the central sector as a canyon by the Poti River, which superimposes itself on the structure, draining toward the Parnaíba River Basin.

From the structural point of view, this relief represents the eastern limit of the Parnaíba sedimentary basin, which constitutes one of the most important Brazilian structural provinces. It composes a vast Paleozoic syneclise (Milani and Zalán 1999; Zalan 1999) formed by subsidence (Almeida et al. 1981; Brito Neves et al. 1984), whose area is about 600.000 km^2 (Vaz et al. 2007; Góes and Feijó 1994) over a basement formed by Brasiliano cratons and faulted orogenic belts (Silva et al. 2003).

The depositional sequences of the Parnaíba Basin have a polycyclic genesis, derived from events of immersion of the platform surface below the regional base level followed by its uplift (Camaca 2015), finding several mapped geological units with varied sedimentary lithologies (sandstones, conglomerates, siltstones, and argillites), in addition to the occurrence of Mesozoic volcanic rocks of the Sardinha

Fig. 4.19 Morphological behavior of the Ibiapaba Glint in its northern sector with a sedimentary cornice (Tianguá Formation) right above the basement rocks. The basin rim was uplifted jointly with the shoulders of the Cariri-Potiguar Rift in the first stage of the Pangea (Cretaceous) split and is today attacked by chemical morphogenesis in its northern segment. Sedimentary rocks support the glint top, while crystalline stones hold part of the slope and the foothill. *Photo* Frederico de Holanda Bastos

Formation, which occur in the form of dikes and sills of Cretaceous age, observed at the eastern edge of the basin (Ferraz et al. 2017).

Despite the diversity of structures, there are only lithologies of the Serra Grande Group (Silurian), which constitutes the basal layer of the basin, divided into, from the oldest to the most recent, the Ipu, Tianguá, and Jaicós Formations (CPRM 2020). The stratigraphic sequences of the Parnaíba Basin in the Ibiapaba Plateau present a monoclinal behavior with a W-shaped inclination toward the center.

The morphology of the immediate reverse of this relief, still in Ceará, is characterized by fluvial dissection of consequent rivers interspersed by tabular interfluves (Souza 1988). The altimetric amplitudes between the talweg and interfluves tend to increase with the reverse slope, near the limits with Piauí. In general, the behavior of this immediate reverse of the Ibiapaba Glint presents a little dissected and slightly inclined plateau in the northern sector of the glint (North of the Sobral/Pedro II Fault) and a greater inclination of the reverse in the southern sectors (South of the Sobral/Pedro II Fault) with more pronounced dissection of the consequent rivers.

In the Ibiapaba Glint, the Serra Grande Group, discordantly superposed over the crystalline basement, sustains a long North–South direction escarpment (>400 km), situated practically in the entire western Ceará but interrupted only by the fluvial superimposition of the Poti River, configuring one of the most significant faults of the Brazilian Northeast. Controlling the escarpment, the Ipu, Tianguá, and Jaicós Formations maintain the cornices in various sectors.

Along its extension, the behavior of the scarp of the Ibiapaba Glint presents variations from the altimetric, structural, and morphological points of view. In the northern sector of the glint (North of the Sobral/Pedro II Fault), the rocks of the crystalline basement reach elevations above 600 m in contact with the sedimentary material of the Tianguá Formation. Thus, the escarpment (cornice) begins just above 600 m and can reach 900 m. Below it, there are pediments dissected into ridges (Fig. 4.19). In the sector situated just South of the Sobral/Pedro II Fault, the glint presents the most expressive vertical scarps associated with the Ipu Formation, as can be seen in the surroundings of Ipu (Fig. 4.20).

From the Poti Fault, following Southward, the glint escarpment is supported by the sandstones of the Jaicós Formation at gradually lower elevations (700 m), compared to the northern sector, previously presented (Fig. 4.21).

Another relevant aspect regarding the morphological behavior of the Ibiapaba Glint refers to the variation of the basement lithologies overlying the sedimentary rocks of the Serra Grande Group. Thus, in the northern sector, the exhumation of the quartzite rocks of the São Joaquim Formation preserves crests higher than 600 m (glintoide relief, which represents a contact of an elevated sedimentary basin with an elevated crystalline one (Claudino-Sales et al. 2020)), besides the Frecheirinha Formation meta-calcars, whose chemical action allowed the development of karstic forms just below the sandstone cornice, highlighting the utmost cave system in Ceará (Ubajara Cave).

The surrounding processes that sculpted the Ibiapaba Glint formed an abrupt topographic scaling that separates Sertaneja surfaces, with 200–300-m elevations from a sedimentary plateau that reaches altitudes from 700 to 900 m. No sedimentary

Fig. 4.20 Drainage (Ipu Waterfall) in one of the largest escarpments of the Ibiapaba Glint, supported by the Ipu Formation. *Photo* Frederico de Holanda Bastos

evidence is noticeable over the surrounding Sertaneja surfaces, except an elongated ridge with a Northeast direction in Santana do Acaraú, constituted by rocks of the Ipu Formation preserved in the Jaibaras Graben (Souza 1988).

4.5.2 Araripe Plateau

The Araripe Basin is the most extensive inland basin of the Brazilian Northeast, whose limit extends beyond the Araripe Plateau, also incorporating the Cariri Valley, totaling approximately 9000 km^2 with an elongated structure in the East–West direction (Assine 1992; Fambrini et al. 2020).

This intracratonic basin presents five stratigraphic sequences: 1. Pre-Rift Sequence (Paleozoic), represented by Cariri Formation (former Mauriti Formation), located in the eastern basin, defining part of the peripheral depression of Araripe; 2. Early Rift Sequence (Neojurassic), represented by the Brejo Santo Formation and the lower part of Missão Velha Formation, also outcropping in the eastern basin; 3. Climax Rift Sequence (Neocomian), constituted by the upper part of Missão Velha and Abaiara Formations, also in the surroundings of the plateau slopes; 4. Post-Rift

Fig. 4.21 Morphological behavior of the Ibiapaba Glint South of the Poti Fault, near Crateús, supported by the Jaicós Formation. *Photo* Frederico de Holanda Bastos

Sequence I (Aptian-Albian), constituted by Barbalha (former Rio Batateiras Formation) and Santana Formations; and 5. Post-Rift Sequence II (Albino-Cenomonanian), composed of the Araripina (former Arajara Formation) and Exu Formations at the top of the Araripe Plateau (Assine 1992; Assine 2007; Fambrini et al. 2020).

The local structural and sculptural influences directly condition the landscapes developed in this geological unit. The typical sedimentary basin reliefs are plateaus, hills, cuestas, glints, and peripheral depressions.

From the geomorphological point of view, the Araripe Basin shows the plateau as the main compartment supported by the sandstones of the Exu Formation, maintaining a structural surface (Fig. 4.22). The plateau topography presents a flattened relief with average altitudes of around 900 m, surrounded by scarps in the most diverse directions (Souza 1988). The porosity and the horizontal disposition of the sandstones of the Exu Formation justify the absence of surface runoff, with practically no fluvial action in terms of notching at the top of the plateau.

This relief evolution relates to the processes of post-Cretaceous topographic inversion as a result of regional uplift, whose surrounding peripheral depressions resulted from regressive erosive processes, with significant evidence of fluvial incisions, washout, and mass gravitational events (Peulvast and Bétard 2015). The circum-denudational retreated the escarpment supported by the Exu Formation. Moreover,

Fig. 4.22 Escarpment in northeastern Araripe Plateau (Crato, Ceará) with the Exu Formation sandstones. *Photo by* Frederico de Holanda Bastos

the only sector with proof of such a formation outside the context of the plateau is the butte, Little Hand Mountain, in Missão Velha.

Although the top of the plateau has practically no surface drainage network, the overlying layers have a slight inclination to the East, favoring the occurrence of river springs and the development of watercourses in the peripheral depressions. Indeed, numerous resurgences occur on the northern slope facing Ceará, which implies the development of humid conditions at the foot of the relief (Souza et al. 1979). The vegetation is mainly characterized by the occurrence of the perennial rainforest, considering the existence of conditions of accentuated humidity, except for some segments.

The area constitutes the principal water disperser of the Salgado River Basin, contributing to the supply of the largest dam in Ceará (Castanhão Dam) and, consequently, to the state water security strategies. The water behavior of this area represents a limiting factor for occupation at the top of the plateau, which is noticeable through the absence of municipal seats and low population density.

4.5.3 Apodi Cuesta

The Potiguar and Araripe Basins are part of a system of Cretaceous rifts of the Northeast whose formation relates to tectonic events of crustal stretching that culminated in the separation of the South American and African plates and the consequent opening of the Atlantic (Bertani 1990; Araripe and Feijó 1994). The lithostratigraphic composition of this basin presents three sequences: (1) Rift Sequence (Lower Cretaceous)—fluvial-deltaic and lacustrine deposits of the Pendência and Pescada Formations; (2) Post-Rift Sequence (Lower Cretaceous)—fluvial-deltaic deposition and the first records of marine ingression with the Alagamar Formation; and (3) Drift Sequence (Tertiary)—transgressive fluviomarine depositions (Açu, Ponta do Mel, Quebradas, Jandaíra, and Ubarana Formations) covered by the regressive depositions (Fm Ubarana, Tibal, and Guamaré) (Pessoa Neto et al. 2007). In the context of Ceará, the outcropping of the Açu and Jandaíra Formations controls the relief.

The sedimentary layers of the Apodi Basin present a slight inclination in the North-Northeast direction, whose monoclinal behavior conditions the formation of a cuestiform relief, with the front preserved by the carbonate rocks of the Jandaíra Formation superposed to the sandstones of the Açu Formation, which constitute a peripheral depression and whose topographic contact with the surrounding hinterland surface occurs without morphological ruptures.

In the evolution of the Apodi escarpment, in the context of Ceará, the role played by the Jaguaribe River, which presents itself as a subsequent river, is noteworthy. In the transition between the middle and lower course, morphological elements indicate a displacement of the bed to the right, contributing to the regressive erosion of the escarpment.

Among the circum-denudational forms analyzed here, the Apodi Cuesta is the one that presents the lowest altimetry (80–150 m) since it is in a sector of post-rift thermal subsidence (Peulvast and Claudino sales 2004). The circum-rifting processes in the surroundings of the Apodi Cuesta exhumed the Cretaceous paleosurfaces fossilized under sediments of the basin, resulting in relief inversion (Peulvast and Claudino Sales 2004). The identification of ancient surfaces in lower terrains (Jaguaribe surface) contributed to questioning the classical models of interpretation of the regional geomorphological evolution based on the pediplanation theory (King 1956), which interpreted the topographic scaling as an evolution of the relief by denudational processes, inferring older ages for higher surfaces.

The reverse of the Apodi cuesta (Fig. 4.23) presents a surface of flat topography with a very gentle slope in the N/NE direction, consisting of limestones of the Jandaíra Formation. In this area, the chemical action allowed the development of outer and inner karsts, whose evolution has a close structural relationship, being more common in the sector in Rio Grande do Norte (Maia et al. 2012). The low dissection results from the permeability presented by the limestone and sandstone terrains that characterize the relief (Souza et al. 1979) and the low altimetry of the area, which does not

Fig. 4.23 Apodi Cuesta in the rainy season with cornice and peripheral depression masked by vegetation. *Photo* Rubson Pinheiro Maia

favor the development of more humid climates, being the environmental conditions similar to the semi-arid planation surfaces.

4.6 Depositional Forms

The depositional forms are features constituted by sedimentation, whose development relates to the erosive processes that modeled the continent, accumulating in lowered sectors. They represent the coastal front (coastal tablelands and plain) and the fluvial plains.

The Atlantic seaboard of the continental margin of Ceará is absent of a great escarpment, thus distinguishing itself from numerous other passive continental margins where such reliefs are present. This geomorphological reality derives from the fact that the final separation of Africa from South America occurred through transformational processes—which do not imply the formation of new terrains or uplift of the coastal zone, such as the rifting that characterizes the separation of southeastern and eastern Brazil.

The fluvial plains are, in turn, characterized by the presence of exorheic rivers. They are intermittent rivers with a fluvial regime commanded by the rainy season, some presenting an ephemeral behavior (wadis). They contain flood valleys in disagreement with the current climatic conditions, which may imply evidence of wetter climates in the past or interdigitations with Quaternary marine transgressions.

The following paragraphs characterize the coastal tablelands, coastal plains, and river plains.

4.6.1 Coastal Tablelands

The coastal zone of Ceará is 573 km long, from the boundary with Rio Grande do Norte in the East to the one with Piauí in the West. In Ceará, the coast is the sedimentary lowland strip, the coastal tablelands, that make the transition between the planation surfaces of the interior and the maritime zone. Its principal morpho-structural characteristic is the presence of azoic sedimentary deposits of continental and coastal origin (Alheiros et al. 1988; Maia 1993; Claudino-Sales 2002) of Miocene age that characterize the Barreiras Formation.

The coastal tablelands are shaped by the action of areal runoff and, close to the sea, by coastal processes commanded by wind and waves and continental processes (deflation, free surface runoff). The fluvial dissection, poorly developed due to the low altitudes and intermittence of watercourses, favors the formation of coastal lakes. However, the modest thickness of sediments in the lower courses makes rivers exhume the lowered basement in which were deposited the Barreiras Formation. Toward the coast, Quaternary sands cover the tablelands, and in contact with the beach strip, the wave action on active cliffs shapes the Barreiras sediments.

In general, coastal tablelands consist of flattened landforms with extensive flat tops. The Barreiras Formation that sustains them presents little lithification with a predominance of pedogenesis processes. They are dissected by a network of channels of low drainage density and dendritic and parallel patterns, forming shallow U-shaped valleys with, at most, 20 m of unevenness. Only West of Camocim and extending to the coast of Piauí, the tablelands are more dissected, frankly carved by a network of channels of moderate drainage density (CPRM 2014).

Represents ancient depositional surfaces, presenting significant gradients toward the coastline. In such a structural feature, relief coincides with the tabular geological structure. They have a continuous original extent and would have been formed in the Neogene, between Miocene and Pleistocene, as sequences of coalescing alluvial fans produced by erosion of the hinterland surface and high massifs and the deposition over a wide range at a marine paleo-level lower than today (Peulvast et al. 2008).

The coastal tableland area presents a sub-humid climate with an average annual precipitation between 1000 and 1400 mm and a significant drought between 4 and 6 months. Thick, sandy and sandy-clayey soils of low natural fertility predominate, such as Quartzarenic Neosol and dystrophic Red-Yellow Acrisol and, subordinately, Yellow Latosol, and Haplic Plinthosol (IBGE-EMBRAPA 2001; CPRM 2014). The caatinga greenery of the surrounding hinterland surface is replaced by lush vegetation, creating the tableland forest (Fig. 4.24).

4.6.2 Coastal Plains

Of current or recent modeling, the coastal plain corresponds to the area of a direct action of the waves, tides, winds, and fluvial flows. From the geomorphological point

Fig. 4.24 Coastal tableland at the rear of a dune field in Beberibe, Ceará. *Photo* Frederico de Holanda Bastos

of view, this unit has a diversity of forms such as inlets, promontories, barriers, fluvial-marine plains, fluvial-lacustrine plains, lagoons, inter-dune ponds, cliffs (Fig. 4.25), and strands of beach rocks (Claudino-Sales 2005), being further characterized by the predominance of sandy materials that demonstrate the action of dry climates, resulting in the formation of barchan, frontal, transverse, and parabolic dune fields.

Two erosive forms predominate in this sector—promontories and cliffs. The first ones, also called headlands, show the influence of Precambrian crystalline materials on the coast associated with the division of Pangea and are represented by Iguape, Mucuripe, Pecém, and Jericoacoara Capes (Claudino-Sales 2005; Costa et al. 2020). The promontories form the zeta (zigzag) coastline, alternating between coastal capes and inlets (Claudino-Sales 2005; Claudino-Sales and Peulvast 2006). The cliffs, in turn, are modeled on the layers of the Barreiras Formation. The headlands, river mouths, and strands of beach rocks create diversification and ruptures throughout the area.

The coastal dynamics receive strong influence from the past and current morpho-structural organization. Currently, the dynamics are under the control of the Southeast and Northeast trade winds (average velocities of 7 and 4 m/s, respectively (Claudino Sales 1993)), Southeast sea waves (1.1 m high and frequency of 5–6 s (MAIA 1998), Northeast swell waves, Northwest-oriented longitudinal currents, and semi-diurnal mesotidal seas (amplitude of about 2.7 m and a level of 1.5 m) (DHN 2022).

Fig. 4.25 7 m cliffs established by sea level variations in the Tertiary sedimentary packages of the Barreiras Formation, Morro Branco Beach, Ceará. *Photo* Frederico de Holanda Bastos

The coastal plain synthesizes the action of climatic, oceanographic, tectonic, and morpho-sedimentary processes (Moreira 1977; Claudino Sales 2005, 2016; Morais et al. 2006; Meireles 2014; Rios et al. 2016; Costa et al. 2020). The zone suffers from marginal flexure and sea level changes, which boost erosion inside the state and deposition on the coast. The area also presents evidence of structural control associated with the Cretaceous evolution of the northeastern continental margin (Claudino-Sales 2016).

The coastal plain (denominated semi-arid coast (Mabesoone 1996)) comprises a narrow strip between the coastline and the coastal tableland. The coastline is rectified and shaped like a wide convex arc of West-Northwest to East-Southeast direction, interrupted by headlands (Fig. 4.26) and river mouths (CPRM 2014). This unit presents a diverse set of depositional relief patterns of wind, fluvial, and marine origins, among which we highlight the dune fields (Fig. 4.27) and the fluviomarine plains (Fig. 4.28).

Wind processes are responsible for the mobile dunes near the coastline, with fixed and vegetated dunes also occurring to the rear, the whole being between 5 and 10 km wide. The dune belt exposes predominant barchan and longitudinal-type forms, which can reach 30 m in height (Claudino-Sales 2007). These eolian formations are built from the reworking of marine sands from the coastal plain and the shallow inner continental shelf, under the dominance of ocean currents and East

Fig. 4.26 Historical photograph (1996) of Pecém Cape, established before the installation of a port, indicating the formation of the littoral barrier on the lower shore, with mobile dune fields within the coastal zone. *Photo* Jean-Pierre Peulvast

Fig. 4.27 Mobile barchan dunes transgress over the mangrove forest of the Jaguaribe River in Aracati. *Photo*: Vanda Claudino-Sales

quadrant trade winds, in a semi-arid climate (CPRM 2014). The fluviomarine plains comprise mangroves, the most developed ones occurring on the banks of the estuaries of the Coreaú, Acaraú, Mundaú, Cocó, Ceará, Piranji, and Jaguaribe Rivers.

The coastal plain has a sub-humid climate with an average annual precipitation between 1000 and 1400 mm and a significant drought between 4 and 6 months (FUNCEME 2022). The rigor of the semi-arid climate does not reach the coast in such an expressive way due to the influence of East trade winds. This unit presents

Fig. 4.28 Fluviomarine plain of Cocó River near its mouth, where it meets the Coaçu River in Fortaleza, Ceará. *Photo* Frederico de Holanda Bastos

an accelerated expansion of urban centers and is subject to the erosive action of both the sea and society.

4.6.3 Fluvial Plains

The fluvial plains cut across all domains—that is, the mountains (crystalline massifs), the hinterlands (planation surfaces), and the coasts (Atlantic seaboard). They are units resulting from the accumulation of alluvial sediments deposited by the action of the drainage network of intermittent character—effectively, there is no perennial river in Ceará. The drainage dividers coincide with the topographic crests represented by the crystalline massifs and sedimentary plateaus.

The fluvial plains follow longitudinally major drainage collectors, especially the Jaguaribe, Acaraú, and Banabuiú Rivers (Fig. 4.29). According to CPRM (2014), they have inexpressive widths upstream. In lower courses, as they carve through the Barreiras Formation, the deposition belt is widened due to the decrease in the fluvial gradient.

The structural control associated with tectonics often conditions the fluvial valleys. Examples are the structural valleys represented by the Jaguaribe and Acaraú Rivers.

Fig. 4.29 Fluvial plain of the Jaguaribe River with the Apodi Cuesta on the horizon. *Photo* Rubson Pinheiro Maia

In the shear zones arranged in the Northeast-Southwest direction, the rivers preferentially follow this direction (Maia et al. 2008). Thus, the regional drainage has a Northeast orientation, normal to the orientation of the coastline, as a response to structural control. According to Souza (1988), these sectors in Ceará constitute areas of regional differentiation in the semi-arid hinterlands for harboring better soil conditions and water availability.

4.7 Final Considerations

The geomorphology of Ceará here presented considered the integration between form and process, classifying the relief according to its denudational, circum-denudational, and depositional characteristics, thus defining the actions of erosion, transport, and sedimentation. In this sense, the present work is based on and extends the geomorphological elements indicated by Costa et al. (2020).

The continental denudation revealed on the surface the structure of the rocks, which suffered attacks from differential erosion. As a result, the depositional forms settled in the lower areas, while the circum-denudational ones retreated, allowing the denudational ones to shape themselves. Despite being based on processes, the examination does not dismiss interpretations about structural and tectonic control. Thus, the analysis presented here shows itself as relevant to understand the landscape of Ceará.

References

Ab'sáber AN (1975) Formas de Relevo. Texto Básico. São Paulo, FUNBEC/Edart, pp 80

Ab'saber AN (1964) Posição das superfícies aplainadas nas paisagens do Nordeste Brasileiro. Notícia Geomorfológica 19:1–38

AGB (1988) Trabalhos de campo. Anais do 3o Congresso Brasileiro de Geografia. Associação dos Geógrafos Brasileiros, Fortaleza

Alheiros MM, Lima Filho MF, Moteiro FAJ, Oliveira Filho JS (1988) Sistemas deposicionais na Formação Barreiras no Nordeste Oriental. In: Anais Do Congresso Brasileiro De Geologia, 35, Belém, PA, Brasil 2, 753–760

Almeida AR, Ulbrich HGJO (2003) Batólito Granítico Quixeramobim-Ce. Aspectos estruturais internos e mecanismo de alojamento crustal. Revista de Geologia UFC, Fortaleza, CE

Almeida FFM, Hasui Y, Brito Neves BBE, Fuck RA (1981) Brazilian structural provinces: an introduction. Earth-Sci Rev 17(1–2):1–29

Arai M, Uesugui N, Rosseti DF, Goes AM (1988) Consideracoes sobre a idade do Grupo Barreiras no Estado do Para. Revista USP 23:17–37

Araripe PT, Feijó FJ (1994) Bacia Potiguar. Boletim de Geociencias da Petrobras 8(1):127–141

Assine ML (1992) Analise Estratigrafia da Bacia do Araripe, Nordeste do Brasil. Revista Brasileira de Geociencias 22(3):289–300

Assine ML (2007) Bacia do Araripe. Boletim de Geociências da Petrobras 15(2)371–389

Bertani J (1990) Sedimentares. Rio de Janeiro: Petrobras. SEREC.CENSUD, pp 291–310

Bastos FH, Cordeiro ANM (2021) Propriedades geomorfologicas das rochas e suas repercussoes no relevo do Nordeste Setentrional Brasileiro. William Morris Davis Revista de Geomorfologia 2(1):1–19

Bétard F, Peulvast JP (2011) Evolução morfoestrutural e morfopedológica do Maciço de Baturité e seu Piemont: do Cretáceo ao Presente. In: Bastos FH (org.) Serra de Baturité: uma visão integrada das Questões ambientais. Fortaleza: Expressão Gráfica e Editora

Betard F, Claudino-Sales F, Claudino-Sales V (2007) Caracterização morfopedólogica de uma serra úmida no semi-árido do Nordeste brasileiro: o caso do maciço de Baturité-CE. Mercator 6:12

Bigarella JJ (2003) Estrutura e Origem das Paisagens Tropicais, vol 3. Florianópolis: Ed. UFSC

Boillot G (1996) Dynamique de la lithosphère. Une introduction à la géologie. Masson, Paris

Brito Neves BB (1999) América do Sul: Quatro fusões, quatro fissõese o processo acrescionário andino. Rev Bras de Geociências 29(3):379–392

Brito Neves BB, Santos EJ, Van Schmus WR (2000) Tectonic history of the Borborema province, northeastern Brazil. In: Cordani et al (Eds) Tectonic evolution of South America, Int. Union of Geol. Sci., Ottawa, Ont., Canada, pp 151–182

Brito Neves BB, Fuck RA, Pimentel MM (2014) A colagem Brasiliana na América do Sul: uma revisão. Brazilian J Geol 4(3):24–39

Brito Neves BB, Fuck RA, Cordani UGE Thomaz-Filho A (1984) Influence of basement structures on the evolution of the major sedimentar basins of Brazil: a case of tectonic heritage. J Geodynam 1:495–510

Caby R, Arthaud MH, Archanjo CJ (1995) Lithostratigraphy and petrostructural caractherization of supracrustais units in the Brasiliano Belt of Northeast Brazil: geodynamics implications. In: Silva Filho AF, Lima ES (eds) Geology of The Borborema Province, pp 235–246

Camaca MSJB (2015) Assinatura Estrutural e Geofísica da Porção Norte (Fronteira Ceará-Piauí) do Lineamento Transbrasiliano: Reativação na Bacia do Parnaíba. Dissertação (Dissertação em Geodinâmica e Geofísica). Centro de Ciências Exatas e da Terra, Universidade Federal do Rio Grande do Norte. Natal, pp 92

Castro DL, Fuck RA, Phillips JD, Vidotti RM, Bezerra FH, Dantas EL (2014) Crustal structure beneath the Paleozoic Parnaíba Basin revealed by airborne gravity and magnetic data Brazil. Tectonophysics 614:128–145

Chang HK, Kowsmann RO, Figueiredo AMF (1992) New concepts on the on the developpment of East Brazilian marginal basins. Episodes 2:194–202

Claudino-Sales V (2016) Megageomorfologia do Nordeste Brasileiro. São Paulo, Novas Edicoes Academicas

Claudino-Sales V (2005) Os Litorais Cearenses. In: Silva JB, Cavalcante TC (Org) ; Dantas, E. W. C. (Org.) . Ceará: um novo olhar geográfico. 1. ed. Fortaleza: Edições Demócrito Rocha

Claudino-Sales V (2002) Les littoraux de l´Etat du Ceara, Bresil: du court terme au long terme. Universite Paris-Sorbonne, Paris, Thesis de Doctorat
Claudino-sales VC, Peulvast JP (2007) Evolução morfoestrutural do relevo da margem continental do Estado do Ceará. Nordeste Do Brasil. Caminhos De Geografia (UFU) 8:2–22
Claudino-Sales V, Peulvast JP (2006) Geomorfologia da zona costeira do Estado do Ceara. In: Dantas EW et al Litoral e Sertão, natureza e sociedade no Nordeste brasileiro. Edições Demócrito Rocha, Fortaleza
Claudino Sales V, Lima EC, Diniz SF, Cunha FSS (2020) Megageomorfologia do Planalto da Ibiapaba, Estado do Ceará: Uma Introdução. William Morris Davis: Revista De Geomorfologia. 1(1):186–209
Costa LRF, Maia RP, Barreto LL, Claudino-Sales V (2020) Geomorfologia do Nordeste Setentrional Brasileiro: uma proposta de classificação. Revista Brasileira De Geomorfologia 21(1):185–208
CPRM (2020) Serviço geológico do Brasil. Mapa Geológico do Estado do Ceará. Escala 1:500.000, Ceará. CPRM
CPRM (2014) Serviço Geológico do Brasil. Geodiversidade do Estado do Ceara. CPRM, Brasilia
CPRM (2003) Serviço geológico do Brasil. Mapa Geológico do Estado do Ceará. Escala 1:500.000, Ceará. CPRM
Cremonini OA (1995) A reativação tectônica da Bacia Potiguar no Cretáceo Superior. V SimpósioNacional de Estudos Tectônicos, Gramado
Cremonini OA, Kender GD (1995) Reativação mesozóica da Bacia Potiguar. *VI Simpósio de Geologia do Nordeste*, Natal
DHN (2002) Tábua de Marés de Fortaleza. Rio de Janeiro: Ministério da Marinha
Fambrini GL, Silvestre DC, Barreto Junior AM, Silva-Filho WF (2020) Estratigrafia da Bacia do Araripe: estado da arte, revisão crítica e resultados novos. Revista do Instituto de Geociências—USP. Série Científica. São Paulo 20(4):196–212
Ferraz NC, Córdoba VC, Sousa DC (2017) Análise estratigráfica da sequência Mesodevoniana-Eocarbonífera da bacia do Parnaíba, Nordeste do Brasil Geociencias. UNESP 36(1):154–172
Fetter A, Van Schmus WR, Santos TJS, Arthaud M, Nogueira Neto JA (1997) Geocronologia e estruturação do Estado do Ceará: NW da Província da Borborema, NE Brasil. XVII Simpósio de Geologia do Nordeste, Fortaleza
Fetter AH ,Van Schmus WR, Santos TS, Nogueira N, José A, Henriarthaud M (2003) U-Pb and Sm-Nd geochronological constraints on the crustal evolution and basement architecture of Ceará State, NW Borborema Province, NE Brazil: implication for the existence of the Paleoproterozoic Supercontinent "Atlantica". Revista Brasileira de Geociências, São Paulo 30(1):102–106
Funceme (2022) Sistema de Informacao. Fortaleza: FUNCEME
Goés AMO, Feijó FJ (1994) Bacia do Parnaíba. Boletim De Geociências Da PETROBRÁS 8:57–67
Gurgel SPP, Bezerra FHR, Corrêa ACB, Marques FO, Maia RP (2013) Cenozoic uplift and erosion of structural landforms in NE Brazil. Geomorphology (Amsterdam) 186:68
IBGE (2001) EMBRAPA SOLOS. Mapa de Solos do Brasil. Rio de Janeiro, 1 mapa color., 107 cm x 100 cm. Escala 1:5.000.000
King LC (1956) A geomorfologia do Brasil oriental. Revista Brasileira De Geociências 18(2):147–265
Lima DL, De S, Bastos F, De H, Cordeiro AMN, Maia RP (2019) Geomorfologia granítica do Maciço de Uruburetama, Ceará, Brasil. Revista Brasileira de Geomorfologia 20(2): 73–395
Mabesoone JM (1996) Relief of Northeastern Brazil and its correlated sediments. Zeif F Geomorph 10:419–453
Mabesoone JM, Castro C (1975) Desenvolvimento geomorfológico do nordeste brasileiro. Boletim Do Núcleo Do Nordeste Da Sociedade Brasileira De Geologia, Recife 3:5–37
Maia LP (1998) Procesos costeros y balance sedimentario a l o largo de Fortaleza (NE-Brasil): Implicaciones para una gestión adecuada de la zona litoral. Universitat de Barcelona, Tesis Doctoral, p 1998
Maia LP (1993) Controle Tectônico e evolução geológica/sedimentar da região da desembocadura do Rio Jaguaribe, Ceará. Dissertação de Mestrado, Departamento de Geologia, UFPE

Maia RP (2012) Geomorfologia e Neotectônica na Bacia Hidrográfica do Rio Apodi-Mossoró RN. Tese de Doutorado, Departamento de Geologia, UFRN

Maia RP, Bezerra FHB (2014) Tópicos de geomorfologia estrutural: nordeste brasileiro. Fortaleza: Edições UFC

Maia RP, Nascimento MJL (2018) Relevos graniticos do Nordeste brasileiro. Revista Brasileira de Geomorfologia 19(2):373–389

Maia RP, Sousa MOL, Bezerra FHR, Neto PX, Lima ENM, Silva CCN, Santos RD (2012) A importância do controle tectônico para a formação da paisagem cárstica na bacia Potiguar, Nordeste do Brasil. Revista Brasileira De Geomorfologia 13(4):451–463

Maia RP, Bezerra FH, Claudino-Sales VC (2008) Vales Fluviais do NE: Considerações Geomorfológicas. Geografi a em Debate (UFPB), Okara

Marinho M, Mascle J (1987) Evolução estrutural do platô marginal da Guiné e sua relação com a formação do Oceano Atlântico Centro-Equatorial. Revista Brasileira De Geociências 17:111–117

Matos RMD (2000) Tectonic evolution of the Equatorial South Atlantic American Geophysical Union. Geophysical Monograph 115:331–354

Matos RMD (1992) The Northeast Brazilian rift system. Tectonics 11(4):766–779

Meireles AJA (2005) Geomorfologia Costeira: funcoes ambientais e sociais. Fortaleza: Editora Universidade Federal do Ceara

Meireles AJA (2005) As unidades morfoestruturais do Ceará. In: Silva JB, Cavalcante TC, Dantas EWC et al (Org). Ceará: um novo olhar geográfico. Fortaleza: Edições Demócrito Rocha

Milani EJ, Zalán PV (1999) An outline of the geology and petroleum systems of the Paleozoic inteiror basins of South American. Episodes 22(3):199–205

Morais JO, Freire GS, Pinheiro LS, Souza MJN, Carvalho AM, Pessoa PRS, Oliveira SHM (2006) Ceará. In: Muehe D (Org) Erosão e Progradação do Litoral Brasileiro. 154 ed. Brasília: Ministério do Meio Ambiente

Moreira AAN (1973) Geologia, Geomorfologia. Atlas do Ceará. Governo do Ceará–IBGE. Rio de Janeiro

Moreira AAN (1977) Relevo. In: IBGE. Geografia do Brasil: região Nordeste. Rio de Janeiro, SERGRAF, 5, p 1–45

Moreira MMM, Gatto LCS (1981) *Geomorfologia.* In: PROJETO RADAMBRASIL. FOLHA FORTALEZA, 21, pp 213–252. DNPM/MME, Brasília

Murphy JB, Nance RD (2013) Origins of the supercontinent cycle. Geosci Front 4(4):439–448

Murphy JB, Nance RD (1995) Montagnes et super-continents. Pour la Science, Dossier hors-série, Paris, pp 46–53

Neves SP (2012) Granitos Orogênicos: da geração dos magmas à intrusão e deformação. Synergia, Rio de Janeiro

Nogueira Neto JA, Santos TJS, Arthaud MH, Parente CV (1991) Condições PT e soerguimento de granulitos (Kingizitos) adjacentes à zona de cisalhamento de Orós-Sistema Orós-Jaguaribe (NE do Brasil). *VAnais do VII Simpósio Nacional de Estudos Tectônicos*, Bahia

Oliveira DC, Mohriak WU (2003) Jaibaras trough: an important element in the early tectonic evolution of the Parnaíba interior sag basin, Northern Brazil. Mar Pet Geol 20:351–383

Pessoa Neto OC, Soares UM, Silva JGF, Roesner EH, Florencio CP, Souza CAV (2007) Bacia Potiguar. Boletim de Geociências da Petrobras 5(2):357–369

Peulvast JP, Bétard F (2015) A history of basin inversion, scarp retreat and shallow denudation: the Araripe basin as a keystone for understanding long-term landscape evolution in NE Brazil Amsterdam. Geomorphology 233:20–40

Peulvast JP, Claudino Sales V (2004) Stepped surfaces and Paleolandforms in the Northern Brasilian <Nordeste>: Constraints on models of morfotectonic evolution. Geomorphology 3:89–122

Peulvast JP, Claudino Sales V (2003) Mapa morfoestrutural do Ceará e áreas adjacentes do Rio Grande do Norte e da Paraíba. In: Brasil. Atlas digital de geologia e recursos minerais do Ceará. Mapa na escala 1:500.000. Fortaleza: Serviço Geológico do Brasil

Peulvast JP, Claudino-Sales V, Betard F, Gunnell Y (2008) Low post-Cenomanian denudation depths across the Brazilian Northeast: implications for long-term landscape evolution at a transform continental margin. Global Planet Change 62(1):39–60

Peulvast JP, Claudino-Sales V, Bezerra FHR, Betard F (2006) Landforms and Neotectonics in the Equatorial Passive Margin of Brazil. Geodinamica Acta 19(1):51–71

Popoff M (1988) Du Gondwana à l'Atlantique sud: les connexions du fossé de la Bénoué avec les bassins du Nord-est brésilien jusqu'à l'ouverture du Golfe de Guiné au Crétacé Inférieur. J Afric Earth Sci 7(2):p. 409–431

Radambrasil (1981) FOLHA SB.23/24 JAGUARIBE/NATAL: geologia, geomorfologia. Rio de Janeiro

Rios VPL, Amaro VE, Vieira MM, Matos FM, Prudêncio MC, Camara MC (2016) Influência neotectônica na morfologia do sistema de ilhas barreiras, Nordeste do Brasil. Revista Brasileira de Geomorfologia 17(3)

Shobbenhaus C, Brito Neves BB (2003) A Geologia do Brasil no contexto da Plataforma Sul-americana. In: Bizzi LA, Shobbenhaus C, Vidotti RM, Gonçalves JH (eds) Geologia, Tectônica e Recursos Minerais no Brasil. CPRM, Brasília

Silva MB, Maia RP (2017) Caracterização morfoestrutural do alto curso da bacia hidrográfica do rio Jaguaribe. Ceará-Brasil Revista Brasileira De Geomorfologia 18(3):637–655

Silva AJP, Lopes RC, Vasconcelos AM, Bahia RBC (2003) Bacias Sedimentares Paleozóicas e Meso-Cenozóicas Interiores. In: Bizzi LA, Schobbenhaus C, Vidotti RM, Gonçalves JH (eds) Geologia, Tectônica e Recursos Minerais do Brasil. Brasília: CPRM

Souza MJN (1997) Geomorfologia. In: Atlas do Ceará, IPLANCE, Fortaleza, pp 14–15

Souza MJN (1988) Contribuição ao estudo das unidades morfoestruturais do Estado do Ceará. Revista de Geologia da UFC. Fortaleza 1:73–91

Souza MJN (1979). Compartimentação Topográfica do Estado do Ceará. Revista Ciência Agronômica, Fortaleza 9:31–52

Souza MJN, Lima FA, Paiva JB (1979) Compartimentacao topográfica do Estado do Ceara. Revista Ciencia Geografica 9(12):77–86

Szatmari P, Françolin JBL, Zanotto O, Wolf S (1987) Evolução Tectônica da margem equatorial brasileira. Revista Brasileira De Geociências 17(2):180–188

Teixeira JBG, Misi A, Silva MG (2007) Supercontinent evolution and the proterozoic metallogeny of South America. Gondwana Res 11:346–361

Trindade IV, Martins Sá J, Macedo MHF (2008) Comportamento de elementos químicos em rochas mineralizadas em ouro na Faixa Seridó, Província Borborema. Revista Brasileira de Geociências 38

Vauches A, Neves S, Caby R, Corsini M, Egydio-Silva M, Arthaud M, Amaro V (1995) The Borborema shear zone system, NE Brazil. J South Am Earth Sci 8:247–266

Vaz PT, Rezende NGAM, Wanderley Filho JR, Travassos WAS (2007) Bacia do Parnaíba. Boletim De Geociências Da Petrobrás 15(2):253–263

Zalán PV (1999) Evolucao fanerozoica das bacias sedimentares brasileiras. Boletim de Geociencias da Petrobras, pp 595–612

Vanda Claudino-Sales She holds a degree in geography from the University of Brasilia (UnB), an specialization in coastal geology at the Federal University of Rio Grande do Sul (UFRGS), a master's degree in physical geography at the University of São Paulo (USP), a Ph.D. in environmental geography at the Université Paris-Sorbonne, and post-doctoral studies in coastal geomorphology at the University of South Florida. She is a professor of the Department of Geography of the Federal University of Ceara (UFC) and was a visiting professor at the University of Florida, at the University Paris-Sorbonne, and the geography master's program of the Acaraú Valley State University (UVA). She is the regional vice president of the Coastal Education and Research Foundation (CERF), and has experience in structural geomorphology, mega-geomorphology, coastal geomorphology, and environmental analysis, and is an environmentalist, having been in the movement for the preservation of the environment in Brazil for over 40 years.

Rubson Pinheiro Maia He is a geographer (UECE), holding a master's degree in physical geography with an emphasis on geomorphology (UFC) and a Ph.D. in geodynamics and geophysics (UFRN). He works in the area of geomorphology with an emphasis on morphotectonics, having worked on morphotectonics of karstic environments, fluvial systems, deformation zones, and crystalline massifs. He participated in the project of geological and geomorphological mapping of the sheet SB-24-X-D-I (Mossoró-RN) funded by CPRM (Geological Service of Brazil) and the project mapping/characterization of karst (Porocarste) in the Potiguar Basin funded by Petrobras. He participates in the Porocarste 3D project funded by Shell and integrates the INCT (National Institute of Science and Technology) in the area of tectonic studies. He taught courses related to geoscience at UFC (2005), UECE (2006/2007), FATEC-CE (2006/2008), and UFRN (2011/2013). He is a geomorphology professor at the Federal University of Ceará and a CNPq research productivity fellow (level 2).

Frederico de Holanda Bastos He has a post-doctorate in physical geography, a Ph.D. in geography, and a master's degree in development and environment from the Federal University of Ceará (UFC). He has a specialization in geoprocessing applied to environmental analysis and water resources and a bachelor's degree in geography from the State University of Ceará (UECE). He is an adjunct professor of the undergraduate courses in geography at UECE, a permanent faculty member of the Graduate Program in Geography (ProPGeo/UECE), and a CNPq research productivity fellow (level 2). He has experience in geoscience, working on geomorphology, environmental geography, and geoprocessing applied to environmental studies.

Chapter 5
Geomorphology of Rio Grande do Norte State

Marco Túlio Mendonça Diniz, George Pereira de Oliveira, Rubson Pinheiro Maia, and Bruno Ferreira

Abstract The geomorphological mapping of the state of Rio Grande do Norte largely followed the methodology adopted by Santos et al. (Revista Brasileira de Geomorfologia 7:03–12, 2006) for the state of Paraná, in order to update knowledge of regional geomorphological mapping. The scale adopted was 1:250,000, which allowed the first three taxons to be used, i.e., the cartographic representation of morphostructural units, morphocultural units, and morphocultural subunits. The first taxon identified the morphostructural units: Brazilian Orogenic Belt; Marginal Sedimentary Basins; Cenozoic Volcanism and/or Plutonism; and Quaternary Sedimentary Covers. The second taxon identified the following morphological and cultural units: Sertaneja Depression; Borborema Plateau; Massifs and Inland Plateaus; Coastal Plateaus and Tablelands; Tectonic Reliefs in Marginal Basins; Reliefs Associated with Neogene Volcanism/Plutonism; and Coastal and River Plains. There were thirty-four morphological and cultural subunits in the third taxon. The aim of this work is to help plan and organize the territory of the state of Rio Grande do Norte.

Keywords Geomorphological mapping · Rio Grande do Norte map · Applied geomorphology

M. T. M. Diniz (✉)
Federal University of Rio Grande do Norte, Caico, Brazil
e-mail: tulio.diniz@ufrn.br

G. P. de Oliveira · B. Ferreira
Federal University of Pernambuco, Recife, Brazil
e-mail: georgesb@bol.com.br

B. Ferreira
e-mail: bruno.ferreira@igdema.ufal.br

R. P. Maia
Federal University of Ceara, Fortaleza, Brazil
e-mail: rubsonpinheiro@yahoo.com.br

V. Claudino-Sales and J. F. Sobrinho (eds.), *Geomorphology of the Northeast Region of Brazil*, The Latin American Studies Book Series,
https://doi.org/10.1007/978-3-031-58874-7_5

5.1 Introduction

The geomorphological studies both on detailed and regional scales are significant contributions of Physical Geography to society. Since they clarify the dynamic and evolutionary possibilities of the landscape through spatialization, hierarchization, and explanation of the landforms and their geomorphic predecessors, they become useful to the scientific community and to agents responsible for planning and territory management (Martins and Rodrigues 2016).

As the relief results from the interaction among the lithostructural framework, tectonic events, and climate, ample knowledge thereof is indispensable when it comes to integrated studies in the academic and technical scopes. Related to the latter, environmental planning is highlighted as an important set of necessary measures and actions for the current land use and occupation activities (Santos 2004).

According to Santos (2004), geomorphological knowledge is crucial to environmental planning, since it allows the interpretation of the relationship among the superficial configurations of the terrain, the spatial distribution of the population center, and the different types of soil use derived from the limitations imposed by the relief. The author also emphasizes the adoption of the relief variable, in most cases, as the reference subject in studies of environmental planning and in determinations of managerial spaces, thus converging other relevant subjects for the elaboration of diagnoses that integrate information derived from Geomorphology.

Amidst the mapping methodologies used in Brazil, the proposal of Ross (1992) stands out. It was based on the methodology created by the team of Getulio Vargas Barbosa, geomorphologist and director of the RADAM Project division, which was later expanded to the RADAMBRASIL project. This methodology obtained greater acceptance by the scientific community of Brazilian geomorphologists, as stems from the understanding of the interactions between exogenous and endogenous factors that mold the terrestrial surface. Ross (1992) proposal creates relief classification levels based on the hierarchy of the *geoforms* assemblages encompassed by each other and adopts the well-discussed concepts of morphostructure and morphosculpture, coined by the Russian School.

This methodology has been applied in several areas and at different scales in the Brazilian territory, as in studies of Ross and Moroz (1996), Santos et al. (2006), Nascimento and Souza (2010), Martins and Rodrigues (2016), to name a few. It consists of an effort to standardize the geomorphological cartographic base of Brazil, and it has been progressively used by governmental institutions of planning, without significant alterations (e.g., map of Relief Units of Brazil (IBGE 2006) on the 1:5.000.000 scale; and Technical Guide of Geomorphology (IBGE 2009)).

Studies of regional geomorphological mapping for the State of Rio Grande do Norte are scarce. Among the few existing works, the publication of Dantas and Ferreira (2010) can be cited; however, the used mapping scale is not specified and the relevance of the morphostructural factors in the actual relief setting is mostly neglected. Another work that deserves attention is Maia et al. (2013), in which a relief classification proposal for the State of Rio Grande do Norte is presented, emphasizing the morphostructural and morphotectonics interpretation of the actual landforms.

It is therefore imperative to update the State of Rio Grande do Norte's geomorphological cartographic base with a systematic methodology whereby the results become a reference for other base mappings and regional planning policies. From this premise, the present study aims to present a geomorphological mapping proposal for the State of Rio Grande do Norte (1:250,000) in accordance with the model of geomorphological compartmentalization (Ross 1992 adapted by Santos et al. 2006).

5.2 Materials and Methods

5.2.1 Study Area

The area of the State of Rio Grande do Norte (RN) is situated between the coordinates 4° 50′ S and 7° S/35° W and 38° 30′ W, bounded in the south by the State of Paraíba, in the west by the State of Ceará, and in the north by the Atlantic Ocean (Fig. 5.1). The climate is dominated by ca. 80% of semi-arid subtypes. Only two drainage basins are of greater importance—Piranhas-Açu and Apodi-Mossoró—both of intermittent regimes.

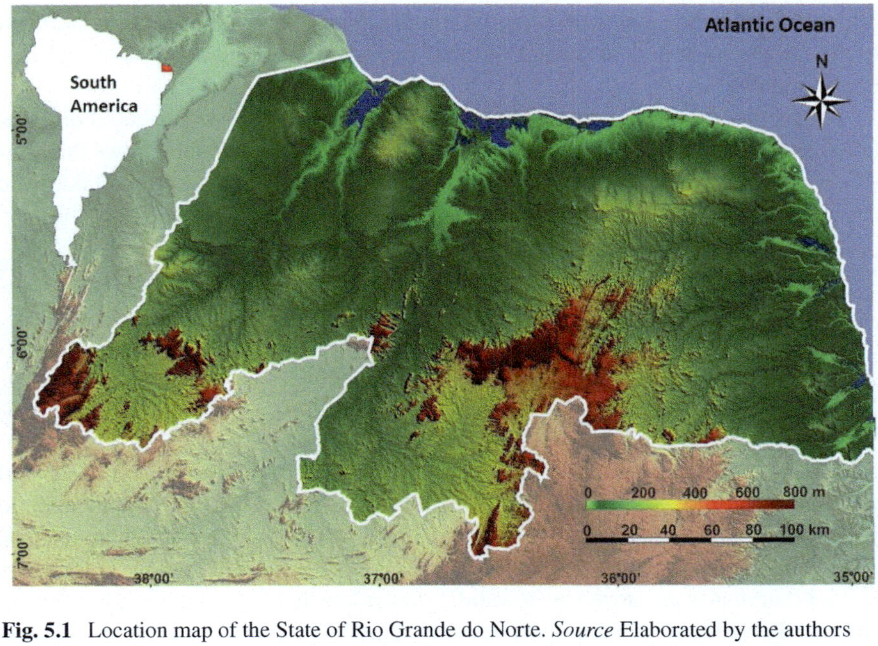

Fig. 5.1 Location map of the State of Rio Grande do Norte. *Source* Elaborated by the authors

5.3 Methodological Procedures

The elaboration of this work encompassed the following procedures: (1) bibliographic and cartographic revision on the landforms of Rio Grande do Norte in North-east of Brazil; (2) elaboration of the geomorphological map in SIG software; (3) fieldwork activities to validate the results obtained by geoprocessing and remote sensing methods and to obtain photographic records of the study area.

The geomorphological mapping was carried out initially by the development of a base map derived from the IBGE digital state mesh of Rio Grande do Norte at the scale of 1:250,000. The geological mesh used is related to the geological mapping of the State of Rio Grande do Norte (1:500,000) performed by Angelim et al. (2006).

The map elaboration involved the acquisition of 90 m-resolution SRTM (Shuttle Radar Topographic Mission) altimetric data, obtained from EMBRAPA (Brazilian Agricultural Research Corporation) through the Project Brasil em Relevo. The following grids (on a 1:25,000 scale) were provided: SB-24-X-A, SB-24-X-B, SB-24-X-C, SB-24-X-D, SB-25-V-C, SB-24-Z-A, SB-24-Z-B e SB-25-Y-A.

The SRTM grids were reprojected from WGS 1984 datum to SIRGAS 2000 datum (Geographic coordinate system), on ArcGIS 10.3 Software. The scenes were mosaicked thereafter, forming a single raster, from which the mapping variables were analyzed. In this regard, topographic contour lines were extracted, relief shading and slope data were extracted, and topographic profiles were traced. These data were layered on the geological mesh to delineate and analyze relief compartments.

For the identification of these compartments, the relation of relief levels in the State of Rio Grande do Norte with different tectonic reactivation episodes, crucial to the arrangement and constraint of their dynamics and evolution, was taken into consideration.

The methodology of hierarchization of the different relief taxa is based on Ross (1992) with nomenclature adaptions by Santos et al. (2006), which allows the classification of six relief taxonomic units. In this study, only the first three taxa were applied, due to the scale adopted (1:250,000). They are hereafter described:

(1) **Morphostructural Units**—defined by megastructures that constrain the regional relief.
(2) **Morphoscultural Units**—resultant from the tectonic-climatic processes acting on the lithostructural framework, directly affecting the current relief's evolution.
(3) **Morphoscultural Subunits**—relief assemblages distinguished within the morphoscultural units on account of their different morphology and texture.

The color scheme used followed the recommendations of the Technical Manual of Geomorphology (IBGE 2009), in which the shades of orange-brown, green, and yellow are applied respectively to Orogenic Belts; Phanerozoic sedimentary basins, and Quaternary sedimentary deposits.

Field surveys were carried out to validate and refine the results obtained by geoprocessing. In addition, photographic records regarding several relief compartments mapped were acquired.

A scale of 1:250,000 was used for the final map, and its layout was designed in ArcGIS 10.3 Software. Additional information obtained by SRTM data analysis, such as the dimension of the mapped units, and the elevation range, were organized in a table, elaborated by the worksheet editor in Microsoft Excel 2010.

5.4 Results and Discussion

According to Maia and Bezerra (2014), the primary elements constraining the geomorphological evolution of the State of Rio Grande do Norte (Fig. 5.2) result from a sequence of three tectonic events: Brasiliano Orogeny, fragmentation of Gondwana megacontinent, and Cenozoic tectonic reactivations. The current morphostructural units identified in this area and related morphoscultural units and subunits are clearly reflected in their morphology pattern and associated evolution.

Four morphostructural units (1st taxon), seven morphoscultural units (2nd taxon), and twenty-nine morphoscultural subunits (3rd taxon) were identified in the State of Rio Grande do Norte. The number of recognized compartments attests to the diversity of the lithostructural and morphoclimatic framework of the region (Table 5.1).

In the following sections, the concepts of morphostructural and morphoscultural units are used in a methodological context. It is important to note that the relief is primarily based on the several tectonic reactivation episodes that affected the geomorphological setting of the study area.

5.5 Brasiliano Mobile Belt

The Brasiliano Mobile Belt in Rio Grande do Norte has an area of ca. 32,818.5 km^2 and occupies 61.12% of the State's territory. It corresponds primarily to Precambrian basement fold belts, resulting from the polyorogenic events that affected the Borborema Province during the Proterozoic, especially the Brasiliano/Pan-African Cycle and the Cretaceous reactivation, which ultimately culminated in the break-up of the Gondwana megacontinent. Further, this unit also exhibits a few Archaean nuclei (Angelim et al. 2006).

The lithological diversity is a prominent attribute of this Belt. Its eastern zone is composed of the Bom Jesus Archaean Nucleus and Paleoproterozoic metamorphic complexes; in the central–western zone, an assemblage of Neoproterozoic shelf turbidite metasedimentary rocks of the Seridó Belt and Paleoproterozoic metavolcanic-sedimentary rocks of the Rio-Piranhas Basement are found. The latter is marked by a series of metaplutonic intrusions. The westernmost area corresponds to the outcrop of the Jaguaribe Belt and is constituted of Paleoproterozoic gneiss-migmatite basement rocks. This belt covers a narrow zone in the RN State but extends to the Ceará State.

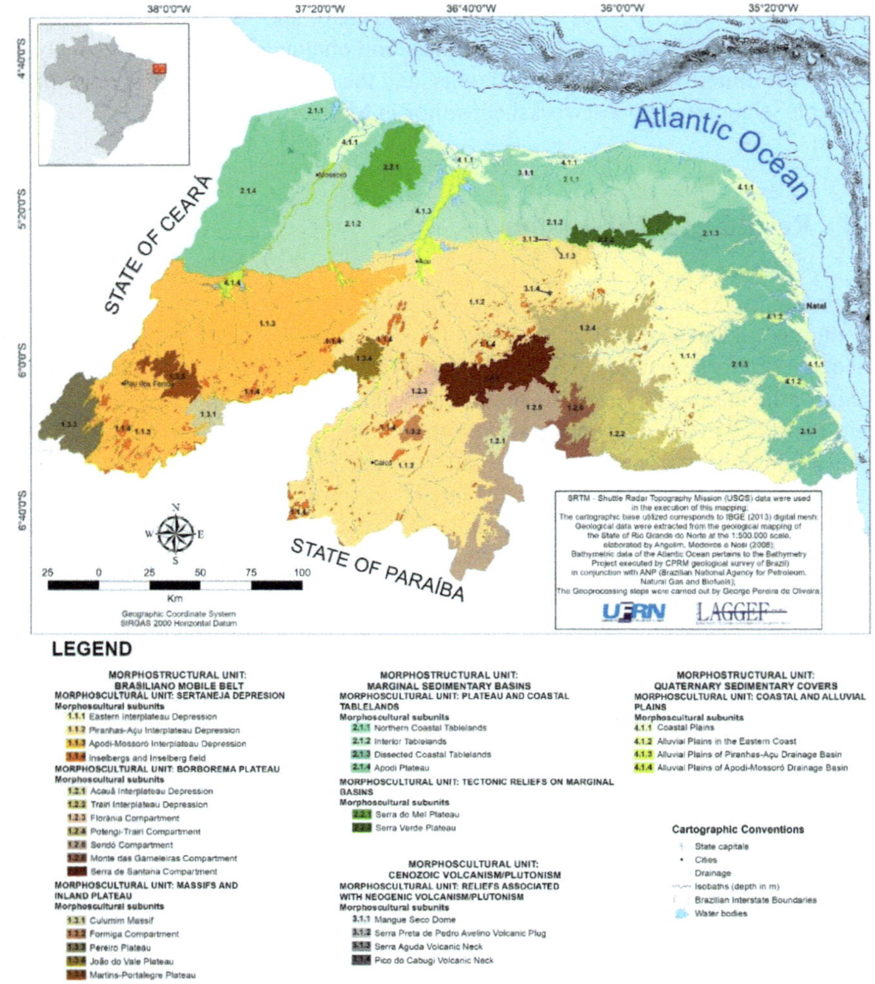

Fig. 5.2 Geomorphological map of the State of Rio Grande do Norte. *Source* Elaborated by the authors

A large extent of this unit is marked by planation surfaces, which correspond to the Sertaneja surface named by Peulvast and Claudino Sales (2004). The Brasiliano Mobile Belt also encompasses a number of Precambrian remobilized massifs that, due to differential erosion processes and tectonic reactivations, have remained in the landscape as high landforms. These massifs reflect Cretaceous tectonic reactivations, related to the Gondwana break-up and the South Atlantic Ocean. Paleoproterozoic and Neoproteroic basement metamorphic supracrustal belts, in addition to the Brasiliano-age plutonic intrusions, are the main rock assemblages supporting these massifs (Angelim et al. 2006).

Table 5.1 Summary of the primary geomorphological aspects of the mapped units and subunits

Morphostructural unit (area Km^2/%)	Morphoscultural unit (area Km^2/%)	Morphoscultural subunit	Elevation (meters a.s.l)			Total area (Km^2/%)
			Min.	Max.	Elevation range	
Brasiliano Mobile Belt (32,818.5 km^2/ 61.12%)	Sertaneja Depression (24,429.5 km^2/ 45.45%)	1.1.1 Eastern Interplateau Depression	40	200	160	5647 (10.5%)
		1.1.2 Piranhas-Açu Interplateau Depression	50	330	280	10,532.5 (19.6%)
		1.1.3 Apodi-Mossoró Interplateau Depression	50	400	350	7413 (13.8%)
		1.1.4 Inselbergs and Inselberg field	50	720	670	837 (1.55%)
	Borborema Plateau (6649 km^2/12.44%)	1.2.1 Acauã Interplateau Depression	290	610	320	144 (0.26%)
		1.2.2 Trairi Interplateau Depression	170	550	380	876 (1.62%)
		1.2.3 Florânia Compartment	110	710	600	270.5 (0.5%)
		1.2.4 Potengi-Trairi Compartment	140	660	520	2302 (4.3%)
		1.2.5 Seridó Compartment	270	800	530	1738 (3.23%)
		1.2.6 Monte das Gameleiras Compartment	320	680	360	366 (0.68%)
		1.2.7 Serra de Santana Compartment	140	760	620	997.5 (1.85%)
	Massifs and Inland Plateaus (1740 km^2/ 3.23%)	1.3.1 Culumim Massif	220	750	530	253 (0.47%)
		1.3.2 Formiga Massif	210	690	480	102 (0.2%)
		1.3.3 Pereiro Plateau	260	860	600	752 (1.4%)
		1.3.4 João do Vale Plateau	80	750	670	268 (0.49%)
		1.3.5 Martins-Portalegre Plateau	170	750	580	365 (0.67%)

(continued)

Table 5.1 (continued)

Morphostructural unit (area Km^2/%)	Morphosculptural unit (area Km^2/%)	Morphosculptural subunit	Elevation (meters a.s.l)			Total area (Km^2/%)
			Min.	Max.	Elevation range	
Marginal Sedimentary Basins (18,150 km^2/ 33.73%)	Plateaus and Coastal Tablelands (16,884 km^2/ 31.38%)	2.1.1 Northern Coastal Tablelands	0	50	50	4768 (8.86%)
		2.1.2 Interior Tablelands	10	200	190	5663 (10.53%)
		2.1.3 Dissected Coastal Tablelands	40	150	110	3745 (6.96%)
		2.1.4 Apodi Plateau	40	270	230	2708 (5.03%)
	Tectonic Reliefs on Marginal Basins (1266 km^2/2.35%)	2.2.1 Serra do Mel Plateau	90	280	190	861 (1.6%)
		2.2.2 Serra Verde Plateau	190	320	130	405 (0.75%)
Cenozoic Volcanism and/or Plutonism (37.5 km^2/0.1%)	Reliefs associated with Neogenic Volcanism/ Plutonism (37.5 km^2/0.1%)	3.1.1 Mangue Seco Dome	50	120	70	28 (0.05%)
		3.1.2 Serra Preta de Pedro Avelino Volcanic Plug	120	240	120	4 (0.02%)
		3.1.3 Serra Aguda Volcanic Neck	130	260	130	1.5 (0.01%)
		3.1.4 Pico do Cabugi Volcanic Neck	190	560	370	4 (0.02%)
Quaternary Sedimentary Covers (2722 km^2/ 5.05%)	Coastal and Alluvial Plains (2722 km^2/ 5.05%)	4.1.1 Coastal Plains	0	120	120	1755 (3.26%)
		4.1.2 Alluvial Plains in the Eastern Coast	0	60	60	282 (0.52%)
		4.1.3 Alluvial Plains of Piranhas-Açu Drainage Basin	0	180	180	437 (0.81%)
		4.1.4 Alluvial Plains of Apodi-Mossoró Drainage Basin	0	80	80	248 (0.46%)

Source Elaborated by the authors

Besides being remnants of ancient geotectonic cycles, recent studies based on geophysical data (Oliveira 2008; Oliveira and Medeiros 2012) reported that the preservation of the massifs in the landscape may be associated with disequilibriums of the lithosphere–asthenosphere and/or crust–asthenosphere systems, resulting from the Cenozoic intraplate magmatic underplating.

The consequential density contrasts may be triggering isostatic compensation mechanisms, responsible for epirogenesis and recent volcanism. Evidence of Cenozoic movements was identified by Gurgel et al. (2013) in the Pereiro Massif, where are seen signs of Cenozoic tectonic processes that generated accommodation spaces to the accumulation of colluvial deposits.

The NE-SW striking shear zones, inherited from the Brasiliano-Pan-African Orogeny, play a key role in the dissection of both planation areas and remobilized massifs, since they control drainage networks and, therefore, erosion patterns (Maia and Bezerra 2014).

In the southern limit of the eastern sector of the Potiguar Basin, an E-W elongated plateau relief was developed, named Chapada da Serra Verde. In this sector, the basin relief rises from 150 to 250 m a.s.l in the most elevated zones. This structural elevation coincides with the contact zone between the basin and the Precambrian basement, specifically where the Picuí-João Câmara Shear Zone (PJCSZ) cuts the Potiguar Basin. In the basement, the relief along the OJCSZ is characterized by a system of N-NE structural ridges that extend to the north until the Ceará-Mirim river valley, which superimposes these structures.

The PCJSZ belongs to one of the most active fault systems in the Brazilian Northeast, and therefore, the effects of its reactivation may have a direct repercussion on sedimentary environments in the form of structural control on the dissection or the formation of elevated segments, associated with tectonic movements.

Three morphoscultural units are distinguished in the Brasiliano Mobile Belt: Sertaneja Depressions, the Borborema Plateau, and the Massifs and Interior Plateaus (Fig. 5.3).

The **Sertaneja Depressions** (24.429.5 km^2/45,45% of the RN state's territory) correspond to the morphoscultural unit that comprises the planation areas of the Brasiliano Belt, where pronounced denudation processes predominate. They are extensive planation surfaces where the monotony of lowered relief is broken by isolated elevations, commonly built of rocks more resistant to erosion than its surroundings (inselbergs), such exhumed as plutonic intrusions (Fig. 5.4). The genesis of these depressions is related to successive episodes of climatic variations during the Cenozoic.

The dissection in the depressions is driven by a drainage network that clearly follows the orientation of the main ductile and brittle structures of the crystalline basement mainly NE-SW.

Four morphoscultural subunits were identified in the area of the Sertaneja Depressions in the RN: 1.1.1 Eastern Interplateau Depressions, 1.1.2 Piranhas-Açu Interplateau Depression, 1.1.3 Apodi-Mossoró Interplateau Depression, and 1.1.4 Inselbergs and inselberg fields.

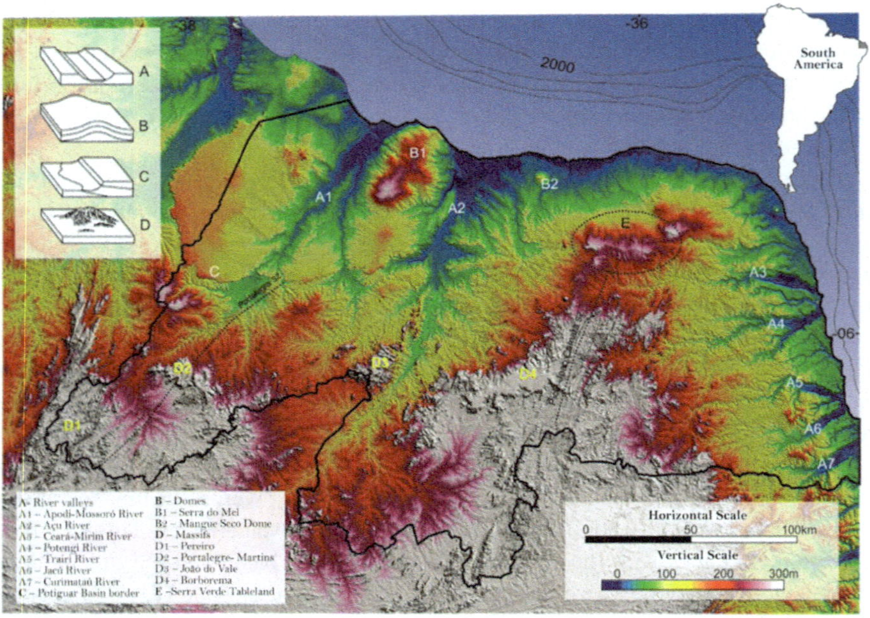

Fig. 5.3 Relief units and major shear zones in Rio Grande do Norte. *Source* Elaborated by the authors

The **Borborema Plateau** (6649 km^2/12.44% of the RN state's territory) comprises an assemblage of highlands located in the easternmost Northeastern of Brazil and whose limits are marked by a series of topographic unevenness, with amplitudes around 100 m in relation to the surrounding areas (Corrêa et al., 2010). The genesis of this plateau is associated with the break-up of the Gondwana Megacontinent ad with the Cenozoic intraplate magmatism (Corrêa et al. 2010; Oliveiras and Medeiros 2012), as aforementioned.

In the State of Rio Grande do Norte, this plateau is primarily composed of Neoproterozoic supracrustal metamorphic rocks of the Seridó Group, in addition to a belt of Paleoproterozoic metamorphic rocks of the Santa Cruz Complex, in its eastern portion. In contrast, a few areas of the Borborema Plateau are capped by a 30 m thick sedimentary cover that supports small tablelands (Fig. 5.5) and residual hills (named Morro testemunho in Portuguese), which are not mapped in this study due to the mapping scale adopted.

This cover corresponds to conglomerate sandstones of the Serra do Martins Formation (SMF), dated to the Paleogene–Neogene interval (Morais Neto et al. 2008; Lima 2008). The most frequently reported SMF incidence in this area in the literature regards to the Serra de Santana Compartment, the longest tableland where altitudes are over 700 m. In fact, the 630 m contour line extends to all the SMF outcrops in other relief units in the State of RN.

Fig. 5.4 Interplateau depressions in the State of RN. **a** Apodi-Mossoró Interplateau Depression; **b** Eastern Interplateau Depression; **c** Piranhas-Açu Interplateau Depression. *Source* Authors' archive

Fig. 5.5 Borborema Plateau in the State of Rio Grande do Norte. **a** In the background, the western hillside of the Borborema in outcrops of the Acari granite with talus deposit; in the foreground is the Gargalheiras reservoir built in the terrains of its small gorges and in the Acauã Interplateau Depression. **b** In the background, Serra de Santana supported by sandstones of the SMF overlaid the basement; in the middle ground, the Acauã Interplateau Depression, where the overlook is a residual of (of SMF) which covers the Acari Granite and is known as "Serra das Antenas" or "Serra da Telern". Both photographs are of Acari, Rio Grande do Norte. *Source* Authors' archive

Overall, the Borborema Plateau is a dynamic relief, as undulated, inclined, and steep slopes occur. The altitudes vary from 350 to 780 m. The plateau summit is characterized by large planation areas, with gentle hills. In the areas where the SFM occurs, the topographic is roughly plain and bordered by steep slopes, almost rectilinear, as in sedimentary tablelands (Fig. 5.5), which allows one to refer to these areas as "chapadas" (Portuguese term for tablelands) overlying crystalline massifs. The shape and the relief dissection are arranged in the orientation NE-SW, as a result of the influence of Brasiliano basement structures, including the shear zones of Portalegre, Picui-Joao Câmara, and Frei Martinho, among others.

In the steep hillsides, such as in the headwaters of the Acauã River (western slope) and the Trairi river (eastern slope), the dissection is incisive and thus interplateau depressions were formed (Fig. 5.5).

In the Borborema Plateau, were identified the following subunits: 1.2.1 Acauã Interplateau Depression, 1.2.2 Trairi Interplateau Depression, 1.2.3 Florânia Compartment, 1.2.4 Seridó Compartment, 1.2.5 Potengi-Trairi Compartment, 1.2.6 Monte das Gameleiras Compartment, and 1.2.7 Serra de Santana Compartment.

The Massifs and Inland Plateaus (1740 km^2/3.23% of the RN state's territory) correspond to isolated elevations of a minor spatial extent that stand out from the surrounding depression. The lithologic supporting these plateaus is diverse and generally characterized by the association of Brasiliano intrusive suites (Itaporanga, Poço da Crux, Dona Inês) with the Paleoproterozoic and Neoproterozoic metamorphic supracrustal belts (Angelim et al. 2006).

In the Martins-Portalegre and João do Vale plateaus, the relief at the summits are relatively plain due to the sedimentary capping of the SMF (Fig. 5.6), similar to the observed in the Borborema Plateau.

The escarpments are mostly dissected. The fact that strongly dissected fluvial-derived sediments are situated in high altitudes (over 700 m a.s.l) suggests a conspicuous relief inversion (Peulvast and Claudino-Sales 2004). Maia et al. (2016), in studies regarding the geomorphology of Martins-Portalegre, state that the inversion and the maintenance of flat-plain surfaces of these sandstone plateaus in the present-day landscape are related to Cenozoic tectonism resultant from the reactivation of basement structures (e.g., shear zones) and the differential erosion of the laterites formed on sandstones under humid climate conditions.

In the others Inland Plateaus, where no sedimentary capping of SMF occurs, the relief is mostly dynamic and exhibits dissected and steep slopes. They are mostly elongated elevations dissected by incisive valleys (Maia et al. 2013). The dissection patterns and drainage array follow the basement structures, including shear zones, faults, fractures, and foliation planes (Maia and Bezerra 2014).

The Inland Plateaus are subdivided into the following morphoscultural subunits: 1.3.1 Formiga Massif, 1.3.3 Pereiro Plateau, 1.3.4 João do Vale Plateau, and 1.3.5 Martins-Portalegre Plateau.

5.6 Marginal Sedimentary Basins

The morphostructural unit regarding the Marginal Sedimentary Basin corresponds to 18,150 km^2 or 33.73% of the territory of the State of RN. These sedimentary terrains have their genesis related to the break-up of the Gondwana Megacontinent, where aborted rifts favored, since the Early Cretaceous, the formation of several basins in

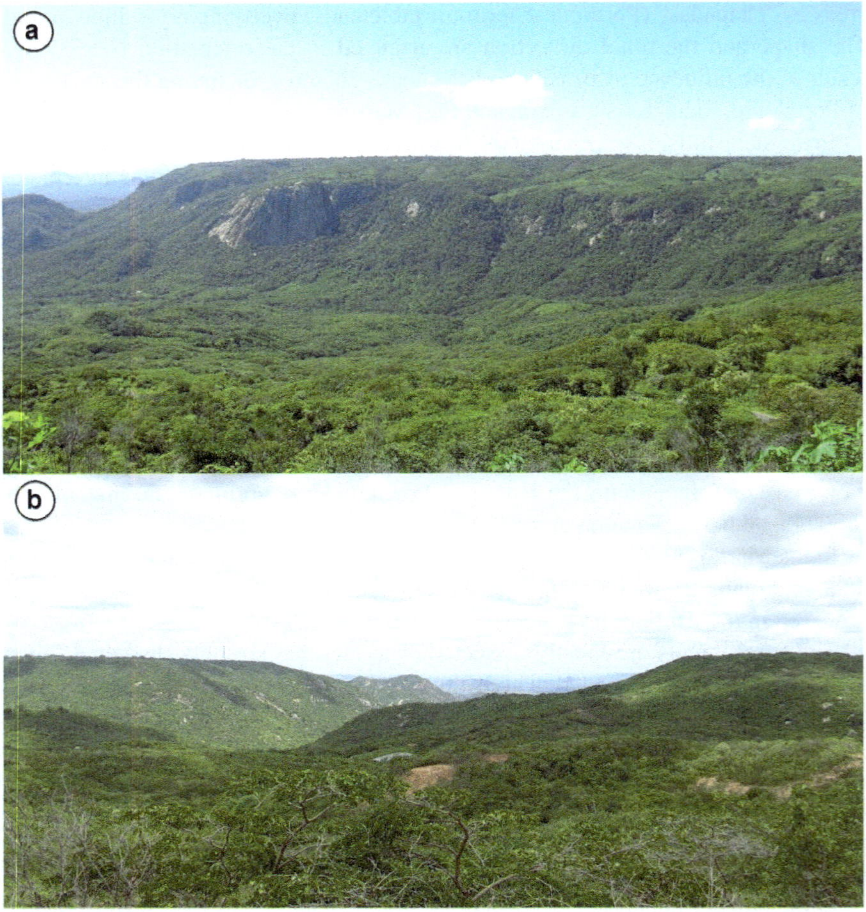

Fig. 5.6 Inland crystalline massifs covered by the Serra do Martins Formation (note the flat-plain aspects of the summit surfaces and the steep slopes). **a** Serra do Martins; **b** Serra de João do Vale. *Source* Authors' archive

the Brazilian continual margin (Matos 2000). Over the Cenozoic, these basins were filled by deposits derived from the dissection of crystalline terrains in the continent's interior (Barreiras Formation) (Mabessone 2000).

In the State of Rio Grande do Norte, this unit is represented by two marginal basins: the Potiguar Basin (northern coast) and Paraíba Basin (eastern coast).

The Potiguar Basin has developed from the Early Cretaceous, with the opening of the Potiguar Rift. This took place in a setting where the differential dextral rotation between South America and Africa culminated in the Borborema Province, in an N-S distensional tectonic regime and E-W compressional regime, accountable for the development of various rift basins under transtensional (Potiguar Rift) and transpressive regimes (Soares et al. 2003).

According to Pessoa Neto et al. (2007), the stratigraphic record of this basin is marked by three sedimentary supersequences: rift supersequence, of Cretaceous age; post-rift supersequence, deposited during Aptian-Albian interval; and drift supersequence, deposited from the Albian up to the present. In the outcrops of its continental portion, this basin exposes sandstones of the Açu Formation and limestones of the Jandaíra Formation, both encompassed by the Apodi Group, besides de basaltic rocks formed during Serra do Cuó and Macau volcanisms (Angleim et al. 2006). Near the coats, these formations become covered by the Barreiras Formation sandstones.

The Paraíba Basin's boundary with the Potiguar basin is the Alto de Touros, (structural crest) situated in an area known as the "continent corner", according to Lima Filho (1998), which considers distinctive the Paraiba and Pernambuco Basins and limited by the Pernambuco Lineament. Due to the lithological resistance, the aforementioned area of the lineament was the last link between South America and Africa during the break-up of Gondwana, which resulted in a late sedimentary deposition, when compared to the other Brazilian rift basins (Matos 1998).

Barbosa et al. (2007) claimed that the sedimentary deposition in the portion of Bacia Paraiba on the State of RN is likely to have occurred after Turonian, as no evidence of refit depositional sequences was found. The carbonate deposits of the drift phase are layered upon the crystalline basement, the oldest (Turonian) correlated to the Jandaíra Formation in the Potiguar Basin, while the youngest (Campanian–Maastrichtian) may be correlated to the Gramame Formation, identified in other sectors of the basin (Barbosa et al. 2007). The Barreiras Formation covers the basin in the coastal zone.

This morphostructural unit comprises two morphoscultural units, the first one being the **Plateaus and Coastal Tablelands** (16,884 km^2/31,38% of the RN state's territory), formed of planted reliefs developed mainly on the sandstones and conglomerates of the Barreiras Formation, sandstones of Açu Formation, and limestones (Jandaira Formation) of Apodi Group. In this unit, there are four morphoscultural subunits: 2.1.1 Northern Coastal Tablelands, 2.1.2 Interior Tablelands, 2.1.3 Dissected Coastal Tablelands, and 2.1.4 Apodi Plateau.

On Potiguar Basin's western coast, the outcropping area of the carbonate platform of the Jandaira Formation stands out from the *cuesta*-like relief, expressed in the landscape by the Apodi Plateau (Chapada). This *cuesta* is characterized by karst morphologies and has an S/SW-facing front and a NE-facing reverse (tilted 0.1°) (Maia et al. 2012).

In the Northern Coastal Tablelands (Fig. 5.7), the relief is mostly flat, with inclinations below 8°. It is situated between 0 and 50 m a.s.l and is predominantly supported by sandstones and conglomerates of the Barreiras Formation. The Dissected Coastal Tablelands occur in the easternmost RN state, in the Paraíba Sedimentary Basin, often in contact with the ocean and elevated up to 100 m a.s.l to the inland.

The Coastal Tablelands stand in altitudes between 50 and 100 m a.s.l., mainly supported by rocks of the Apodi Group (post-rift units of the Potiguar Basin). Among them, the Serra Verde Plateau stands out in the landscape. It consists of a small Tableland in the boundary of the Potiguar basin with the Borborema Province, where the relief is elevated above the interior tablelands situated at 100 m a.s.l. Such unit is

Fig. 5.7 Northern Coastal Tablelands. **a** In the foreground, a gentle ramp sloping to the Atlantic, interrupted only by Tectonic Reliefs, as the Serra do Mel Dome in the back. **b** Coastal Cliff of the Barreiras Group in the São Cristóvão Beach (Areia Branca, RN). *Source* Authors' archive

situated in the prolongation of the Picui-Joao Camara Shear zone over the Potiguar Basin post-rift. This morphostructural layout suggests a genetic association of the unit with Cenozoic tectonic reactivations of Precambrian shear zones, showing the geomorphological repercussion of the reactivations on sedimentary environments.

The other morphoscultural unit corresponds to the **Tectonic Reliefs on Marginal Basins** (1266 km^2/2.35% of state territory), forms whose origin is associated with the Cenozoic reactivation of basement structures. Among others, the domical elevation, known as Serra do Mel (Fig. 5.7), is genetically related to the Cenozoic reactivation of former distensional faults in a compressional setting.

Therefore, two morphoscultural subunits can be defined for the Tectonic Reliefs in sedimentary basins: 2.2.1 Serra do Mel Plateau and 2.2.2 Serra Verde Plateau.

5.7 Cenozoic Volcanism

The Cenozoic Volcanism in the State of Rio Grande do Norte corresponds to an area of 1266 km^2, or 2.35% of the state territory. This event is represented by small volumes of mafic and ultramafic volcanic rocks originating from the Macau Magmatism aligned in an N-S trending belt in the central portion of RN, from the Potiguar Basin, in its emerged and submerged areas, to the Precambrian basement in the continent interior (Silveira 2006). The lithologies constituent of these structures consists of predominantly olivine-basalts and basanites dated of Paleogene (Angelimet al. 2006).

Formerly associated with mantle plume activity, the Macau magmatism has been recently attributed to an edge-driven convection mechanism, due to the absence of spatial–temporal correlation between the two events (Knesel et al. 2011; Oliveira and Medeiros 2012).

The morphoscultural unit in which occurrences of the Macau magmatism were found is named **Relief associated with Neogene Volcanism/Plutonism**. They crop out as necks, plugs, and laccoliths, with varied shapes and altitudes. Pico do Cabugi Volcanic Neck (Fig. 5.8) was exhumed among the crystalline basement and is elevated up to 560 m a.s.l, with a conical shape and medium-angled slopes. The Mangue Seco Dome, in turn, (Fig. 5.8) is a domical feature outcropping amid the Cenozoic sedimentary lithologies of the Potiguar Basin. Its altitudes reach ca. 120 m, creating a rounded topographic contrast with slopes gently inclined to the Coastal Tablelands. Outcrops in form of flow* also occur; although not mapped in this study due to their small sizes, and thus not encompassed by the scale used.

The morphoscultural subunits identified were: 5.1.1 Mangue Seco Dome, 5.1.2 Serra Preta de Pedro Avelino Volcanic Plug, 5.1.3 Serra Aguda Volcanic Neck, and 5.1.4 Pico do Cabugi Volcanic Neck.

5.8 Quaternary Sedimentary Covers

The Quaternary Sedimentary Covers represent an area of 2722 km^2, or 5.05% of the state territory, and comprise two recent sedimentary deposits, located on specific parts of the other morphostructural units. They are mostly composed of clays, sands, and gravel resulting from recent erosion cycles over the Quaternary.

These sediments may be derived from fluvial (deposited along the major river courses crossing the State of Rio Grande do Norte), aeolian (forming extensive dune fields characteristic of the RN coast), lake, transitional, and marine depositional systems (Fig. 5.9). They fill the lowered relief, where aggradational processes become predominant over the dissection, namely, the plains. Hence, the morphoscultural unit formed by these covers is the Coastal and Alluvial Plains.

The plains' unit is subdivided into four morphoscultural subunits: 4.1.1 Coastal Plains, generally Holocene and composed of marine sediments with sand fraction;

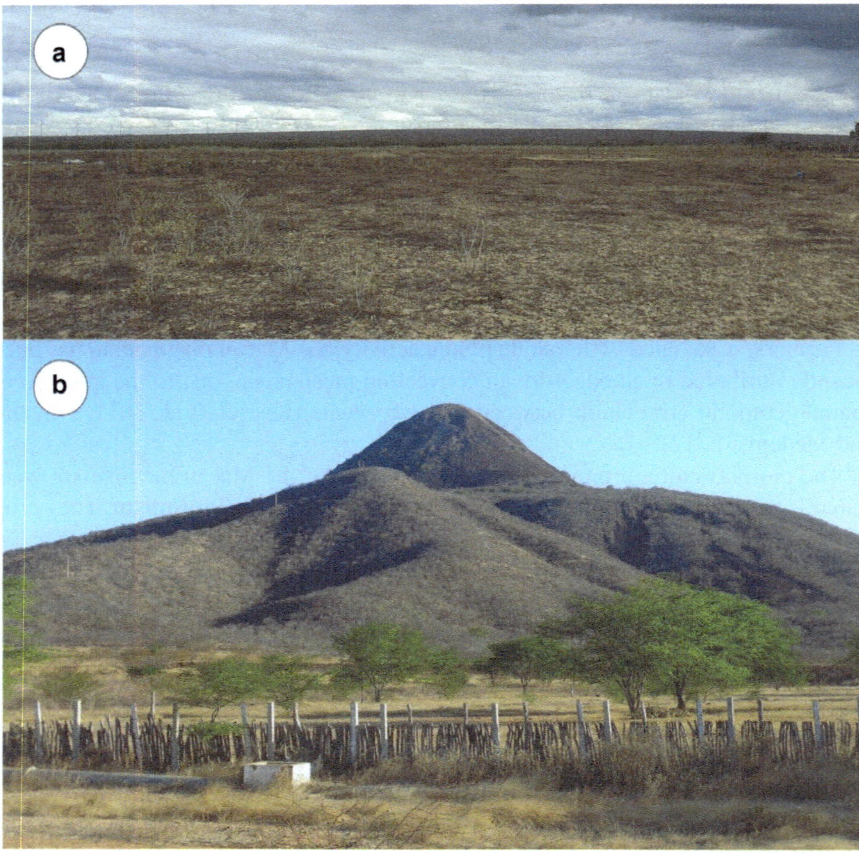

Fig. 5.8 **a** Mangue Seco Dome and **b** Pico do Cabugi Volcanic Neck, both features originating from the Cenozoic magmatism. *Source* Authors' archive

4.2.1 Alluvial Plains in the Eastern Coast, formed of sands and clays, in which the latter are related to areas with a humid and semi-humid climate of the rivers in the eastern sector that are born in the Borborema Plateau; 4.1.3 Alluvial Plains of the Piranhas-Açu Drainage Basin, and 4.1.4 Alluvial Plains of the Apodi-Mossoró Drainage Basin. These two latter units are composed of sand and gravel carried by large rivers draining the semi-arid areas in the RN.

Fig. 5.9 Quaternary depositional features of the Coastal Plain in the RN. **a** Beach range with the occurrence of beach rocks (Extremoz—RN). **b** Ramp of aeolian deposits covering the Coastal Cliff (Serra do Mel—RN). **c** Semi-fixed dunes (Extremoz—RN). **d** Beach range with structures for limiting coastal erosion (São Bento do Norte—RN). **e** Mobile dunes with an aeolian park (Galinhos—RN). *Source* Authors' archive

5.9 Final Considerations

The mapping undertaken in this study allowed the identification of four morphostructural units, seven morphoscultural units, and 30 morphoscultural subunits, which reveal the complexity of the geomorphology of the State of Rio Grande do Norte, considering the interplay of lithology, tectonic events, and morphogenetic dynamic.

The geomorphological map elaborated provides support for territory planning and ordering in the state of RN and may serve as an example to other states in the Northeast of Brazil.

In regards to previous works, we consider that the major contribution of this work is the possibility of comparison between the official mapping of the relief units of Brazil, as well as of other states since the same system is used and widely known by the geomorphological community around the country.

In the State of Rio Grande do Norte, the largest morphostructural unit is the Brasiliano Mobile Belt, which occupies 61.12% of the state territory. The most extensive morphoscultural unit is the Sertaneja Depression, which dominates the landscape of the state (45.45% of the territory).

The second largest morphostructural unit is the Marginal Sedimentary Basins (33.73% of the state territory), almost fully dominated by the morphoscultural unit of the plateaus and Coastal Tablelands (31.38% of the state territory).

The other morphostructural units are related to Cenozoic Volcanism/Plutonism (37.5 km^2/0.1%) and Quaternary Sedimentary Covers (2722 km^2/5.05%).

References

Angelim LAA, Medeiros VC, Nesi JR (2006) Programa Geologia do Brasil—PGB. Projeto Mapa Geológico e de Recursos Minerais do Estado do Rio Grande do Norte. Mapa Geológico do Estado do Rio Grande do Norte. Escala 1:500,000. CPRM/FAPERN, Recife

Barbosa JA, Neumann VH, Lima Filho M, Souza EM, Moraes MA (2007) Estratigrafia da faixa costeira Recife-Natal (Bacia da Paraíba e Plataforma de Natal), NE Brasil. Estudos Geológicos 17(2):3–30

Corrêa ACB, Tavares BAC, Monteiro KA, Cavalcanti LCS, LIRA DR (2010) Megageomorfologia e morfoestrutura do Planalto da Borborema. Revista do Instituto Geológico 31(1–2):35–52

Dantas ME, Ferreira RV, Relevo. In: Pfaltzgraff PAS, Torres FSM (2010) Geodiversidade do estado do Rio Grande do Norte. CPRM, Recife, pp 77–92

Gurgel SPP, Bezerra FHR, Corrêa ACB, Marques FO, Maia RP (2013) Cenozoic uplift and erosion of structural landforms in NE Brazil. Geomorphology 186(1):68–84

IBGE (2006) Instituto Brasileiro de Geografia e Estatística. Mapa de unidades de relevo do Brasil. IBGE, Rio de Janeiro. 1 mapa, color. Escala 1:5.000.000

IBGE (2009) Instituto Brasileiro de Geografia e Estatística. Manual Técnico de Geomorfologia. 2nd edn. IBGE, Rio de Janeiro, 182 p

Knesel KM, Souza ZS, Vasconcelos PM, Cohen BE, Silveira FV (2011) Young volcanism in the Borborema Province, NE Brazil, shows no evidence for a trace of the Fernando de Noronha plume on the continent. Earth Planet Sci Lett 302(1–2):38–50

Lima Filho MF (1998) Análise estratigráfica e estrutural da Bacia Pernambuco. 200 f. Tese (Doutorado em Geoquímica e Geotectônica)—Universidade de São Paulo, Instituto de Geociências, São Paulo

Lima MG (2008) A história do intemperismo na Província Borborema Oriental, Nordeste do Brasil: implicações tectônicas e paleoclimáticas. 594 f. Tese (Doutorado em Geodinâmica e Geofísica)—Universidade Federal do Rio Grande do Norte, Centro de Ciências Exatas e da Terra, Natal

Mabesoone JM (2000) Abertura do Oceano Atlântico Sul: como, onde, quando? In: Simpósio de Geologia do Nordeste, 18., Recife. Resumos. SBG, Recife, p 62

Maia RP, Bezerra FHR (2014) Condicionamento estrutural do relevo no Nordeste Setentrional Brasileiro. Mercator 13(1):127–141

Maia RP, Bétard F, Bezerra FHR (2016) Geomorfologia dos Maciços de Portalegre e Martins – NE do Brasil: Inversão do Relevo em Análise. Revista Brasileira De Geomorfologia 17(2):273–285

Maia RP, Sousa MOL, Bezerra FHR, Xavier Neto P, Lima ENM, Silva CCN, Santos RD (2012)A importância do controle tectônico para a formação da paisagem cárstica na Bacia Potiguar, Nordeste do Brasil. Revista Brasileira de Geomorfologia 13(4):431–456

Maia RP, Amaral RF, Gurgel SPP (2013) Geomorfologia do Rio Grande do Norte. In: Albano GP, Ferreira LS, Alves AM (Orgs.). Capítulos de Geografia do Rio Grande do Norte. Fundação José Augusto, Natal, pp 19–59

Martins TIS, Rodrigues SC (2016) Compartimentação geomorfológica da Folha Piumhi, região do alto São Francisco, Minas Gerais. Revista Brasileira De Geomorfologia 17(1):145–162

Matos RMD (1998) The transversal zone: a key feature between NE Brazil and W Africa. In: Mello MR, Yilmaz PO (eds) AAPG international conference & exhibition, Rio de Janeiro - Brazil, Proceedings, vol 1, pp 426–427

Matos RMD (2000) Tectonic evolution of the equatorial south Atlantic. In: Mohriak WU, Talwani M (eds) Atlantic rift in continental margins. American Geophysical Union, Washington, DC, pp 331–351

Morais Neto JM, Green PF, Garner GD, Alckmim FF (2008) Age of the Serra do Martins Formation, Borborema Plateau, northeastern Brazil: constraints from apatite and zircon fission track analysis. Boletim de Geociências da Petrobrás 16(1):23–52

Nascimento MD, Souza BSP (2010) Mapeamento geomorfológico da área abrangida pela carta topográfica de Santa Maria – RS como subsídio ao planejamento ambiental. Revista Brasileira De Geomorfologia 11(2):83–90

Oliveira RG, Medeiros WE (2012) Evidences of buried loads in the base of the crust of Borborema Plateau (NE Brazil) from Bouguer admittance estimates. J S Am Earth Sci 37:60–76

Oliveira RG (2008) Arcabouço geofísico, isostasia e causas do magmatismo cenozoico da Província da Borborema e de sua margem continental (Nordeste do Brasil). Tese (Doutorado em Geodinâmica e Geofísica)—Programa de Pós-Graduação em Geodinâmica e Geofísica, Universidade Federal do Rio Grande do Norte, Natal

Pessoa Neto OC, Soares UM, Silva JGF, Roesner EH, Florencio CP, Souza CAV (2007) Bacia Potiguar. B. Geoci. Petrobrás 5(2):357–369

Peulvast JP, Claudino Sales V (2004) Stepped surfaces and Paleolandforms in the Northern Brasilian <<Nordeste>>: Constraints on models of morfotectonic evolution. Geomorphology 62(1–2):89–122

Ross JLS, Moroz IC (1996) Mapa geomorfológico do estado de São Paulo. Rev. do Depto. Geografia 10(1):41–58

Ross JLS (1992) O registro cartográfico dos fatos geomorfológicos e a questão da taxonomia do relevo. Rev. do Depto. Geografia 6(1):17–29

Santos LJC, Oka-Fiori C, Canali NE, Fiori AP, Silveira CT, Silva JMF, Ross JLS (2006) Mapeamento Geomorfológico do estado do Paraná. Revista Brasileira De Geomorfologia 7(2):03–12

Santos RF (2004) Planejamento ambiental: teoria e prática. Oficina de Textos, São Paulo, 184 p

Silveira FV (2006) Magmatismo Cenozoico na porção central do Rio Grande do Norte, NE do Brasil. 194 f. Tese (Doutorado em Geodinâmica e Geofísica), Universidade Federal, Natal

Soares UM, Rosseti EL, Cassab RCT (2003) Bacias sedimentares brasileiras: Bacia Potiguar. Phoenix 5(56):1–6

Marco Túlio Mendonça Diniz He holds a degree (2005), a master's degree (2008), a Ph.D. (2013) in geography from the State University of Ceará (UFC), and a post-doctorate (2018) in the graduate program in geography at the Federal University of Sergipe (UFS). He is an associate professor at the Federal University of Rio Grande do Norte (UFRN), a CNPq research productivity fellow (level 2), the editor-in-chief of the Journal of Geoscience of the Northeast, and a researcher of the Geoprocessing and Physical Geography (LAGGEF) at the UFRN (leader), Integrated Coastal Zone Management (LAGIZC) at the CNPq/UECE, Geoecology and Territorial Planning (GEOPLAN) at the CNPq/UFS, Environment and Sustainable Development (MADES) at the CNPq/IFPI, and Research Group on Risk Management and Natural Disasters (GENAT) at the CNPq/UFRN. He received an award as an outstanding researcher from the UFRN (2020) in human and social science, arts, and letters. He is a member of the internal committee of PIBIC/CNPq/UFRN and the external committee of PIBIC/CNPq/UFPI, an external partner of the UECE scholarship system. He has experience in geography with an emphasis on physical geography, acting mainly on coastal geography, integrated landscape, and geomorphological heritage.

George Pereira de Oliveira He holds a degree in geography (2017) from the Federal University of Rio Grande do Norte (UFRN), Seridó Higher Education Center (Caicó-RN), a bachelor's degree in geography (2022) from the Federal University of Pernambuco (UFPE), a master's degree in geography (2019) from the graduate program in geography at the Federal University of Pernambuco, where he is a Ph.D. candidate in geography. He has experience in physical geography with an emphasis on structural and tectonic geomorphology, geomorphometry, geomorphological mapping, and integrated landscape analysis. He is a member of the Quaternary Study Group of the Brazilian Northeast (GEQUA) and the Study Group on Archeology, Geoscience, and Technology (GEAGT).

Rubson Pinheiro Maia He is a geographer (UECE), holding a master's degree in physical geography with an emphasis on geomorphology (UFC) and a Ph.D. in geodynamics and geophysics (UFRN). He works in the area of geomorphology with an emphasis on morphotectonics, having worked on morphotectonics of karstic environments, fluvial systems, deformation zones, and crystalline massifs. He participated in the project of geological mapping geomorphological sheet SB-24-X-D-I (Mossoró-RN) funded by CPRM (Geological Service of Brazil) and the project mapping/characterization of karst (Porocarste) in the Potiguar Basin funded by Petrobras. He participates in the Porocarste 3D project funded by Shell and integrates the INCT (National Institute of Science and Technology) in the area of tectonic studies. He taught courses related to geoscience at UFC (2005), UECE (2006/2007), FATEC-CE (2006/2008), and UFRN (2011/2013). He is a geomorphology professor at the Federal University of Ceará and a CNPq research productivity fellow (level 2).

Bruno Ferreira He is a professor with a Ph.D. in geoscience from the Federal University of Pernambuco (UFPE) and has experience in geoscience with an emphasis on physical geography and applied physical geography. He holds a master's degree in geoscience (CAPES 4) (2010) and a degree in geography (2007), both from the Federal University of Pernambuco.

Chapter 6
Geomorphology of Paraíba State

Bruno de Azevedo Cavalcanti Tavares, Antônio Carlos de Barros Corrêa, George Pereira de Oliveira, and Daniel Rodrigues de Lira

Abstract Paraiba is located at the easternmost corner of South America; about 80% of its territory is found on Proterozoic rocks of the Borborema Geotectonic Province. The remaining area is structured by Mesozoic continental or marginal rift basins, Cretaceous volcanic rocks, remnants of Paleogene/Neogene platform covers, and Quaternary accumulation units. The state comprises four morphostructural compartments corresponding to structural massifs, intermontane lowlands, coastal piedmonts, and Quaternary plains. These landforms are subject to two well-marked tropical morphogenetic regimes, the humid and the semi-arid. The humid domain occurs along the coastal strip and is associated with an annual rainfall varying from 1200 to 1700 mm. The prevailing landforms of the domain are the coastal tablelands and pluriconvex hills covered in argisols, latosols, and deep-weathering mantles. However, most state landform compartments, including the Borborema Highlands and the Sertaneja lowlands, are dominated by the drier climatic conditions associated with the semi-arid domain, with year-round high temperatures and precipitation averaging 400–800 mm. This scenario is associated with thinner soils covered with rocky pavements and extensive pediments.

Keywords Coastal tablelands · Structural massifs · Semi-arid lowlands

B. de A. C. Tavares (✉) · A. C. de B. Corrêa · G. P. de Oliveira · D. R. de Lira
Federal University of Pernambuco (UFPE), Recife, Brazil
e-mail: bruno.tavares@ufpe.br

A. C. de B. Corrêa
e-mail: antonio.correa@ufpe.br

G. P. de Oliveira
e-mail: george.oliveira@ufpe.br

D. R. de Lira
e-mail: daniel.rlira@ufpe.br

V. Claudino-Sales and J. F. Sobrinho (eds.), *Geomorphology of the Northeast Region of Brazil*, The Latin American Studies Book Series,
https://doi.org/10.1007/978-3-031-58874-7_6

6.1 Introduction: The Geotectonic Context

Located at the easternmost corner of South America, the state of Paraíba has a coastline facing the South Atlantic Ocean to the east with a total of 140 km in length. Its land limits are formed by the State of Pernambuco to the south, Ceará to the west, and Rio Grande do Norte to the north. It occupies an area of 56,468.427 km^2 and extends between parallels 6°1′33″ S and 7°9′18″ S and meridians 34°47′34″ W and 38°45′56″ W.

Of the five tectonostratigraphic units traditionally considered in Paraíba (Santos et al. 2002), about 80% of its territory is found on Proterozoic rocks belonging to the Borborema Geotectonic Province. The others are represented by the Mesozoic continental or marginal rift basins, Cretaceous volcanic rocks, remnants of Paleogene/Neogene platform cover, and Quaternary accumulation forms, especially along the coastline and following the course of the main drainages in the form of plains and fluvial terrace levels.

The Borborema Province constitutes a mosaic of basement fragments, micro-continents, and fold belts ranging from the Archean to the Neoproterozoic, separated by transcurrent shear zones and thrust faults. Its geological organization culminated with the apex of the Brasiliano Cycle with the collage of the West Gondwana continent. The primary shear zones, called lineaments, divide the Borborema Province into sub-provinces: Southern, Transverse, and Northern (Brito Neves et al. 2014). In addition, the Borborema Province is limited to the west by the Parnaíba basin, to the south by the São Francisco Craton, and to the north and east by the marginal rift basins formed after the breakup of the Western Gondwana continent and the opening of the equatorial and south Atlantic Ocean (Fig. 6.1).

Despite the reactivation of Proterozoic structures closer to the eastern seaboard, the State of Paraiba lacks the set of inland uplifted phanerozoic sedimentary basins that form table-like and homoclinal plateaus in other sectors of the Borborema Province (Correa et al. 2019). In fact, except for the Paleogene and Neogene unconsolidated covers that suffered relief inversion due to uplift along the coast (Barreiras Fm.) and topped a few summits within the Borborema Highlands (Serra dos Martins Fm.), other Cretaceous continental rift basins (Rio do Peixe and Boa Vista basins) remain at relatively lower elevations, with little topographic differentiation from the adjacent lowlands structured on crystalline Proterozoic lithologies.

The coastline of Paraíba is bordered by a marginal rift basin, originated during the Cretaceous continental rifting, and opening of the Atlantic Ocean (Matos 1992). This Mesozoic basin along with the Neogene sediments that cap it underwent tectonic uplift, relief inversion, and partial exhumation of the unconsolidated covers during Cenozoic (Maia and Bezerra 2014; Marques et al. 2014) (Figs. 6.2 and 6.3).

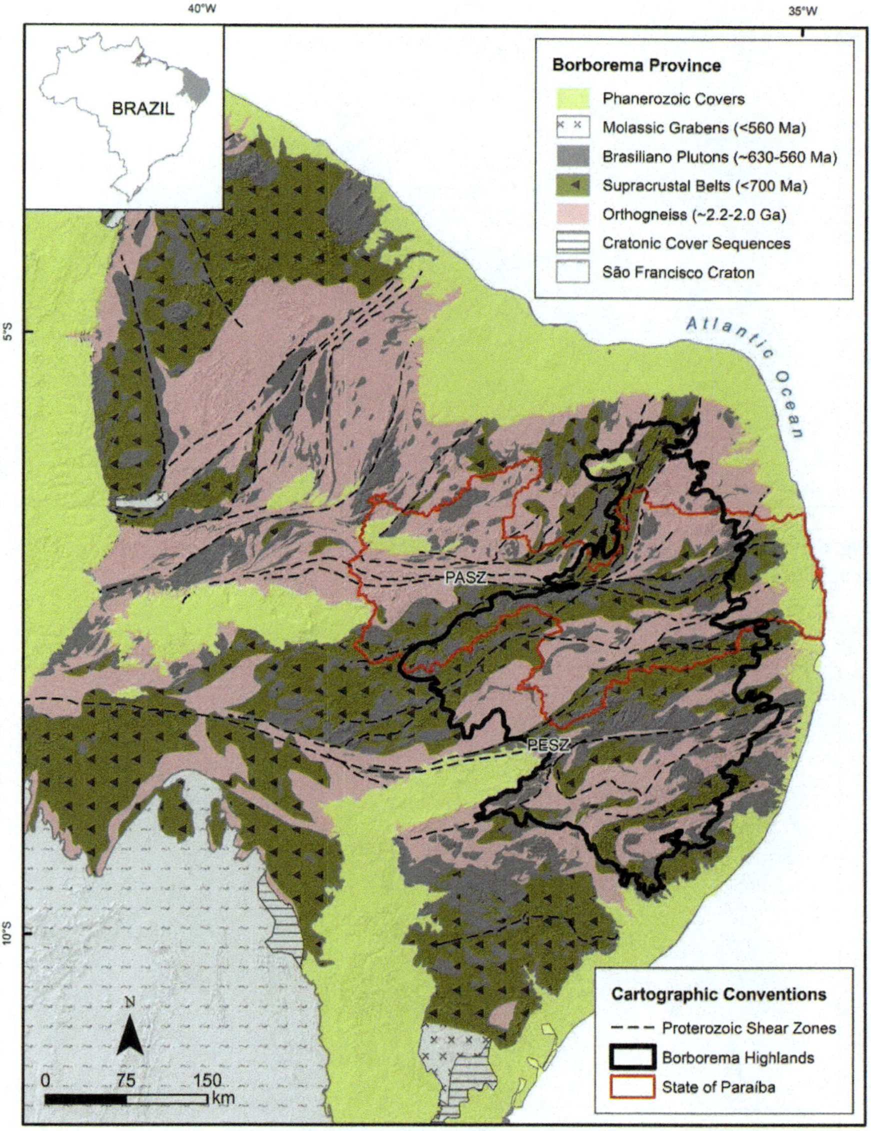

Fig. 6.1 State of Paraíba within the Borborema Province. PASZ—Patos Shear Zone; PESZ—Pernambuco Shear Zone. *Modified from* Santos et al. (2014) and Santos et al. (2021)

6.2 The Borborema Highlands

As in the other three coastal states in the east of the Northeast of Brazil where this morphostructure occurs, the Borborema Highlands in the state of Paraíba corresponds to the continuous set of landforms elevated above the altitude of 200 m, whose limits

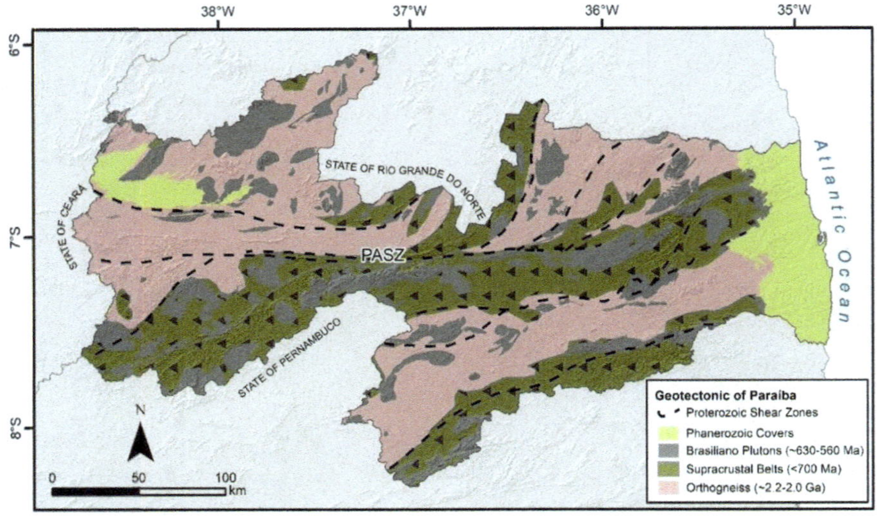

Fig. 6.2 Geotectonic map of the State of Paraíba. PASZ—Patos Shear Zone. *Modified from* Santos et al. (2014, 2021)

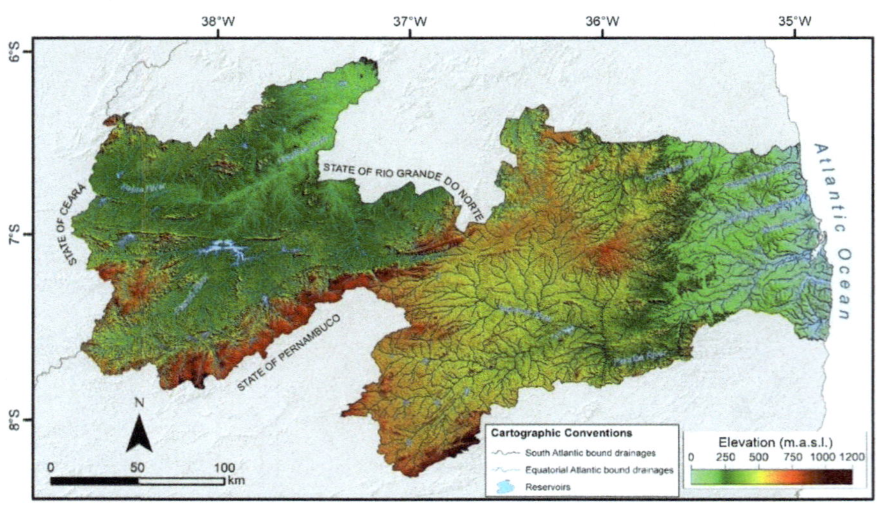

Fig. 6.3 Hypsometric map of Paraíba with the drainage network, highlighting Equatorial and South Atlantic bound watersheds. *Source* the authors

with the adjacent units are marked by topographic breaks in the form of escarpments that can reach more than 500 m, as in the Teixeira range in the center of the State (Correa et al. 2010). The origin of the Highlands was initially linked to the overall readjustment of the regional landforms and the establishment of a new general base level following the opening of the South Atlantic Ocean in the Cretaceous. Later this event was superseded by the consortium action of tectonically conditioned epirogenesis and isostatic readjustments throughout the Cenozoic. In the state of Paraíba and Rio Grande do Norte, a secondary control on regional uplift was played by intraplate magmatism (Oliveira et al. 2023) during the Miocene, which resulted in basaltic flows in some sectors of the Highlands, such as in the Boa Vista basin.

The territory of the state of Paraíba lies between the northern and transverse sub-provinces. The first is characterized by the almost continuous exposure of the Paleoproterozoic basement, with some Archean cores reworked during the Brasiliana Orogeny (900–570 Ma). The ancient cores form massifs such as the Piranhas River, and São José do Campestre, interspersed by supracrustal belts composed of volcanic-sedimentary and metasedimentary sequences, especially from the Meso and Neoproterozoic. The Transversal sub-province is delimited by the E-W trending Patos Lineament to the north and Pernambuco Lineament to the south. These 1000 km long shear zones cut through the center of the Borborema Province. This sector is characterized by the presence of a series of Meso- and Neoproterozoic fold belts, as well as Neoproterozoic plutons.

Some areas behave as text-book examples of a complete Wilson cycle (Stage I) remnants, as in the case of the outcrops associated with the Cariris Velhos Orogeny (~1 Ga) in the conspicuously level surface at an average elevation of 500 m west of the city of Campina Grande approximately 120 km wide. According to Crandall (1910), in one of the earliest geomorphological descriptions of the territory of Paraíba, the extreme flatness of this summit surface corresponded to a rejuvenated granitic peneplain which the author referred to as part of the Borborema Massif (Fig. 6.4).

Syn and post-orogenic alkaline granites of Brasiliana age are placed amongst the shear zones of the Transversal sub-province. The differential erosion triggered by the Cenozoic planation cycles has topographically highlighted these plutons, which nowadays correspond to the regional summit surface, reaching elevations close to 1200 m, as in Pico do Jabre, at the border with the State of Pernambuco. Unlike the dynamics observed in the transforming margin to the north of the Borborema Province (Peulvast and Claudino-Sales 2004), Corrêa et al. (2010), drawing from the proposal first put forward by Czajka (1958) point to the prevalence of a distinct behavior in the eastern sector of the Province which affected the coastal façade of Paraíba as far inland as the highest summits of the Borborema Highlands. This morphological pattern is derived from the generalized and repeated up-warping of the continental margin (Monteiro and Corrêa 2020). The authors recognized two sectors of regional uplift corresponding to relief lineaments of ENE-WSW and E-W trend, the latter coinciding with a sector of the western escarpment of the Borborema Highlands in Paraíba, the Teixeira range, above which the highest elevation of the state, Pico do Jabre, a Neoproterozoic syenitic intrusion, at 1197 m occur.

Such differential erosion landforms structured on mostly Neoproterozoic plutons placed along ductile shear zones are common structures in denudated orogens. Their origin is linked to thermo-tectonic processes occurring in parallel bands that follow the structural trend of the original orogen. Their outcropping areas are in excess of 10^2 Km^2, resulting in the erosive exposure of plutonic cores sided by mylonites and gneisses (Fig. 6.5).

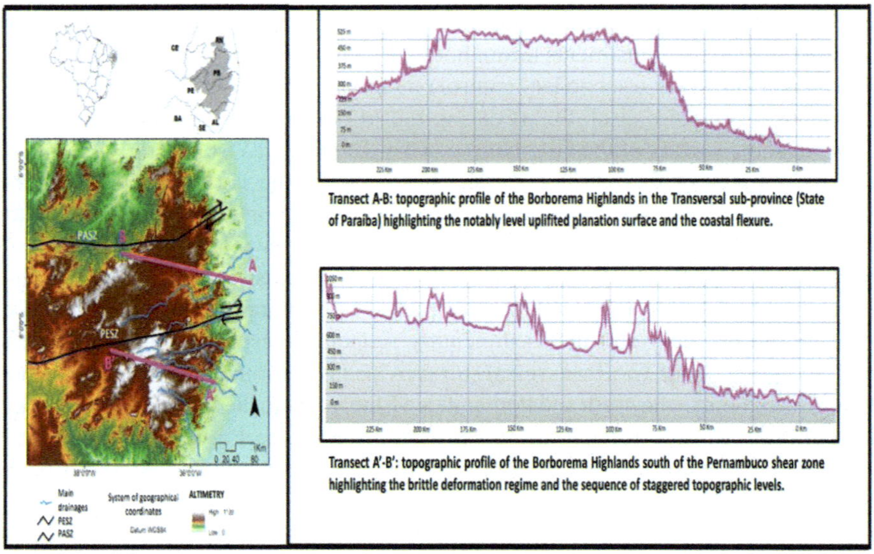

Fig. 6.4 Topographic transects north and south of the Pernambuco shear zone highlighting the distinct morphological responses of the Borborema Highlands to a flexure dominated (State of Paraíba) versus brittle tectonics dominated scenario. *Source* the authors

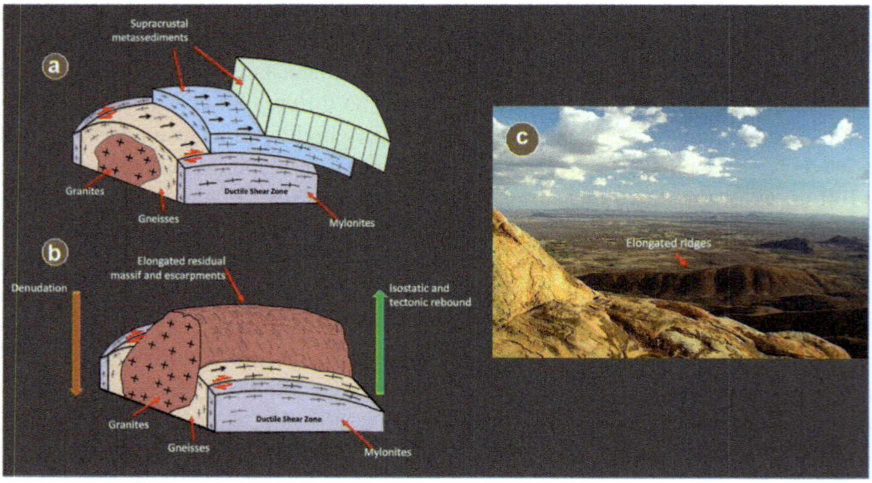

Fig. 6.5 Evolution of plutonic intrusions into structurally controlled residual massifs. **a** Pluton placement within ductile shear zone during the Brasiliana Orogeny; **b** differential erosion following tectonic and isostatic uplift; **c** elongated residual ridges and escarpment controlled by Patos shear Zone. *Source* the authors

Fig. 6.6 Set of granitic ridges controlled by the Patos shear zone at Teixeira range escarpment

The Cenozoic reactivation of the relief in the Transversal sub-province followed preferential pathways along the main shear zones (Fig. 6.6), such as the Patos Lineament. The repeated episodes of regional uplift during the Cenozoic, associated with differential erosion, created interior escarpments, defining edges of crystalline highlands, or delimiting structurally controlled residual massifs. Uplift created accommodation spaces for Quaternary sedimentation, but the lack of thick weathering mantles, probably since the end of the Mesozoic, did not produce the necessary sedimentation to fill these structural depressions, nor did it generate significant remaining covers of typical tropical soils.

The Borborema Highlands in the state of Paraíba present three morphostructural sub-compartments as identified by Corrêa et al. (2010), from north to south: São José do Campestre structural surface; Paraiba intermontane depression and the Remobilized Massifs of the Transversal Zone (Figs. 6.7 and 6.8).

6.2.1 São José do Campestre Structural Surface

This sector of the Highlands corresponds to its northern end, occurring in the State of Paraíba and Rio Grande do Norte. Minor structural highs characterize the unit, sometimes covered by Paleogene continental deposits (Serra dos Martins Formation)

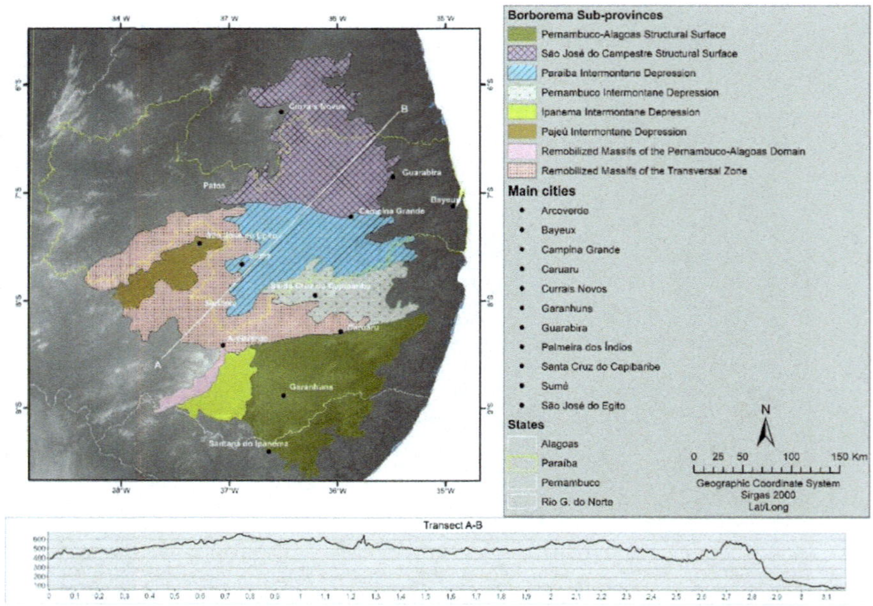

Fig. 6.7 Morphostructural sub-compartments of the Borborema Highlands in Paraíba. *Source* the authors

cut at their base by Miocene basaltic dykes. The dating of the Miocene volcanism that cut through the sediments permitted constraining the time of uplifting of these fluvial sandstone/conglomerate covers to elevations above 700 m. Despite being capped by sediments, these summit surfaces are considered part of the Borborema Highlands since they are circumscribed by the larger morphostructure and share the exact uplifting mechanisms (Fig. 6.9).

6.2.2 Paraiba Intermontane Depression

This sub-compartment is identified in the central portion of Borborema Highlands in Paraíba and delimited east and west by the scarp compartments. The lack of significant structural and lithological differences allowed this area to develop relatively flat morphologies bounded to the south by the more rugged terrains of the Remobilized Massifs of the Transversal Zone. In addition, the prevailing severe semi-arid climate (Cabaceiras records only 275 mm/year and Soledade 350 mm/year) does not allow the development of thick regoliths, which exposes the gneisses and migmatites directly to the surface. Crandall (1910), in his classic work regarding the geology of the semi-arid Northeast, had already referred to this area as a granitic peneplain (Fig. 6.10).

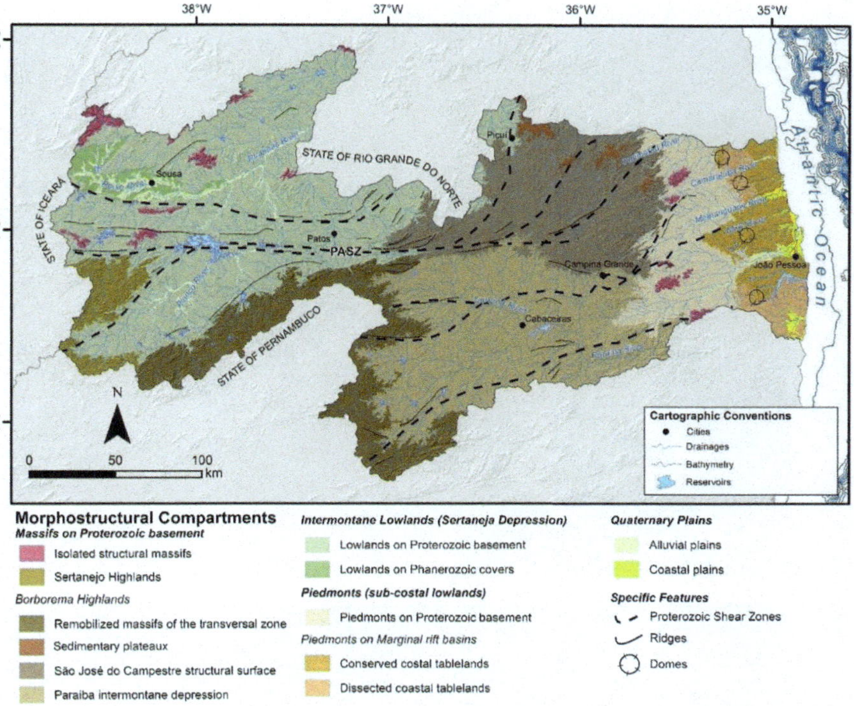

Fig. 6.8 Morphostructural compartments of the State of Paraíba. PASZ—Patos Shear Zone. *Source* the authors

Fig. 6.9 Tablelands on the sediments of Serra dos Martins Formation capping the Borborema Highlands summit in the municipality of Picui. The Paleogene/Neogene covers structure the table-like morphologies in sectors of the São José do Campestre structural surface. *Source* the authors

Fig. 6.10 Landscape aspect of Bravo pluton as it outcrops in Lajedo do Pai Mateus (**a**). The granitic structures are preferably E-W oriented and represent the highest landforms within the Paraibana Intermontane (**b**). *Source* the authors

6.2.3 Remobilized Massifs of the Transversal Zone

The Transversal zone was the most affected area by the tectonic up-warping within the Highlands, hence displaying the highest summits and the most rugged topography. Located in an approximately E-W strip along the border with Pernambuco, this zone is characterized by an aligned succession of isolated plutonic massifs, ridges, and narrow intermontane lowlands. The Teixeira batholith is the most conspicuous intrusion within this unit, and the exposure of its northern edge by differential erosion results in the highest escarpment of the entire Highlands (Fig. 6.11).

This sector of the Highlands is strongly conditioned by Brasiliana age intrusions of different dimensions that result in a series of residual massifs, some of which have a significant outcropping area. The batholiths of Serra da Baixa Verde and Teixeira stand out in this unit for their elevation, reaching 1185 and 1197 m, respectively. Both are located on the limits of the states of Pernambuco and Paraíba. These mountain ranges are conditioned by sinistral shear zones subsidiary to the Pernambuco Lineament. The escarpment of the Teixeira range is part of the western escarpment of the Highlands. Its rocky surface is adapted to reactivated fault lines and fractures which, under the semi-arid climate, does not undergo substantial chemical weathering. The elevation range in excess of 700 m between the top of Pico do Jabre (Peak of Jabre)

Fig. 6.11 Serra do Teixeira (Teixeira Range) delimits the highest sectors of the Borborema Highlands in Paraíba. The granitic massifs represent notably rejuvenated landforms controlled by the area's major shear zones. *Source* the authors

Fig. 6.12 Contact between the Serra do Teixeira escarpment and the pediments of Patos depression. *Source* the authors

and the bottom of Patos depression to the north grants this landform the classification as a "relief of dissected mountains" (Fig. 6.12) (Mabesoone and Neumann 1995; Corrêa 1997).

6.3 Intermontane Semi-arid Depressions

To the west of the Borborema Highlands, under semi-arid conditions, landforms were less affected by the doming processes that operated along the eastern seaboard of the region; thus, the intermontane depressions prevail, usually bounded by higher relief forms and escarpments. These low-lying landscapes display various levels of dissection, according to the dominant lithology and rheological framework, varying from moderately hilly topographies to nearly level pediment surfaces. Such morphostructural units, with notably level interfluves, have been traditionally interpreted as

Fig. 6.13 Patos depression highlighting the E-W oriented ridges lineaments associated to Patos shear zone. Inselbergs rise above the flat lowlands. *Source* the authors

remnants of Neogene planation surfaces and are established primarily on the metasedimentary lithologies of the Brasiliana fold belts or high-grade metamorphics of the Meso and Paleo-Proterozoic complexes.

The low-lying surfaces do not show continuous sequences of surface cover or coalescent levels of sedimentation coming from the mountain fronts, as in the case of the foothills of Serra de Teixeira south of Patos. Most of the time, the pediment surfaces of the depressions present sparse and thin residual alteration mantles and vast rocky outcrops (Fig. 6.13).

6.3.1 Depressions in Continental Rift Basins

In the extreme northwest of Paraíba, on the border with the state of Ceará, there is a lowland level located about 50 m below the general planation surface structured in Proterozoic crystalline rocks of the surroundings. This rhombohedral-shaped depression with an area of approximately 1500 km^2 structurally coincides with the continental rift basin of the Rio do Peixe, delimited by normal and oblique faults that control the lines of receding scarps that served as source areas for sediments (Fig. 6.14).

This Eo-Cretaceous basin is part of a set of basins located in the interior of the Borborema Province, north of the Pernambuco Lineament and south of the limit of

Fig. 6.14 Rio do Peixe depression structured on eo-Createaceous sediments of the Rio do Peixe rift basin in the vicinity of the town of São João do Rio do Peixe. The local base level lies approximately 50 m below the surrounding planation surface. *Source* the authors

the Potiguar basin, formed by grabens and semi-grabens controlled by reactivated structures of the Proterozoic basement during the separation between Africa and South America. Unlike other continental rift basins, the Rio do Peixe basin contains only a sequence of Neo-Comian syn-rift phase deposits, in the form of the Antenor Navarro, Souza, and Rio Piranhas Formations, from base to top, respectively. Its spatial delimitation follows the crossing zone between the Patos Lineament (E-W) and the subsidiary Portalegre shear zone.

After its filling, the Rio do Peixe basin remained at a relatively low topographic level, suggesting the area's tectonic stability between the Cretaceous and Early-Cenozoic. On the contrary, sectors to the E and NE of the basin, like the Borborema Highlands, underwent significant base level readjustments, which at some point in the Neogene must have led the headwaters of the paleo-Rio Piranhas to capture the broad topographical hollow west of the current confluence with the Pianco River. At that time, the end of the basin's sedimentation, likely dominated by post-rift alluvial-colluvial covers, coincided topographically with the surface of the surrounding rocky pediments. It is assumed that until then, the closed depression delimiting the Rio do Peixe sedimentary basin had been isolated from the newly developed exorheic drainage network.

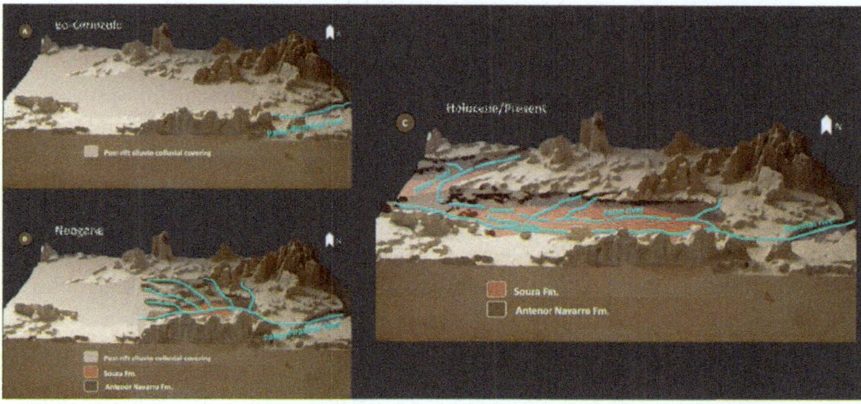

Fig. 6.15 Geomorphic evolution of the Rio do Peixe Depression following the capture and dissection of the original basin by the headwater erosion of the paleo-Piranhas River. *Source* the authors

Following the capture, the Piranhas River would have established its new subsequent course, controlled by the Patos Lineament at the southern edge of the sedimentary basin, while the tributaries originating in the flexural north margin of the basin led to the gradual exhumation of the syn-rift sequences, and established the course of the current Peixe River along the basin's central axis. Nonetheless, this process has yet to develop a longitudinal equilibrium profile for the Peixe River since its channel still lies about 50 m above the local base level of the original capture zone (Fig. 6.15).

Other inland basins that did not experience the relief inversion process becoming sedimentary plateaus occur in the State of Paraíba. Among these, due to its larger area and economic importance, the Boa Vista basin stands out. It is an elongated depression in the WNW-ESE direction located on the Borborema Highlands, 60 km southwest of Campina Grande, in the municipality of Boa Vista. Geologically, the basin is inserted in the Transversal sub-province, having Proterozoic migmatites and supracrustals as the geological basement. The paleo depression results from the extensional reactivation of the Serra do Monte Shear Zone in the Miocene (Santos et al. 2002). The sedimentary fill is dominated by sandy-arcosian to arkose lithic sediments and bentonite clays exploited commercially (Campos Novos Formation). Basaltic flows overlie the sediments. Geomorphologically, the basin behaves as a level surface, topographically continuous, and undifferentiated of the adjoining rocky pediments.

6.4 The Coastal Tablelands

Along the coastal zone of Paraíba, the difference in tectonic behavior between the sector corresponding to the Transversal sub-province, between the Patos and Pernambuco shear zones, and areas further north of the former, resulted in distinct topographical expressions as well as fluvial dissection patterns. According to Lima Filho et al. (2005) the model of tectonic evolution of the Paraíba marginal rift basin, also affecting its Neogene unconsolidated covers (Barreiras Fm. of Miocene age), exhibited a remarkable flexural response that have also affected the morphological aspect of the landforms (Monteiro and Correa 2020).

The coastal tablelands of Paraíba correspond to a low-lying homoclinal plateau with elevations ranging from 50 to 200 m. This morphostructure is established on the lithologies of the Paraíba basin and, above all, on its Neogene covers, associated with the unconsolidated continental sediments of the Barreiras Formation (Fig. 6.16). Morphologically, this low plateau displays a homoclinal attitude, accompanying the flexure of the Paraíba basin toward the coast (Barbosa and Lima Filho 2006; Lima et al. 2017).

The morphogenesis of this unit is driven by the regressive erosion of headwaters within a short distance from the general base level. The rivers vertically carve the incohesive, sandy-clay sediments of the Barreiras Formation, reaching, in the case of some valleys closer to the coastline, the underlying Mesozoic limestones (Gramame

Fig. 6.16 Tabular structures of the Miocene sediments of Barreiras Formation, Coqueirinho beach. *Source* the authors

Formation) and sandstones (Beberibe-Itamaracá Formation). Broad and flat interfluves characterize the relief, deeply dissected by parallel consequent first-order streams that typically converge to a primary collector adapted to the linear structures associated with the E-W trending Proterozoic shear zones that, propagating from the interior through the basement of the Paraiba basin, intersect the coastline.

Further to the west, the tablelands assume a more distinctive homoclinal attitude, with summit surfaces gently tilting eastwards. In this sector, the flat interfluves reach the highest elevation, above 200 m at some places, with dip-slip slopes and cornices structured on the Cretaceous sandstones of the Beberibe-Itamaracá Formation. The higher elevation and the outcropping of Mesozoic strata of the Paraiba rift basin lacking the Neogene cover of the Barreiras Formation attest to the late Cenozoic uplifting of the area (Bezerra et al. 2008). These cuesta-like landforms present west-facing front-slopes and gentle dip-slips tilted toward the coast (Fig. 6.17).

Fig. 6.17 Homoclinal aspect of the coastal tablelands at their westernmost limit. **a** Elements of a cuesta-like landform to the west of the coastal tablelands north of the town of Pedras de Fogo; **b** outcrop of the Beberibe-Itamaracá Formation sandstone at the top of the interfluve. *Source* the authors

6.5 The Coastal Forms

To the east and along the final stretches of the main rivers (Paraíba, Mamanguape, Miriri, Camaratuba), the coastal tablelands are limited by marine, fluvial, and fluvial-marine plains. The glacial-eustatic variations of the Quaternary controlled the incision of the valleys, resulting in a deeply incised parallel drainage pattern, with wide interfluves in the form of sedimentary tables with steep hillslopes, susceptible to the intense gullying and headward erosion. Morphogenesis is primarily controlled by the rainy monsoonal climate, the parent material's sandy-clayey character, and the high slope of the terrain (Mabesoone and Cunha Silva 1991). According to the degree of dissection and the regional pattern of subjacent fractures, the average interfluvial distances vary from 2 to more than 10 km.

The parallel drainages, conditioned by the structures that underly the sediments of the Paraíba basin and propagate into the Neogene covers, are characterized by broad estuarine plains in their final reaches under the influence of the tides. Structural highs and troughs control the N-S morphological alternation between sedimentary mesas and tide-dominated estuaries (horsts and grabens). It is believed that those morphotectonic controls were reactivated numerous times during the Neogene, thus accounting for maintaining the relatively high topography along the coastline.

The coast of Paraíba presents discontinuous plains, especially following the estuarine sector at the mouths of the main rivers. Along the coastline, the tablelands often reach the sea, resulting in linear cliffs structured in the sediments of the Barreiras Formation that reach 80 m in height (Fig. 6.18). In some areas, a narrow sandy beach is interposed between the base of the cliffs and the sea, whereas in others, they are directly reached by the surf zone. The absence of continuous and more developed coastal plains is attributed to the low input of sediments during the Pleistocene/Holocene due to the predominantly semi-arid climate further inland (Santos et al. 2002).

The marine deposits occur as sandy beach ridges that, when coalesced, result in low terraces from 2 to 4 m above sea level. Old strips of barrier islands lithified by carbonate cement of corals and algae form lines of sandstone reefs parallel to the coastline, some of which remain exposed during low tides. In the internal estuarine sector, the plains are dominated by clay deposits and colonized by mangroves. In the most important rivers, such as the Paraíba, up to two levels of sandy terraces are recognized above the current intertidal plains, generally at elevations ranging from 4 to 10 m (Mabesoone and Cunha Silva 1991).

Fixed coastal dunes occur from Baia da Traição to the mouth of the Guaju river on the border with Rio Grande do Norte. In addition to the active frontal dunes, there are strips of up to 3 km in width of stabilized sand mantles, generally made up of reddish-colored sands and levels of concentration of organic matter from plant decomposition and illuviation processes. Parabolic shapes are still discernible on some stretches of the coast, such as Mataraca, with vegetated ridges reaching elevations of over 60 m.

Fig. 6.18 Cliffs on the sediments of Barreiras Formation. Along the coast those landforms may reach up to 80 m in elevation. *Source* the authors

6.6 Piedmont of the Borborema Highlands and the Cariatá Graben

The regional morphostructural unit interposed between the coastal tablelands and the Borborema Highlands to the west was regionally called "the Paraiba pre-coastal depression" (Carvalho 1982). This unit corresponds to the piedmont of the Borborema Highlands, as defined by other authors (Mabesoone and Castro 1975; Corrêa et al. 2010) for the set of hilly forms with altitudes ranging from 50 to 200 m, interspersed with occasional topographically prominent Proterozoic residual massifs, which mark the transition to the eastern escarpment of the Highlands. The geological basement of this morphostructure is dominated by paleo- and meso-Proterozoic metamorphic complexes, crisscrossed by Brasiliana orogeny ductile shear zones. In Paraíba, the piedmont occurs immediately at the rear of the coastal tablelands. It is characterized by broad topographic hollows that follow the channels of the main rivers, such as the Paraíba, which acted as a local base level for the exhumation of former platform covers that previously capped the landscape (Fig. 6.19).

From east to west, this unit is affected by morphostructural controls resulting from the rejuvenation of the "macro-dome" of the Borborema Highlands, which in the area are expressed as topographical steps that separate altimetric levels in different

Fig. 6.19 Low-lying tablelands within the Cariatá graben structured on Pleistocene sediments. The Highlands of the Manuel Matos horst lie in the background. *Source* the authors

stages of dissection. As a result, the hillier pluriconvex landforms occur to the east in areas of higher rainfall totals. Further inland, the watershed divides become sharper forming linear ridges separated by pediment slopes displaying various degrees of fluvial entrenchment.

Southeast of Piedmont, along the course of the Paraíba River, there is an elongated topographically depressed area delimited by extensional and oblique faults reactivated in the Neogene, probably as late as in the Pleistocene/Holocene limit (Bezerra et al. 2008). This relief unit, called Cariatá graben (Brito Neves et al. 2001a, 2001b), is located east of the municipality of Itabaiana and is characterized by morphological elements derived from the action of morphotectonic controls, acting on the creation of accommodation spaces, and morphoclimatic factors responsible for the production and transport of unconsolidated surface covers. The main feature of the graben's landform is the relief inversion which turned the Neogene sedimentary fill into low tablelands along the channel of the Paraíba River. The inversion was caused by the continuous lowering of the base level within the graben, leading to the dissection of the deposits by first-order consequent streams (Figs. 6.20 and 6.21).

The graben presents a sequence of depositional forms lodged within a 20 km long and 10 km wide depression controlled by reactivated NE-SW trending Proterozoic faults. From the top of the interfluves to the banks of the Paraíba River channel, one can observe a succession of depositional geomorphic units: low-lying tabular plateaus structured in Neogenic unconsolidated sediments (debris flows and mud flows sequences), colluvium hillslopes, alluvial plains, and levels of fluvial terraces

Fig. 6.20 The Paraiba River constitutes the local base level within the Cariatá graben. The valley bottom is characterized by pockets of fluvial deposits including the contemporary floodplain and levels of terraces, some of Pleistocene age (Tavares 2010). *Source* the authors

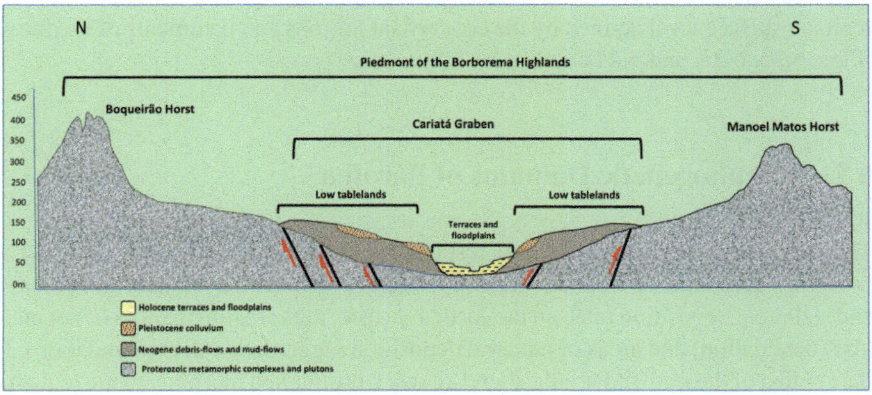

Fig. 6.21 North–South topographic transect of the Cariatá graben highlighting the main geomorphic features lodged within the Piedmont of the Borborema Highlands morphostructure. *Source* the authors

Fig. 6.22 Tablelands with zero-order catchments attesting to the initiation of dissection processes. The Boqueirão horst lies in the background marked by ENE-WSW trending ridges and inselgebirges. *Source* the authors

along the Paraíba River (Tavares et al. 2014). This scenario is laterally delimited by crystalline horsts, in the form of inselberg or inselgebirge residuals that rise above the regional surface level defined by the concordant hilltops and pediments of Piedmont (Figs. 6.22, 6.23, and 6.24).

6.7 Morphogenetic Domains of Paraíba

The morphogenetic domains directly respond to climatic events in a given region. The processes associated with the climate operate and control the weathering, erosion, and sediment deposition rates on the earth's surface, thus creating sectors of net mass loss (denudation) and areas of material deposition (aggradation or accumulation). In the context of the state of Paraíba, there are two well-marked morphogenetic regimes, the humid and the semi-arid (Fig. 6.25).

The humid domain is associated with high precipitation rates and a humid tropical climate, with annual rainfall varying from 1200 to 1700 mm. Thus, morphogenesis is characterized by the formation of thick weathering mantles, with the presence of soils of the Latosol and Argisol types. The coastal tablelands and pluriconvex hills characterize the relief in this sector of Paraiba. These landforms are dissected by a dense drainage network that becomes perennial ass the streams enter the humid

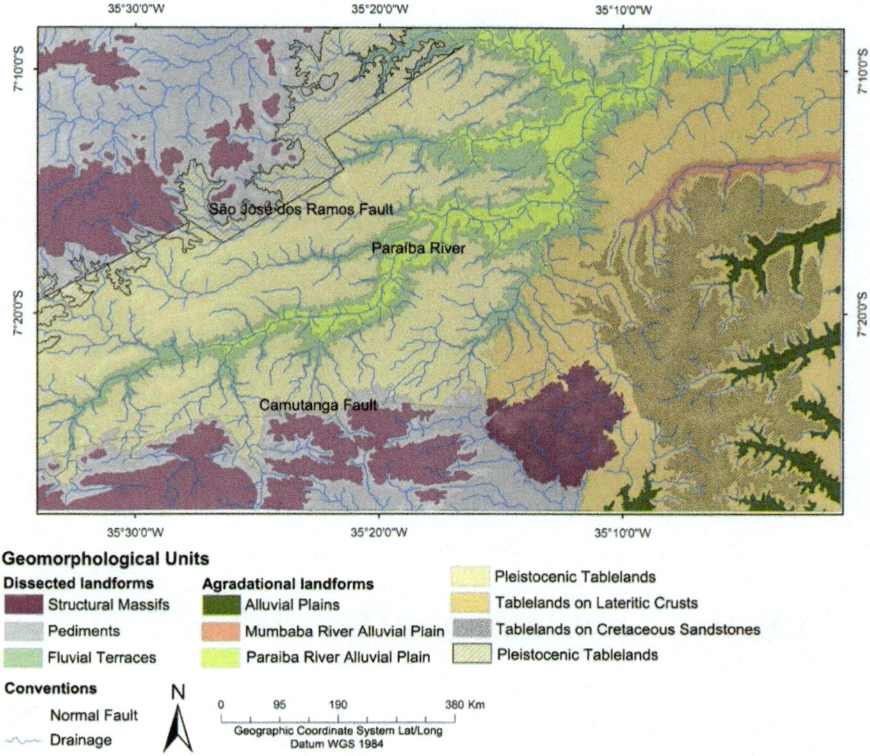

Fig. 6.23 Geomorphic units of the Cariatá graben delimited by the dissected surfaces of the Piedmont of the Borborema Highlands (north and south) and the uplifted westernmost sector of the coastal tablelands to the east. *Source* the authors

domain. The morphogenesis is notably marked by high rates of chemical alteration, which, associated with climatic conditions, trigger the remobilization of weathering mantles, generating mass movements—the humid tropical landscape is characterized by hillslope colluviation and sediment-filled valley bottoms.

The Semi-Arid Domain of Paraíba covers the entire sector of the Sertaneja Depression and part of the mega-morphostructure of the Borborema Highlands. Morphogenesis is associated with drier climatic conditions, with high temperatures and precipitation averaging 400–800 mm. This scenario is associated with extreme climatic events of great magnitude and low recurrence. Consequently, the landscape is dominated by thinner soil covers, and sediment storage is often limited to some sectors of the hillslopes and valley bottoms. Under the tropical semi-arid conditions of Paraiba, potential vegetation is represented by several arrangements of caatinga (thorn-brush vegetation and dry woodlands). Vegetation cover typically provides a buffer to surface processes; hence under the sparse caatinga, thin soils covered with debris pavements or vast expanses of rocky surfaces prevail. As a result, landforms are dominated by barren hillslopes and escarpments. Rock outcrops dominate the rugged summit

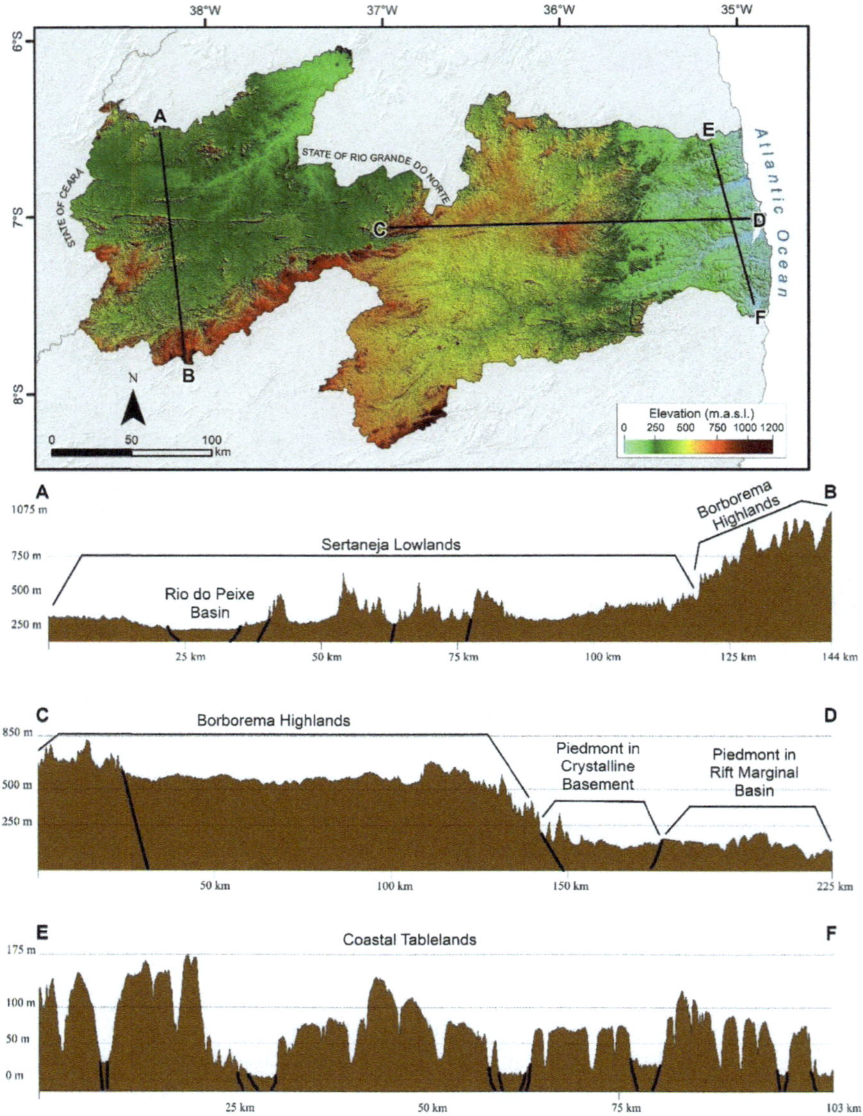

Fig. 6.24 Topographic transects indicating the topographic macro-features of the state of Paraíba and their main tectonic-structural controls (Shear Zones—black lines in topographic transects). *Source* the authors

surfaces in sectors of the Borborema Highlands. In the semi-arid Sertaneja depression, either well-preserved or dissected, pediments are typical. These level surfaces also provide particular topographic prominence to Inselberges and Inselgebirges residuals.

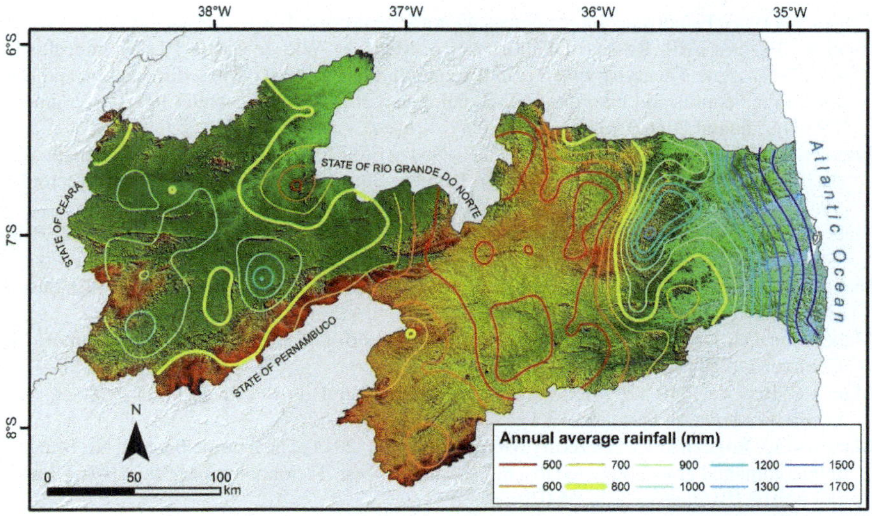

Fig. 6.25 Annual rainfall averages for the State of Paraíba highlighting the 800 mm threshold. *Source* the authors

References

Barbosa JA, Lima Filho MF (2006) Aspectos estruturais e estratigráficos da faixa costeira Recife–Natal: observações em dados de poços. Boletim De Geociências Da Petrobras 14(1):287–306

Bezerra FHR, Britoneves BB, Corrêa ACB, Barreto AMF, Suguio K (2008) Late Pleistocene tectonic-geomorphological development within a passive margin—the Cariatá trough, northeastern Brazil. Geomorphology 97(3/4):555–582. https://doi.org/10.1016/j.geomorph.2007.09.008

Brito Neves BB, Campos Neto MC, Van Schmus WR, Santos EJ (2001a) O sistema Pajeú-Paraíba e o maciço São José do Campestre no leste da Borborema. Revista Brasileira De Geociências 31(2):1–15

Brito Neves BB, Campos Neto MC, Van Schmus WR, Fernandes TMG, Souza S (2001b) O terreno alto Moxotó no leste da Paraíba (maciço Caldas Brandão). Revista Brasileira De Geociências 31(2):16–31

Brito Neves BB, Fuck RA, Pimentel MM (2014) The Brasiliano collage in South America: a review. Brazilian J Geol 44(3):493–518. https://doi.org/10.5327/Z2317-4889201400030010

Carvalho MGDF (1982) Estado da Paraíba: classificação geomorfológica. Editora da UFPB, João Pessoa

Corrêa ACB (1997) Mapeamento Geomorfológico de Detalhe do Maciço da Serra da Baixa Verde, Triunfo—PE: estudo da relação entre a distribuição dos sistemas geoambientais e a compartimentação geomorfológica. Dissertação de Mestrado. UFPE, CFCH, Recife, 183 p

Corrêa ACB, Tavares BAC, Monteiro KA, Cavalcanti LCS, Lira DR (2010) Megageomorfologia e morfoestrutura do Planalto da Borborema. Revista Do Instituto Geológico, São Paulo 31(1/2):35–52. https://doi.org/10.5935/0100-929X.20100003

Corrêa ACB, Tavares BAC, Lira DR, Mutzenberg DS, Cavalcanti LCS (2019) The Semi-arid domain of the Northeast of Brazil. In: Salgado AAR, Santos LJC, Paisani JC (eds) The physical geography of Brazil. Springer, Dordrecht

Crandall R (1910) Geografia, Geologia, Suprimento d'água, transportes e açudagens. I.F.O.S., Série 1BD (Publicação, 10), Rio de Janeiro

Czajka W (1958) Estudos geomorfológicos no nordeste do Brasil. Rev Bras Geogr 20(2):135–180
Lima JC, Bezerra FHR, Rossetti DF, Barbosa JA, Medeiros WE, de Castro DL, Vasconcelos DL (2017) Neogene–Quaternary fault reactivation influences coastal basin sedimentation and landform in the continental margin of NE Brazil. Quatern Int 438(Part A):92–107. https://doi.org/10.1016/j.quaint.2016.03.026
Lima Filho MF, Barbosa JA, Neumann VHML, Souza EM (2005) Evolução estrutural comparativa da Bacia de Pernambuco e da Bacia da Paraíba. Simpósio Nacional De Estudos Tectônicos 10:45–47
Mabesoone JM, Castro C (1975) Desenvolvimento geomorfológico do Nordeste Brasileiro. B Soc Geol Núcleo Nordeste 3:5–35
Mabessone JM, Silva JC (1991) Aspectos Geomorfológicos. Estudos Geológicos (UFPE). Série b, Estudos e Pesquisas 10:117–132
Mabesoone JM, Neumann VHML (1995) Mapa geomorfológico de Pernambuco. Simpósio De Geologia Do Nordeste 16:250–253
Maia RP, Bezerra FHR (2014) Condicionamento estrutural do relevo no nordeste setentrional brasileiro Mercator. Fortaleza 13(1):127–141
Marques FO, Nogueira FCC, Bezerra FHR, de Castro DL (2014) The Araripe Basin in NE Brazil: an intracontinental graben inverted to a high-standing horst. Tectonophysics 630:251–264. https://doi.org/10.1016/j.tecto.2014.05.029
Matos RMD (1992) The northeast Brazilian rift system. Tectonics 11(4):766–791. https://doi.org/10.1029/91TC03092
Monteiro KA, Corrêa ACB (2020) Application of morphometric techniques for the delimitation of Borborema Highlands, northeast of Brazil, eastern escarpment from drainage knick-points. J S Am Earth Sci 103:102729. https://doi.org/10.1016/j.jsames.2020.102729
Oliveira GP, Corrêa ACB, Tavares BAC, Monteiro KA (2023) The influence of cenozoic magmatism on drainage rearrangement processes of the northeast sector of the Borborema Highlands, northeastern Brazil. J S Am Earth Sci 121:104124. https://doi.org/10.1016/j.jsames.2022.104124
Peulvast JP, Claudino-Sales V (2004) Stepped surfaces and Paleolandforms in the Northern Brasilian Nordeste: constraints on models of morfotectonic evolution. Geomorphology 3(1/2):89–122. https://doi.org/10.1016/j.geomorph.2004.02.006
Santos EJ, Ferreira CA, Silva Jr. JMF (2002) Geologia e recursos minerais do estado da Paraíba. CPRM, Recife, pp 234
Santos EJ, Souza Neto JA, Silva MRR, Beurlen H, Cavalcanti JAD, Silva MG, Dias VM, Costa AF, Santos LCML, Santos RB (2014) Metalogênese das porções norte e central da Província Borborema. In: Silva MG, Rocha Neto MB, Jost H, Kuyumjian RM (eds) Metalogênese das províncias tectônicas brasileiras. CPRM, Belo Horizonte
Santos FG, Pinéo TRG, Medeiros VC, Santana JS, Morais DMF, Vale JAR, Wanderley AA (2021) Mapa geológico da província Borborema: projeto geologia e potencial mineral da província Borborema. CPRM, Recife
Tavares BAC, Corrêa ACB, Lira DR, Cavalcanti LCS (2014) Compartimentação geomorfológica e morfotectônica do Gráben do Cariatá, Paraíba, a partir de imageamento remoto. Revista Brasileira de Geomorfologia 15(4):523–538. https://doi.org/10.20502/rbg.v15i4.531
Tavares BAC (2010) A participação da morfoestrutura na gênese da compartimentação geomorfológica do gráben do Cariatá, Paraíba. Recife, PE. 137f. 2010. Dissertação de Mestrado, Universidade Federal de Pernambuco. https://repositorio.ufpe.br/handle/123456789/6407.

Bruno de Azevedo Cavalcanti Tavares He holds a bachelor's degree (2007), a master's degree (2010), and a Ph.D. in geography (2015) from the Federal University of Pernambuco (UFPE), with all research focused on structural geomorphology and morphotectonics. He is a member of the Quaternary study group of the Brazilian Northeast (GEQUA). Since May 2016, he has been an adjunct professor at UFPE in the archaeology department, where he works with research focused on geoarchaeology and spatial archaeology. He is the vice-coordinator of the geoarchaeology laboratory (LABGEOARQ) of UFPE and vice-leader of the group for the study of archaeology, geoscience, and technology (GEAGT). He has experience in geoscience with an emphasis on geomorphology, working mainly on structural geomorphology, morphotectonics, morpho-stratigraphy, paleoenvironmental reconstruction, and platform relief evolution.

Antônio Carlos de Barros Corrêa He holds a degree in geography from the Federal University of Pernambuco (UFPE) (1994), an undergraduate exchange program at Radford University, Virginia, USA, a master's degree in geography (1997) from the Federal University of Pernambuco (UFPE), a Ph.D. in geography (2001) from the Paulista Júlio de Mesquita Filho State University with a Sandwich Internship at Durham University, UK, and a post-doctoral fellowship in geomorphology (2003) from the Júlio de Mesquita Filho Paulista State University. He is a professor and researcher at the Federal University of Pernambuco, a CNPq research productivity fellow (level 2), a permanent member of the graduate programs in geography and archeology, and the leader of the Quaternary study group of the Brazilian Northeast (GEQUA) and the Quaternary geomorphology laboratory at UFPE. He has experience in geoscience with an emphasis on geomorphology, acting mainly on Quaternary geomorphology, geomorphology of semi-arid environments, historical and structural geomorphology, application of morphometric indexes in geomorphological analysis, and geosystemic analysis.

George Pereira de Oliveira He holds a bachelor's degree in geography (2017) from the Federal University of Rio Grande do Norte (UFRN), Seridó Higher Education Center (Caicó-RN), a teaching degree in geography (2022) from the Federal University of Pernambuco (UFPE), a master's degree in geography (2019) from the graduate program in geography at the Federal University of Pernambuco, where he is a Ph.D. candidate in geography. He has experience in physical geography with an emphasis on structural and tectonic geomorphology, geomorphometry, geomorphological mapping, and integrated landscape analysis. He is a member of the Quaternary Study Group of the Brazilian Northeast (GEQUA) and the Study Group on Archeology, Geoscience, and Technology (GEAGT).

Daniel Rodrigues de Lira He holds a degree (2008), a master's degree (2010), and a Ph.D. in geography (2014) from the Federal University of Pernambuco (UFPE), having experience in physical geography with an emphasis on geomorphology, pedology, and geo-technology, working mainly on geomorphology, paleoenvironmental reconstruction, pedology (soil × landscape), geochemistry, geoprocessing and remote sensing, and teaching of physical geography.

Part IV
Eastern Northeast

Chapter 7
Geomorphology of Pernambuco State

Antônio Carlos de Barros Corrêa, Bruno de Azevedo Cavalcanti Tavares, Drielly Naamma Fonsêca da Silva, and Daniel Rodrigues de Lira

Abstract Pernambuco lies on bedrocks of the eastern sector of the Borborema Province, which consists of a collage of metamorphic terrains and plutonic intrusions related to the Brasiliana Orogeny (900–570 Ma) and the formation of the West Gondwana continent. The area is characterized by the predominance of crystalline lithologies separated by NE- SW and E-W shear zones. Phanerozoic sedimentary basins overlie the Proterozoic basement as a result of the platformization that followed the Brasiliana Orogeny and the opening of the South Atlantic in the upper Cretaceous. The up-warping of the passive oceanic margin that preceded the opening of the Atlantic and continued during the Paleogene and part of the Neogene resulted in the distribution of the morphostructures that make up the relief of Pernambuco. The accumulation landforms reflect the morphogenetic processes of the coastal humid tropical climate and the inland semi-arid, corresponding to continuous or fragmented alluvial plains and terraces according to the discharge regime, that varies from perennial to ephemeral. Overall, the morphological units are distributed as a succession of "stepped" surfaces starting from the coast, increasing in elevation to 1200 m on the line that establishes the regional divide between the drainages that run to the Atlantic and those that flow west and southwest toward to the São Francisco River about 300 km inland.

Keywords Regional uplift · Proterozoic shear zones · Stepped landforms

---

A. C. de B. Corrêa (✉) · B. de A. C. Tavares · D. R. de Lira
Federal University of Pernambuco, Recife, Brazil
e-mail: antonio.correa@ufpe.br

B. de A. C. Tavares
e-mail: bruno.tavares@ufpe.br

D. R. de Lira
e-mail: daniel.rlira@ufpe.br

D. N. F. da Silva
Metropolitan University of Technology, Santiago, Chile
e-mail: dfonseca@utem.cl

V. Claudino-Sales and J. F. Sobrinho (eds.), *Geomorphology of the Northeast Region of Brazil*, The Latin American Studies Book Series,
https://doi.org/10.1007/978-3-031-58874-7_7

7.1 Introduction

To propose a genetic and typological approach to the landforms of Pernambuco, one must start by addressing the morphostructural contexts, given the region's peculiar geological evolution. The hierarchy of relief forms currently observed in the landscape translates a chain of tectonic and denudational processes whose dynamics were first established at the end of the Mesozoic. However, the interaction between surface and deformational agents operating on a mosaic of lithologies that resulted in the current arrangement of regional physical settings dates back to the Paleoproterozoic. Understanding these formative events is essential for proposing a compartmentalization of landforms that maintains a desirable level of evolutionary coherence.

Almost all of Pernambuco's territory lies on the geological structures of the eastern sector of the Borborema Province, which consists of a collage of metamorphic terrains and plutonic intrusions directly related to the Brasiliana Orogeny (900–570 Ma) and the formation of the West Gondwana continent (Hasui 2012; Carneiro et al. 2012). The area is characterized by the predominance of crystalline lithologies separated by NE- SW and E-W shear zones (Brito Neves et al. 2000). Paleozoic and Mesozoic sedimentary basins, of lesser spatial expression, lie discordantly on the Proterozoic basement as a result of both the platformization processes after the end of the Brasiliana Orogeny and the stretching and rifting of portions of the continental crust that resulted in the final opening of the South Atlantic at the end of the Cretaceous (Fig. 7.1).

The phases of deformation and uplift of the passive oceanic margin that immediately preceded the opening of the Atlantic and continued during the Paleogene and

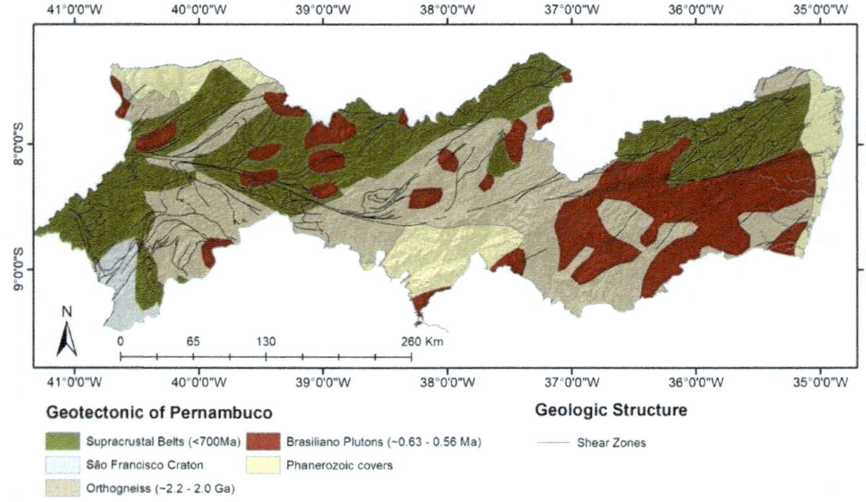

Fig. 7.1 Geotectonic map of the State of Pernambuco. Modified from Santos et al. (2021)

part of the Neogene were directly responsible for the distribution of the larger sets of forms that make up the relief of Pernambuco. The reactivation of the platform, from the north of the state of Rio Grande do Norte to the north of Alagoas, has triggered the operation of recurrent surface uplift mechanisms along the continental margin. The tectonic deformation resulted in the elevation of a regional dome, further enhancing the role of differential erosion, which engendered a mosaic of structurally controlled landforms at various spatial scales.

Furthermore, the presence of this conspicuous topographic bulge (the Borborema Highlands), less than a hundred kilometers from the coast, and its influence on the drainage network and the distribution of regional summit surfaces were addressed by some of the early scientific contributions regarding the evolution of the Pernambuco's physiography (Andrade 1958, 1968; Ab'Saber 1957; Czajka 1959). Thus, the early studies had already acknowledged the interplay between landform dissection, promoted by the fluvial network largely adapted to the rheological grid, and the geological framework, resulting in classical contributions concerning stream superposition and adaptation (Dresch 1957). Other perspectives focused on the dynamics of base levels (general and local), especially regarding the evolution of the oceanic margin—from the Upper Cretaceous onwards—and the role of glacio-eustatic levels on coastal terrace morphology (Andrade 1956).

The general topographic configuration of the eastern margin of the Borborema Province corresponds to a regional macro-dome (basin and swell topography) characteristic of passive platform margins—such as West Africa, marked by a regional up-warped morphology alternating highlands and lowlands within intervals of up to 1000 km. Besides, the regional dome is dissected by semi-radial drainage networks, whose evolution and continued post-rift uplift result from the action of thermal anomalies at the lithosphere-mantle boundary and their geodynamic responses on the surface (Klöcking et al. 2020).

Thus, the morphological units that make up the Pernambuco territory are distributed as a succession of "stepped" surfaces starting from the coast, increasing in elevation to the line that establishes the regional divide between the drainages that run to the Atlantic and those that flow west and southwest toward to the São Francisco River about 300 km inland (Fig. 7.2). From this uplifted dome—the Borborema Highlands—whose most elevated summits, between 1000 and 1200 m asl, correspond to outcrops of Brasiliano plutons, topographically raised by the Cenozoic reactivation and further highlighted in the landscape by differential erosion, the relief drops abruptly inland. Finally, on the banks of the São Francisco River, there is a vast low-lying surface structured on metamorphic complexes, Proterozoic fold-belts, and non-uplifted intracratonic and continental rift basins. This flat and poorly dissected compartment is subject to the semi-arid morphoclimatic system, which emphasizes the role of lithological controls on landforms, especially along scarp lines and the occurrence of isolated residual reliefs such as inselbergs and buttes.

In this drylands scenario, which makes up 85% of the State's surface, granite intrusions, igneous batholiths, and stocks stand out in elevation—as residual massifs. In contrast, fragments of sedimentary basins composed of asymmetrically tilted blocks

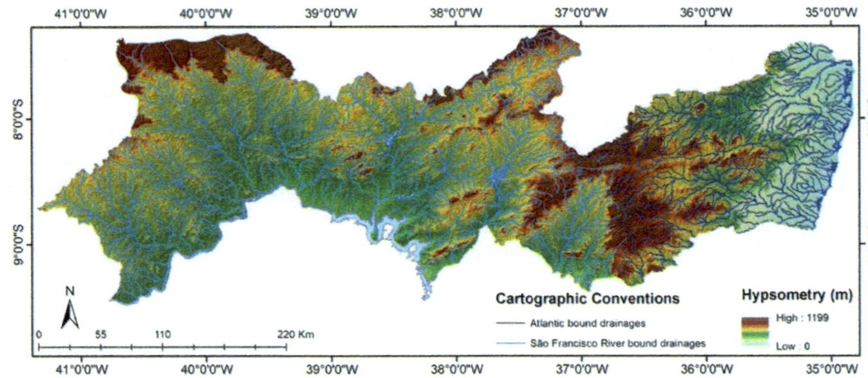

Fig. 7.2 Hypsometric map of Pernambuco with the drainage network, highlighting Atlantic and São Francisco bound watersheds. *Source* the authors

form small homoclinal tablelands. In the extreme northwest of Pernambuco, establishing the border zone with the State of Ceará, the table-like Araripe sedimentary plateau rises above the lowlands as a result of the relief inversion of the homonymous Cretaceous continental rift basin from the reactivation of inverse faults in the course of the Cenozoic (Marques et al. 2014).

7.2 Geology and Morphogenesis; the Morphostructural Approach

The territory of the State of Pernambuco lies within two major geotectonic domains of the Borborema Province, coinciding in its northern sector, roughly above the 8o S parallel with the Transversal Zone Domain, characterized by a pair of dextral strike- slip faults of E-W direction, the Patos and Pernambuco Shear Zones (Fig. 7.3), attributed to the final stages of the Brasiliana orogeny in the Neoproterozoic, and to the placement of numerous granitoid plutons related to that orogeny (Van Schmus et al. 2008; Castellan et al. 2020).

Although the structural architecture and the kinematic interpretation are complex and demand their studies of structural geology, the placement of Neoproterozoic granites along the shear zones and the respective mylonitization bands and migmatites associated with the surroundings of these intrusions result in a pattern of dissection in ridges and valleys. Further away from the deformation zones, the landscape alternates with a sequence of granitic and gneissic domes. Nevertheless, it is essential to emphasize that the role of the sub-vertical foliation of mylonites and gneisses associated with fractures controlled by this arrangement creates an elongated topography of ridges, hillslopes, and valleys adapted to the fracture network. Both in the Borborema macro-dome and to the west of it, the low rates of pedogenesis collaborate to emphasize this structural control, resulting in embedded rivers with long

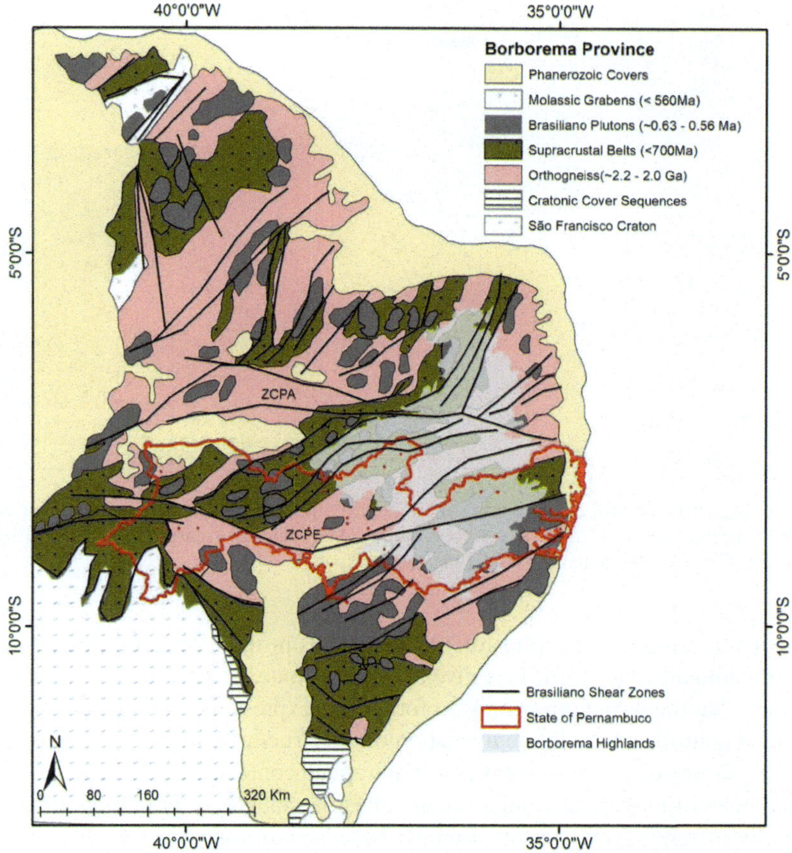

Fig. 7.3 Domains of the Borborema Province indicating the State of Pernambuco, the major shear zones and the outline of the Borborema Highlands. Modified from Santos et al. (2021)

straight stretches installed between remnants of rocky summits in different stages of dissection (Fig. 7.4).

The approach adopted for the subdivision of Pernambuco landforms is based on the proposal by Geramisov and Mescheirov (1968). They suggested using the term morphostructure for a landscape in which the current morphology of the crust surface reflects a combination of pre-existing geological structure and the work of exogenous agents. Cotton (1968) and Saadi (1991) propose that the term morphostructure be used in approaches of a regional nature, which emphasize the control of the geological structure—lithology and structure—upon the arrangement of the overall surface morphology.

The morphostructural units, according to Bates and Jackson (1987), correspond to the morphological elements of greater expression in the landscape, which configure a direct reflection of the geological structure on the surface, and may derive from the tectonic activity and its interaction with the external dynamics. Thus, the guiding

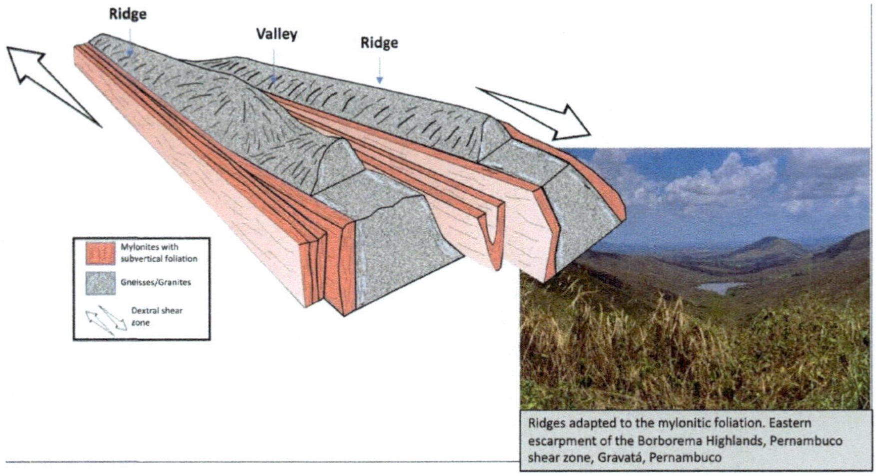

Fig. 7.4 Structural control of ridge and valley relief by the sub-vertical foliation of shear zones' mylonites in the eastern sector of the Borborema Province. Eastern escarpment of the Borborema Highlands, Gravatá, Pernambuco. *Source* the authors

element for defining a morphostructure results from this correlation expressed as the predominant relief form in a given regional context. Additionally, the authors point out that the term corresponds to forms that express the dominance of passive structural controls due to the rejuvenation of old structures, as usually occurs in the passive margins of platform areas and in intraplate contexts.

The geomorphological compartments of Pernambuco are presented based on a hierarchy of relief-form clusters, initially based on a proposal of geotectonic and, secondarily, lithostratigraphic subdivision. The largest regional units are linked to the hypsometrically defined sets of high and lowlands, from an east–west transect profile, from the coastline to the western end of the state, about 700 km inland. This choice points to the fundamental role that the Borborema Highlands macro-dome exerts on the delimitation of the most significant relief components of the State (Fig. 7.5).

Secondarily, on a smaller spatial scale, the morphological units were defined by the lithostructural control expressed by the exposure of metamorphic complexes and Proterozoic plutonic intrusions, in addition to the sedimentary sequences—mostly siliciclastic—of the marginal coastal rift and continental rift sedimentary basins (Fig. 7.5).

The lithostructural elements also exert control over the morphological units, reflected in either lowered or locally elevated forms. Differential erosion leads to the partial conservation of summit surfaces, which present different degrees of dissection and fluvial entrenchment. The latter is directly influenced by the trend and intensity of the fracture framework. Moreover, bedrock foliation associated with the shear zones and other anisotropic features of mylonites, migmatites, and gneisses collaborates to the high degree of adaptation of the drainage network to the geological structure.

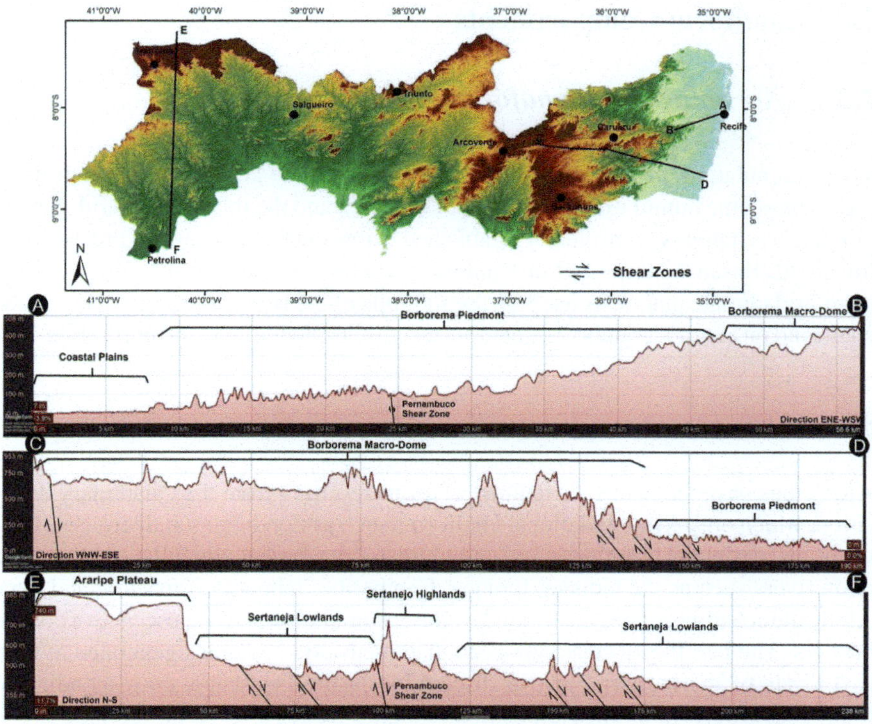

Fig. 7.5 Topographic transects indicating the topographic macro-features of the state of Pernambuco and their main tectonic-structural controls (Shear Zones). *Source* the authors

According to the morphostructural approach described above, digital terrain modeling was carried out. The 3D analysis of the landforms relied on elevation data from ALOS PALSAR. These data have been resampled from SRTM resulting in a spatial resolution of 12.5 m, which allowed the construction of shaded relief block diagrams with a good level of detail. The interposition of contour lines and the elaboration of the shaded relief model were performed using ArcGIS 10.1 software. The terrain modeling sought to highlight the morphostructures and regionally delimit the geomorphic features of each studied area.

7.3 Landform Compartments

7.3.1 Accumulation Landforms

The accumulation landforms in the state of Pernambuco are subdivided into those representing the humid tropical domain of the coast and the inland semi-arid tropical domain. The latter is more clearly established below the 800 mm annual precipitation threshold. Based on a rainfall distribution gradient from the coast to the west, the semi-arid climate manifests itself on the foothills of the Borborema Highlands in the south and on the moderately dissected surfaces of its piedmont to the north (Fig. 7.6).

7.3.1.1 The Humid Tropical Domain

Along the coast, the plains are defined as forms resulting from the Quaternary deposition, which created aggradational landforms such as floodplains and terraces. They occur discontinuously following a longitudinal arrangement along the course of the main rivers such as the Goiana, Ipojuca, Capibaribe, Jaboatão, Tapacurá, and Una. Along the coastline, plains are formed by the accumulation and coalescence of coastal barriers. The late Pleistocene/Holocene glacio-eustatic dynamics generated at least two levels of marine terraces still discernible in the landscape, corresponding to higher sea levels, in altimetric positions ranging from 10 to 12 m, the oldest, and from 0 to 6 m, the most recent ones (Suguio et al. 2011).

In the interior of the estuaries, the plains evolve by fixing tidal flats, deltaic bars, and mangroves or filling small lagoons whose original outlets were intercepted by advancing coastal bars. Areas of active lagoon accumulation occur in the coastal

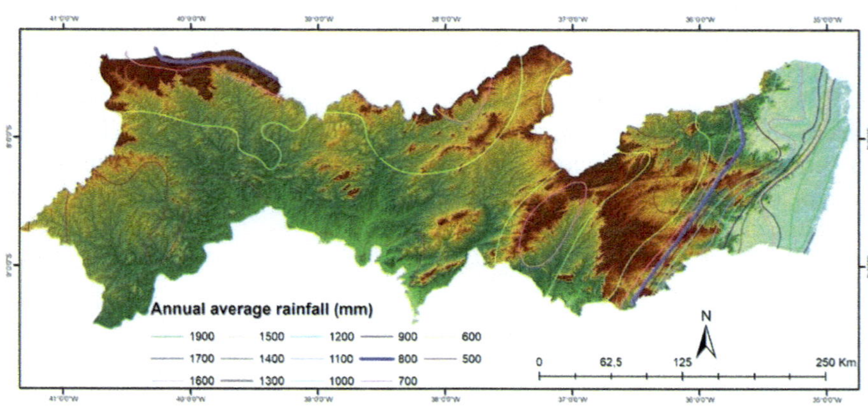

Fig. 7.6 Annual rainfall averages for the State of Pernambuco highlighting the 800 mm threshold. *Source* the authors

environment, to the rear of beach ridges and barriers, as in the case of the Araçá and Olho D'água lagoons in the Recife Metropolitan Region.

The sedimentation dynamics of the estuaries, from the last marine transgression between 6 and 5 Ka AP onward (Suguio et al. 2011), resulted in the formation of the most expressive fluvial-marine plains in the state. The growth of coastal barriers from south to north resulted in the partial closure of estuaries' mouths, most of the "wave-dominated" type (Santos et al. 2011). The rise in base level inside the lagoons led to the creation of small deltas and intertidal plains colonized by mangroves. Among the plains formed by the filling of estuaries are those of the Capibaribe/Beberibe river, where the original urban site of Recife is located; the Ipojuca river south of the capital, the Sirinhaém river, and the Rio Formoso estuary further south (Fig. 7.7).

On the perennial stretches of rivers that flow to the coast, segments of plains occur, sometimes flanked by one or two levels of terraces. The final courses of the river carve the Neogene sediments of the Barreiras Formation, to the north of Recife, or of the Pernambuco basin, to the south, resulting in more extensive fluvio-marine

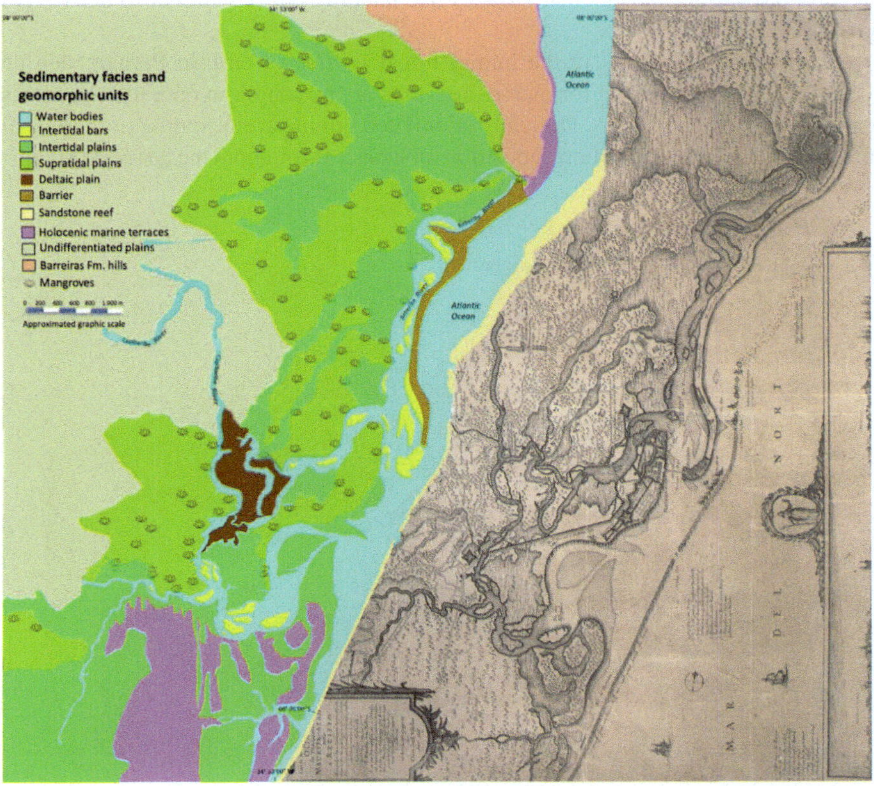

Fig. 7.7 Original sedimentary facies and geomorphic units of the Capibaribe River estuary from the interpretation of Cornelius Goliath (1648) map. *Sources* modified from http://objdigital.bn.br/objdigital2/acervo_digital/div_cartografia/cart535284/cart535284.html

plains with well-marked channel entrenchments. Upstream, on crystalline rocks, a regional "fall-line" can be observed over rocky sills, making it difficult to deepen the channel incision and giving rise to rapids and small waterfalls. It is assumed that the depth of vertical dissection that resulted in suspended fluvial terraces along the lower segments of the rivers was driven by headward erosion following the maximum lowering of the general base level ca. 18 Ka AP (Fonsêca et al. 2020). However, the incision came to a halt as the streams reached the underlying crystalline bedrock on the boundary between the aggradational coastal landforms and the Piedmont of the Borborema Highlands, where denudation forms prevail.

Despite occurring at a smaller spatial scale, as compared to the morphostructural compartments, hillslope deposits along the humid eastern seaboard of the State play a significant role in the evolution of hilly dissected topography and escarpments. They occur as coarse colluvium derived from the accumulation of mass movements—shallow translational slides—high energy gravitational flows, or as colluvial-alluvium ramps accompanying the narrow fluvial plains that delimit the hillslope compartments from those dominated by alluvial deposition. In addition, at several locations, abandoned terraces rise laterally above entrenched active river channels (Fig. 7.8).

Most of such deposits in the humid forested zone of eastern Pernambuco are stabilized, and their initial formative processes are no longer in operation under the current climatic conditions. Optically stimulated luminescence dating of these sediments indicates that colluvium deposition took place throughout the last 50,000 years.

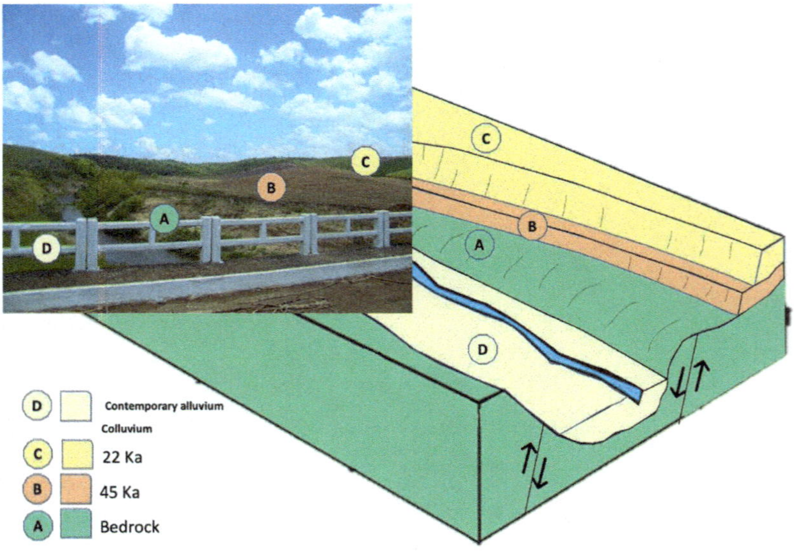

Fig. 7.8 Sequence of alluvio-colluvial sediments disconnected from the contemporary channel upstream of a shear zone, Capibaribe-Mirim River, Borborema Piedmont Modified from Fonsêca et al. (2020)

Nonetheless, hillslope deposition ceased around 5000 years BP with the onset of late Holocene climates (Fonsêca et al. 2020).

7.3.2 The Semi-Arid Tropical Domain

Spatially fragmented, the aggradational landforms characteristic of the semi-arid domain of Pernambuco reflect the varied combinations of rainfall regimes and dry woodland (caatingas) vegetation cover, from the transition to moister coastal environments to the severely dry landscapes to the west. In all cases, the interannual variation in precipitation is only comparable to Earth's subtropical deserts. Consequently, three types of accumulation landforms are identified: hillslopes—subdivided into upper, middle, and lower (debris-covered pediments); fluvial (ephemeral and intermittent) and lacustrine. The latter corresponds to the pelitic facies of deposits accumulating in rock basins (weathering pits), sometimes occupying the entire accommodation spaces of such morphological features.

Hillslope deposits made up of coarse colluvium are the most common in the semi-arid domain of Pernambuco. They derive from the accumulation of high energy gravitational flows: debris flows and mudflows, with a secondary contribution of thin composite alluvial fans. The majority of these deposits are stable under current climatic conditions. Optically stimulated luminescence dating indicates that colluvium accumulated preferably along the last 30,000 years (Corrêa et al. 2015); however, hillslope sedimentation after 5000 years BP, with the onset of late Holocene climatic conditions, is not spatially representative.

Weathering pits situated mainly on pediments display a mixed evolution, combining saprolitic and sub-aerial processes (Silva et al. 2017). They are also filled up by gravitational flows, alternating gravel, mud, and lacustrine sedimentation layers. In Pernambuco, these landforms are restricted to semi-arid landscapes, as in the rest of the northeast of Brazil. The sedimentary filling of the weathering pits, as in the occurrences at the municipalities of Salgueiro (Lagoa do Uri) and Brejo da Madre de Deus (Lagoa do Icó), spans the interval between 60,000 and approximately 4000 years BP. They also constitute invaluable archeological and paleontological geo-archives, as they preserve fossils of the Pleistocene megafauna and, in the upper layers, artifacts of the peoples that inhabited their surroundings throughout the Holocene.

The semi-arid fluvial environment of Pernambuco testifies to the existence of three types of drainage regimens: the exotic perennial of the São Francisco River, the ephemeral, with occasional flow coinciding with the length of the rainfall events, and the intermittent, which may present a flow during a few weeks or even months along the rainy season.

Fluvial deposits dominated by bed load originate asymmetrical mid-channel bars, with an anastomosed pattern and little channel incision. Shallow, gravelly channels are predominant, with cut-and-fill structures along river banks (Mabesoone 1984; Almeida and Corrêa 2020). Significant flood deposits and terraces are restricted to intermittent streams of larger size, such as the Pajeú River and along the margins of the São Francisco River.

The São Francisco River fluvial plain in its sub-middle stretch in Pernambuco has a polygenetic origin. Sediments observed along terrace cuts, and excavated trenches start with gravelly bars at the bottom, dating back to the onset of the Last Glacial Maximum, continuing until the late Holocene with sandy flood and lacustrine deposits capping the sequence (Lira 2014). After 30,000 years BP eolian sedimentation also starts to occur along the river stretch between the present towns of Petrolina and Petrolandia, leading to the accumulation of dune fields composed of parabolic, barchanoid, and nebkha types (Fig. 7.9). Some sand sheets display evidence of eolian reworking as late as 700 years BP attesting to the occurrence of intervals of more intense continental aridity (Lira 2014).

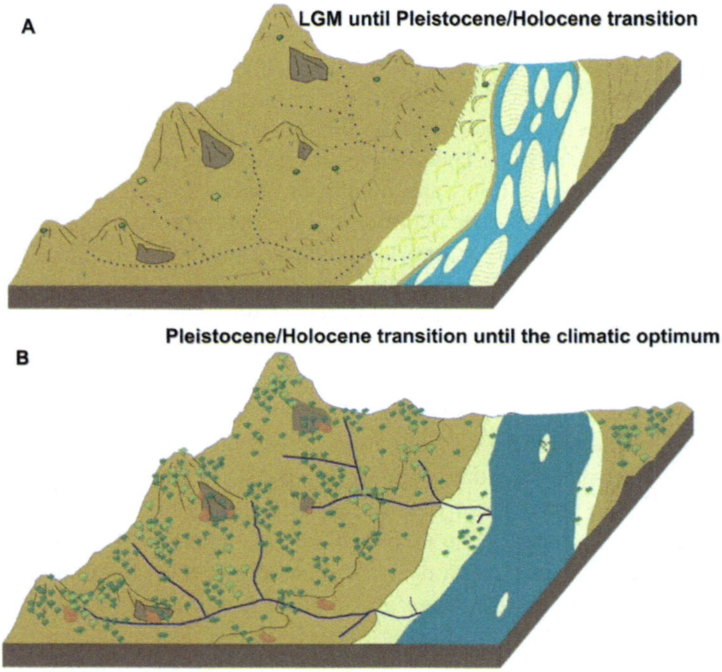

Fig. 7.9 Landscape evolution of the Petrolina region, sub-middle São Francisco Modified from Lira (2014), indicating the polygenetic evidence of a fluvial plain accumulation along the late Pleistocene/Holocene

7.3.3 *Denudation Landforms*

The landform classification proposed for Pernambuco is based primarily on morphostructural controls considering the interaction between the geological framework lithostratigraphy and tectonic—and the overall morphoclimatic setting. The geomorphic nomenclature followed the proposal of Corrêa et al. (2019) for Northeast Brazil, adjusted to the scale and specific characteristics of the state's geography. The morphoclimatic aspects are considered subordinate to the geological ones. For this contribution, they define the role of different regional morphogenetic systems in determining large-scale morphological relief aspects rather than rates of surface process operation. The applied typology aims at emphasizing the contrast between the landforms of the narrow, humid tropical coastal facade from those of the prevailing semi-arid tropical domain inland. The main denudation morphostructural units of Pernambuco are listed below:

i. Piedmonts in Neoproterozoic fold-belts and marginal rift basins
ii. Highlands in Neoproterozoic fold-belts
iii. Depressions in Neoproterozoic fold-belts
iv. Plateaus in Phanerozoic basins
v. Depressions in Phanerozoic basins.

7.3.3.1 Piedmonts in Neoproterozoic Fold-Belts and Marginal Rift Basins the Piedmont of the Borborema Highlands

Starting from the Atlantic façade, the geomorphological compartments of the State of Pernambuco can be defined, in a first approximation scale, based on their geotectonic subdivision and, secondarily, lithostratigraphic. The landforms thus defined are delimited by the emerged sectors of the marginal rift basins and the lithologies of the Proterozoic metamorphic complexes and Brasiliano plutons at altitudes below 200 m. In addition, geological structures, such as shear zones, faults, and networks of joints and fractures, are also manifested in the relief through the delimitation of sets of forms subordinated to lithology, sometimes encompassing more than one geological domain under the same morphostructure.

In this work, the terminology "piedmont" follows the same connotation implied by Allaby (2008) to designate the lowlands in front of an area of prominent relief, highlands, plateau, or mountain. In the case of the Borborema Highlands, this corresponds to its original escarpment, facing the passive oceanic margin, inherited from the rifting process at the end of the Mesozoic. The establishment of the ocean provided the base-level control conditions for the evolution of the escarpment and its Piedmont by denudation and headward erosion. Furthermore, the terminology is classically used for this sector of the state's relief vis-a-vis the contributions of Mabesoone and Castro (1975), Brasil (1983).

The landforms that antecede the escarpment of the Borborema Highlands are generally hilly, with convex summits rarely exceeding the altitude of 200 m. Further to the west, as the coast moisture diminishes, the hilly topography gives place to elongated ridges of E-W and NE-SW trends controlled by the mylonite bands. To the north of the Pernambuco Shear Zone, the piedmont presents relics of level summit surfaces, with vertical drainage dissection in excess of 100 m. These flat divides elevate like ramps from the coastal zone to the foothills of residual massifs situated ahead of the Borborema Highlands escarpment (Fig. 7.10). In this sector, disjunct fragments of Neogene unconsolidated sedimentary covers are also found due to past hillslope processes or preserved as remnants of elevated gravelly fluvial terraces.

The Borborema Piedmont can be morphologically divided according to the underlying lithological controls. On lithologies corresponding to the metamorphic complexes and plutonic intrusions, the relief evolves into multi-convex morphologies, with increasingly sharp summits further away from the shoreline. Inland altitudes can reach up to 300 m in elevation on the top of the residual massifs and ridges that characterize the transition zone to the eastern escarpment of the highlands under drier tropical sub-humid conditions. This domain is noticeable for its hilly topography under different degrees of dissection, presenting isolated elevations or patches of continuous multi-convex landforms. Most of the time, the area's elevation lies below 200 m.

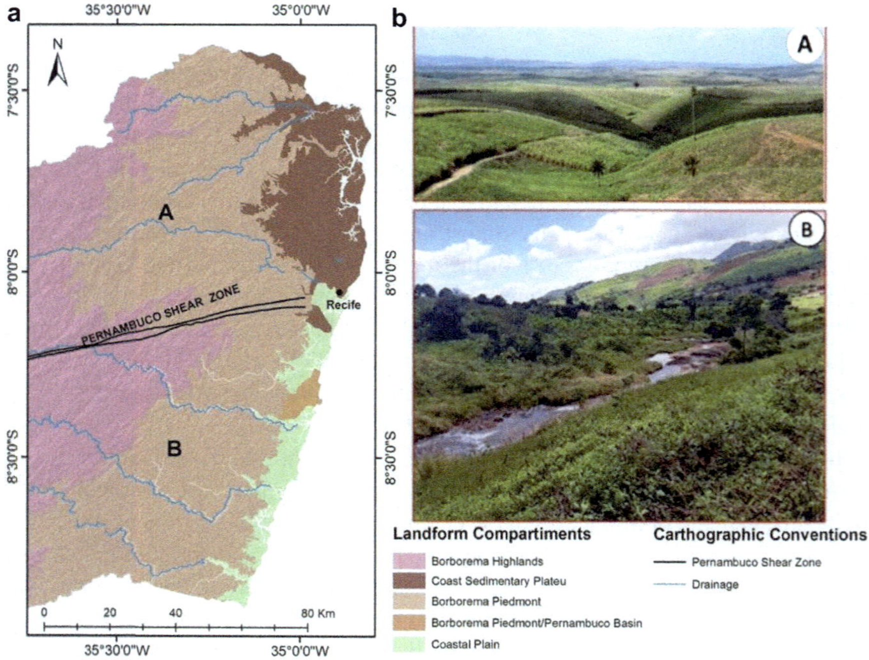

Fig. 7.10 Aspects of landscape morphology and dissection in the Borborema Piedmont, State of Pernambuco, north (**a**) and south (**b**) of the Pernambuco shear zone. Modified from Monteiro and Correa (2020)

The unit is crisscrossed by a dense perennial drainage network of dendritic pattern, with rivers presenting stretches of straight channels, right-angle inflections, and knick-points associated with rapids and small waterfalls. The valleys are well-fitted with a "V" cross-section adapted to the underlying linear structures. Alluvial sedimentation reaches the base of the hills creating flat valley bottoms and small narrow strips of plains. At the western limit of the unit, in the transition to semi-arid conditions, small segments of pediment ramps are interposed between the fluvial accumulation sectors and the base of the hills.

To the south of the Pernambuco Shear Zone and eastward of the crystalline basement, on the siliciclastic sequences of the Pernambuco basin, the hilly surface presents less dissected landforms with broader summits without a clear-cut altimetric distinction between the two geological contexts. Slope segments are shorter and vary from convex to straight. In this sector, the Borborema Piedmont is directly structured on strongly weathered Cretaceous sediments and vulcanites of the rift basin. The influence of the humid tropical climate, dominated by the continuous influx of the moist southeasterly trade winds, favored the development of a thick kaolinitic weathering mantle. On the other hand, the uplift of the marginal basin, together with the adjacent metamorphic complexes and igneous intrusions, resulted in a homogeneous pattern of fluvial incision on both crystalline and sedimentary terrains. However, the main rivers flowing to the Atlantic tend to form rapids and sequences of falls along the contact zone between the crystalline rocks to the west and the sedimentary basin closer to the coast, thus developing a regional "fall-line".

7.3.3.2 The Low Coastal Sedimentary Plateau

To the north of the Pernambuco Shear Zone, the Neogene covers associated with the Barreiras Formation, overlying the Cretaceous sediments of the Paraíba basin, have been uplifted to altimetric levels oscillating 50 m, in the coastal cliffs, to nearly 200 m inland. The drainage is incised into V-shaped valleys, separating flat interfluves of varied widths (tabuleiros, lit. table-like), representing the most conspicuous morphology of this low-lying sedimentary plateau, gently tilted toward the coast (Fonsêca et al. 2016).

In this sector, the notably flat summits, with average interfluvial distances varying from a few hundred meters, close to the coastline, to tens of kilometers further to the northwest near the border with the State of Paraíba, have rendered this landscape the epithet "Superfície dos Tabuleiros (lit. Table-like surface)" by Mabesoone and Castro (1975). The authors have coined this term based on the morphostratigraphic relations, i.e., flat tops structured on unconsolidated Neogene sediments. They chose this terminology to differentiate this sector from the flat-topped hills that, further to the west, are established directly above the crystalline lithologies of the Proterozoic metamorphic complexes and plutonic intrusions referred to as "Superfície das Chãs (lit. Flat-topped surface)." In some circumstances, as in the Itamaracá Island, the tilting of blocks coupled with the deepening of the fluvial incision have exposed the underlying Mesozoic sediments along the valley sides.

7.3.3.3 Highlands in Neoproterozoic Fold-Belts the Borborema Highlands

The area of crystalline highlands that elevate approximately 60–80 km from the coastline, the Borborema Highlands, constitute the most representative and topographically prominent landform of the State. The dome-shaped highlands operate as a regional drainage divide and interfere directly with the intensity of the inland aridity as its eastern escarpment intercepts the moist southeasterly trade winds inducing orographic precipitation along the east-facing summit surfaces.

The highlands are structured on plutons inherited from the Brasiliana Orogeny, eroded fold-belts, and remains of Paleoproterozoic metamorphic terrains. This uplifted regional morphostructure, of dome-like shape, whose most elevated summits barely reach the altitude of 1200 m, originated from a series of apyrogenic pulses initially associated with the breakup of the western Gondwana continent. Later, in Cenozoic times, the uplift mechanisms responded to types of localized intraplate magmatism between 45 and 7 Ma BP, accompanied by a deep-seated thermal anomaly (Klöckling et al. 2020).

Notwithstanding, geochronological evidence points to the Borborema Highlands being uplifted only during the Cenozoic (Corrêa et al. 2010; Oliveira and Medeiros 2012). Gravimetric and magnetometric data suggest that the elevation derives from the placement of oceanic crust magma in the base of the continental crust, thus promoting an "underplating" process that pushed the continental margin upward. Such processes, operating on ancient inherited Proterozoic structures, created a mosaic of relief compartments, later reshaped by differential erosion processes (Corrêa et al. 2010).

The highlands are limited by a series of breaks in the topographic gradient (knicks), characterized by escarpments ranging on average from 100 to 200 m in elevation, reaching an excess of 500 m or higher at some places. These ruptures separate the highlands from the surrounding low-lying surfaces. The lithological control plays a minor role in delimiting the external escarpments of the highlands. However, the regional summit surfaces result, up to a certain level, from a lithological response to differential erosion. Hence, granites, syenites, and quartzites form domes (bornhardts and sugar loaves) or the highest regional level summits, reaching 1200 m in elevation (Tavares 2015). Besides lithology, the most elevated sectors of the highlands in Pernambuco result from a direct response to the tectonic control exerted by reactivated Neoproterozoic shear zones, especially the E-W trending Pernambuco shear zone, and those that branch out of NE-SW direction. These geologic structures delimit the inner escarpments of mountain ranges and lineaments of ridges (Fig. 7.11).

Inside the elevated areas, the remnants of Neoproterozoic fold-belts and metamorphic rocks with sub-vertical foliation plans and schistosity structure the low-lying areas which form intermontane depressions where the main drainage courses are installed, such as the Capibaribe and Ipojuca Rivers, draining directly to the Atlantic, or the Moxotó and Pajeú Rivers that run toward the São Francisco River to the west. In general, the surface of the intermontane depressions coincides with the

Fig. 7.11 DTM of the eastern escarpment of the Borborema Highlands. Boa Vista range, pluton inserted in dextral shear zone forming bornhardt, elevation 1201 m. *Source* the authors

500 m contour line. However, the surface level of the depression elevates from east to west up to 650 m on the foothills of the inner E-W escarpment that follows the Pernambuco shear zone. This geological structure delimits the southern boundary of the loftier sector of the highlands—the remobilized massifs of the "Transversal Zone" (Corrêa et al. 2010), an E-W trending relief compartment contained between the Pernambuco Shear Zone and the Patos Shear Zone to the north in the State of Paraíba.

In many aspects, the Borborema Highlands behave like a rejuvenated anteclise shaped like a regional crystalline macro-dome surrounded to the west by uplifted sedimentary basins that form table-like (mesas) and homoclinal plateaus. Another peculiar evidence of the relief's dome morphology derives from the drainage network dispersal pattern, roughly centrifuge radial, with catchments located around the elevated core of the macro-dome.

The complexity of landforms, inherited from a topographic swell morphology along a passive continental margin, added by the internal differentiation due to lithological, structural, and rheological traits, provide the highlands with the aspect of a regional rejuvenated massif. In terms of morphometry, the Borobrema Highlands can be classified as a "dissected mountainous relief" (Murphy et al. 1968), with local altimetric differences between valley bottoms and summit surfaces exceeding the

600 m threshold, as in the Baixa Verde and Ororubá ranges escarpments along the Pernambuco Shear Zone.

7.3.3.4 The Sertanejo Highlands

This morphostructure lies immediately west of the Borborema Highlands, forming a lower topographic level adjacent to the latter, whose summits do not exceed 750–800 m in elevation. To the west of the Borborema, the Sertanejo Highlands borders the southern escarpment of the Araripe Plateau, creating a moderately dissected topographic step between the uplifted sedimentary basin and the remarkably level surface of the Sertanejo depression further south. This landform also delimits the westernmost border of Pernambuco, with the State of Piaui, through a series of elongated flat-topped divides locally known as the Dois Irmãos range, structured on the metasediments of the Neoproterozoic Riacho do Pontal fold-belt (Caxito and Uhlein 2013; Tavares 2015).

This geomorphic compartment exhibits various degrees of differential dissection, from a "ridge and valley" pattern to a morphology dominated by clusters of convex hills, which rise above the regional base level. A low-lying flat surface characterizes the latter, the Sertanejo Depression, structured on coalescing pediment ramps under severe semi-arid environmental conditions (Fig. 7.12).

The Sertanejo Highlands and its sub-compartments are generally structured on Neoproterozoic meta-volcano-sedimentary and metasedimentary lithologies, constituting the country rocks for Brasiliano plutons which support the highest elevations of the Borborema Highlands to the east. The schistosity of these uplifted metamorphic

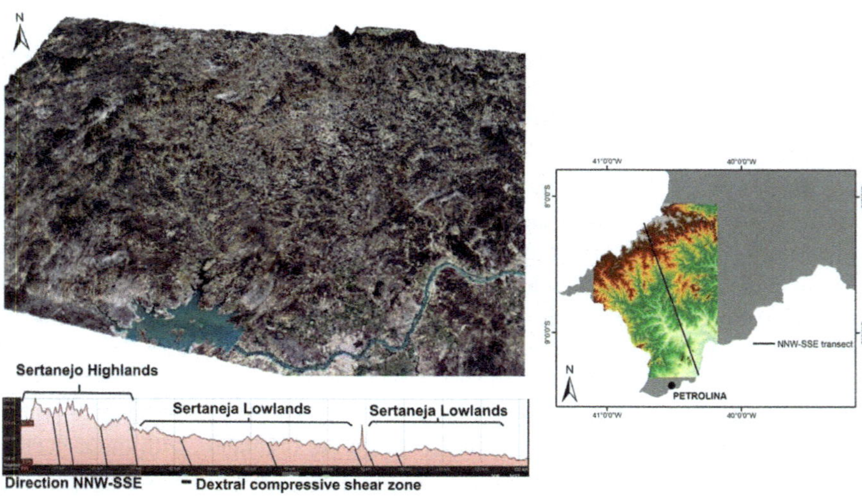

Fig. 7.12 DTM of the Sertanejo Highlands highlighting the lithological control of the Riacho do Pontal fold-belt and dextral shear zones on the dissected topography. *Source* the authors

terrains accounts partially for the dissection patterns of the landforms, with the occurrence of incised gorges, water gaps, and the occasional subordination of the drainage to foliation trends. In short, the Sertanejo Highlands presents a sequence of staggered surfaces, separating landforms with varying degrees of differential dissection according to elevation and lithological characteristics.

7.3.3.5 The Isolated Structural Massifs

Among the areas of prominent relief developed on Neoproterozoic fold-belts in Pernambuco, the isolated structural massifs constitute a separate category, differentiated from other spatially continuous larger geomorphic units. These morphologies are primarily associated with Brasiliano plutons placed along the Neoproterozoic shear zones that cut through the Borborema Province. They stand out in the regional landscape as isolated ranges and massifs, emerging from the bottom of the otherwise flat semi-arid depressions yet disconnected from the main continuous elevated areas such as the Borborema and the Sertanejo Highlands.

The morphology of these massifs is partly inherited from the emplacement of plutons within ductile shear zones. As a consequence of the original geotectonic setting, the resulting intrusions are flanked by mylonitization bands that, after being exposed to the surface by long-term denudation, structure the steep escarpments of the massifs. Compared to the surroundings, the topographic prominence of these landforms is also a response to differential erosion under the prevailing semi-arid conditions, principally when the igneous intrusions cut through the metasediments of the Neoproterozoic fold-belts (Fig. 7.13).

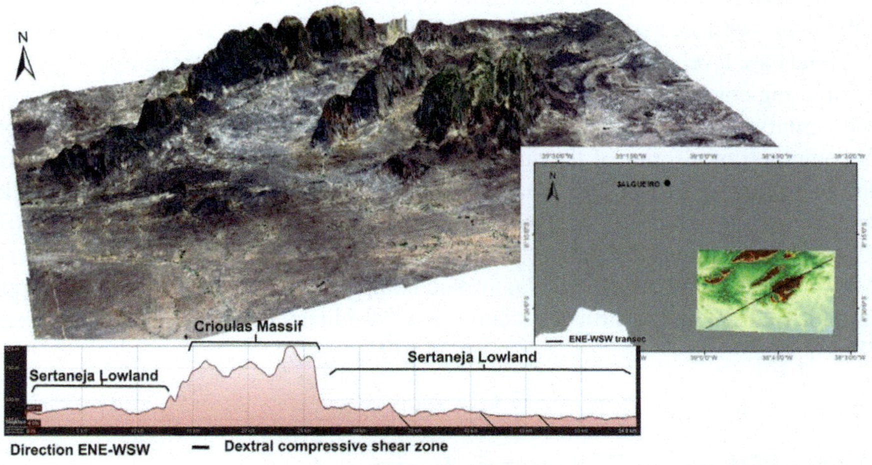

Fig. 7.13 DTM of the Crioulas massif highlighting its dissected morphology and residual character amidst the Sertanejo lowlands. *Source* the authors

For lacking spatial continuity with larger regional geomorphic units, the structural massifs occur as isolated elevations of dome-like or elongated aspects varying in size from a few hectares to hundreds of square kilometers. However, despite the geographical isolation, their summits may reach a wide range of elevations, often surpassing 1000 m of altitude. The maintenance of such lofty morphologies depends on a twofold process that combines the interaction of a pulsatile Cenozoic uplift operating along reactivated fault lines and the work of differential erosion that lowered the surrounding base levels structured on less resistant metamorphic host rocks (Gurgel et al. 2013; Maia and Bezerra 2014; Maia et al. 2015).

7.3.4 Depressions in Neoproterozoic Fold-Belts

To the west of the Borborema Highlands macro-dome, low-lying flat surfaces at elevations varying from 300 to 600 m extend from the margins of the São Francisco River to the foothills of the residual massifs and plateaus. This geomorphic unit occurs either as remnants of semi-arid continental planation surfaces, or intermontane depressions, generally structured on metamorphic complexes, supracrustal metasediments, or yet eroded intrusive suites dominated by porphyritic igneous rocks. These low topography areas present either a moderately dissected hilly morphology or vast expanses of remarkable level surfaces dominated by sequences of staggered pediment ramps with little dissection. Pediments are structured directly on the underlying bedrock and are often capped by thin gravel pavements. They extend from the base of the higher elevations, with which they form quite distinctive slope knicks, to the local base levels, dominated by fluvial accumulation zones (Corrêa 2001).

The surface lowering and denudation of the semi-arid depression on crystalline lithologies west and southwest of the Borborema Highlands are controlled by the regional base level defined by the São Francisco River. Variations in morphology and weathering covers are directly associated with underlying lithological controls and the vicinity of regionally uplifted areas to the north and east. Thus, the depression is composed of a series of flat surfaces that elevate gradually from the margins of the São Francisco River and are occasionally separated from each other by subtle breaks in slope that reflect the presence of lithological contacts (Corrêa et al. 2010). Changes in schistosity, foliation, or gneissic banding, especially within enclaves of Meso- and Paleoproterozoic metamorphic complexes, respond to changes in topographic rugosity and control the occurrence of residual landforms such as inselbergs (Fig. 7.14).

7.3.5 Plateaus in Phanerozoic Basins

The history of the sedimentary Plateaus in semi-arid Pernambuco dates back to the origins of the depositional basins from the Silurian to the opening of the Atlantic

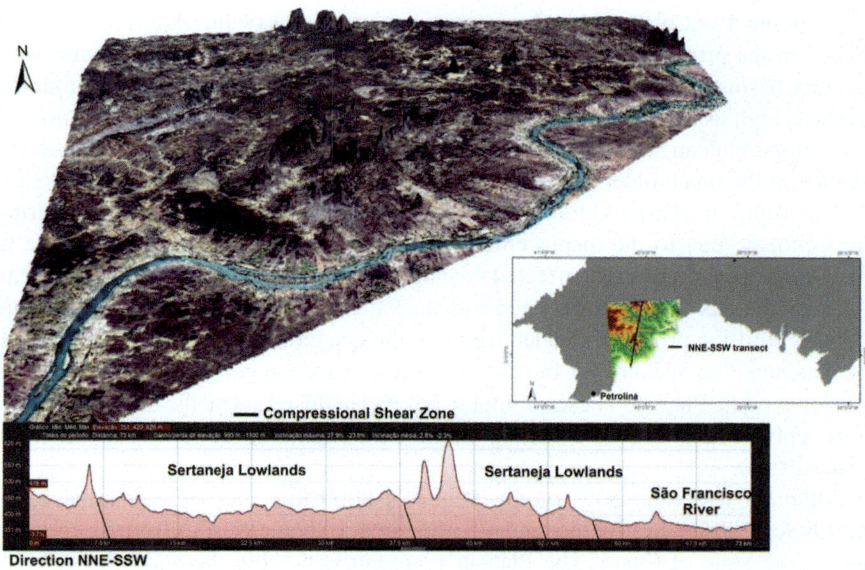

Fig. 7.14 Aspect of the Sertanejo lowlands bordering the São Francisco River, dotted with inselbergs and residual massifs. *Source* the authors

Ocean in the late Cretaceous, following the breakup of the West Gondwana continent (Schobbenhaus and Brito Neves 2003). Later, throughout the Cenozoic, some of these basins suffered topographic inversion due to the tectonic reactivation that affected the Borborema Province and the continental margin. Thus, as it occurred in other elevated areas of the State, the basins subjected to uplifting originated table-like and homoclinal (cuesta-like) plateaus, as in the case of the Araripe and Jatobá basins, respectively.

7.3.5.1 The Araripe Plateau

Located in the extreme northwest of Pernambuco, the Araripe Plateau originated from the opening of the transforming equatorial margin of Brazil. Brittle structures that affected the region's Proterozoic basement created the elongated geometric arrangement that characterizes the Araripe basin. Along the Cenozoic and as late as the Quaternary, these same structures were tectonically reactivated, forming the Plateau itself. The particular spatial disposition of the basin is controlled by geological faults that branch out from the Pernambuco shear zone to the south and the Patos shear zone to the north in the State of Ceará. These structures were reactivated during the rifting phase that opened the equatorial Atlantic, creating accommodation spaces that, despite being distant from the continental margin, allowed for marine ingression, converting the original depression into a continental rift basin with a peculiar infill as compared to other basins in the region.

As pointed out above, the characteristic morphology of the Araripe Plateau is related to the process of relief inversion resulting from the uplifting of some sedimentary basins in the region (Morales and Assine 2015), in many cases to altimetric levels as high as those of the source areas of the basin's sediments. Thus, sectors of the Araripe plateau summit surface exceed 1000 m in elevation, comparable to the altitude of the much older crystalline rocks of the Borborema Highlands to the east.

According to Marques et al. (2014), the basin's relief inversion occurred during the Cenozoic due to the displacement of inverted normal faults as a response to the compressive regime that affected the northeastern margin of the South American plate and the sets of faults associated with the Patos shear zone. The authors postulate that this uplifting process extended well into the Quaternary, hence, being one of the mechanisms that account for the elevation of Pleistocene colluvial deposits on the hillslopes of the Plateau. These colluvium covers are raised above the local base levels and deeply entrenched by the drainage along the northern façade of the Plateau (Lima et al. (2021).

Although less than 50% of the Plateau's 8000 km^2 lies in Pernambuco, its morphology differs in structure and morphoclimatic characteristics from its northern half in the State of Ceará. The Plateau's summit is notably flat and lithologically controlled. It coincides with the outcrop of the late Cretaceous Exu Formation, constituted of post-rift sandstones. On the northern escarpment, facing the State of Ceará, the siliciclastic sediments of the Exu Formation overly the pelites of the Santana Group, resulting in a series of topographical shoulders adjusted to the sediments' erodibility. To the south, in Pernambuco, under semi-arid conditions, the escarpment is dominated by a 200–250 m high cliff structured on the sandstones of the Exu Formation. On that side, the Santana Group (Romualdo Formation)'s outcrops are restricted to topographic hollows encountered on horst blocks.

On the southern escarpment, in the rain-shadow of the moisture brought along from the equatorial coast by the Intertropical Convergence Zone (ITCZ), a severe semi- arid climate and thorn-brush vegetation (caatinga) prevail. Most of the time, the steep sandstone cliffs (Exu Formation) lie discordantly over a narrow topographic step of the Sertanejo Highlands, which is structured on Brasiliano granodiorites. These highlands are bound to the south by an escarpment line adjusted to the SW-NE trending Ouricuri- Bodocó shear zone, a subsidiary of the Pernambuco shear zone. The topographic step is characteristic of an exhumed surface, which has lost part of its Cretaceous sedimentary cover following the area's uplift. This type of morphostructural control, however, is restricted to the southern half of the Plateau (Fig. 7.15).

7.3.5.2 The Jatobá Basin Plateau

This morphostructural unit is situated within the semi-arid domain, and its evolution is linked to processes that anteceded the establishment of the passive Atlantic margin of Northeast Brazil. Initially, in the Silurian, a N-S oriented elongate depression was formed, on which the conglomeratic sandstones of the Tacaratu and Santa

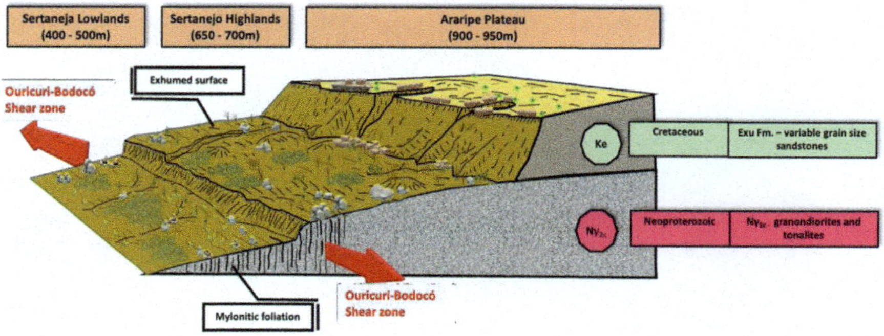

Fig. 7.15 Illustrative diagram of the morphostructural units along the southern escarpment of the Araripe Plateau in Pernambuco, under semi-arid conditions, highlighting the lithological and structural controls upon the slope breaks. *Source* the authors

Brígida Formations were deposited. Then, this original "sag" basin underwent a phase of rupture induced by the rifting processes that led to the opening of the South Atlantic, creating uneven accommodation spaces along a sequence of semi-grabens that controlled the deposition of Cretaceous units (Carvalho 2005; Corrêa et al. 2015).

The rifting process associated with the opening of the basin, initially in the shape of a proto-gulf along the recently opened eastern coast of South America, was halted further to the north as the crustal breakup reached the vicinities of the Pernambuco shear zone. This structural obstacle shifted the rifting direction eastward, abandoning the original inlet and defining the contemporary outline of the eastern coast of the Northeast approximately between the latitudes of 8° and 13° south. As the original rifting ceased to operate, the aborted arm of the triple junction turned into an aulacogen, whose accommodation spaces were filled up by the sedimentary sequences that make up the Recôncavo-Tucano-Jatobá basin's complex.

In Pernambuco, the morphostructural pattern of the Plateau that resulted from the uplift of the Jatobá basin (Fig. 7.16) presents a continuous altimetric lowering from E to W, following the original geometry of the aulacogen, made up of a sequence of grabens and horsts bounded by extensional faults. Along its S-SE edge, the Plateau presents a conspicuous homoclinal structure, with steep escarpments at the "front" and gentle dip-slopes that merge with pediments (glacis) to the west as they reach the local base levels along the main drainages.

The Plateau's elevation decreases from ca. 1100 m near the Borborema Highlands to the east to less than 500 m in the semi-arid depression to the west. The elevated sectors are controlled by horsts capped by the remnants of the basal depositional sequence of the Jatobá basin, the Tacaratu Formation of the Silurian. To the W and SW of the uplifted sector, the limit between the Plateau and the surrounding pedimented semi-arid surfaces are marked by discrete knicks accompanying extensional fault lines. In this sector, the Cretaceous sequences outcrops at the surface of the lowlands, structurally controlled by grabens that remained relatively unaffected by the overall Cenozoic reactivation to the east.

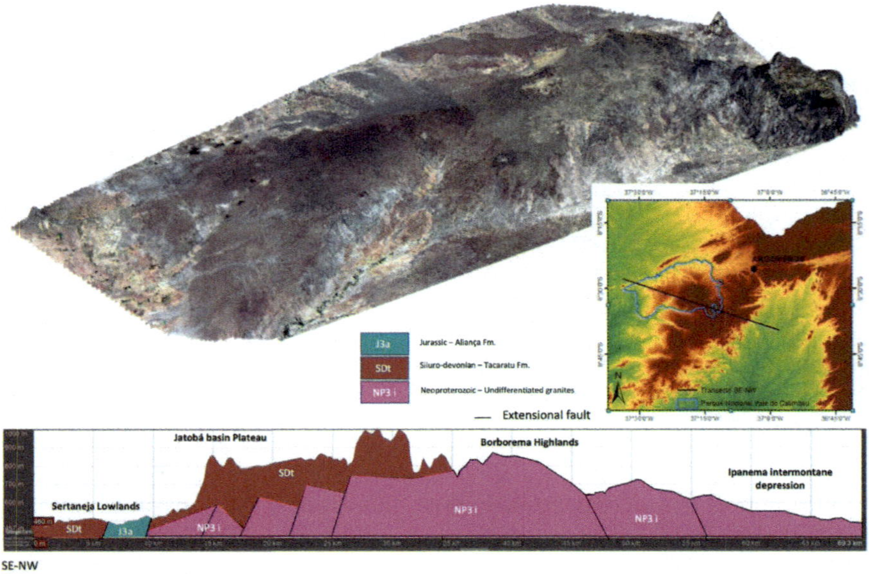

Fig. 7.16 Jatobá basin Plateau, northern sector, topographic transect and lithostructural conditioning. *Source* the authors

The eastern and southeastern limits of the Plateau are arranged in structurally controlled staggered topographic levels with flat summits. Deeply incised drainages subordinated to the network of fractures and joints, flanked by vertical to sub-vertical sandstone cliffs exceeding 200 m in height, separate homoclinal asymmetric mesas. The presence of residual landforms such as buttes, mesetas, and pinnacles ahead of the Plateau's outer escarpments attest to the operation of long-term exhumation and slope retreat by headward erosion.

7.3.5.3 Homoclinal Plateaus in Isolated Paleozoic Basins

Lying above the semi-arid Sertanejo Depression—primarily structured in the eroded remnants of Proterozoic fold-belts and metamorphic complexes—the sedimentary basins of Fátima, Betânia, and Carnaubeira form a fragmented semi-circle of asymmetric homoclinal tablelands immediately to the southwest of Borborema Highlands' western escarpment. These basins are the remnants of a previously more extensive Siluro-Devonian siliciclastic cover inherited from the original sag-basin phase of the Recôncavo- Tucano-Jatobá complex. The sediments inside them were most likely trapped by a system of grabens developed on the crystalline basement and later uplifted at least twice, once still at the end of the Cretaceous (Gomes et al. 2001).

In Pernambuco, the smaller basins of Cedro, São José do Belmonte and Mirandiba are still genetically correlated to the above. Nonetheless, in terms of geomorphology, they integrate the low-lying surface of the Sertanejo Depression, adjacent and in topographic continuity to the lowlands elaborated on Proterozoic crystalline lithologies.

Among the basins that form the isolated homoclinal tablelands of Pernambuco, the Fátima basin (Fig. 7.17) is the most spatially representative one. It is formed by a sequence of siliciclastic deposits of the Tacaratu Formation, characterized by fine to medium sandstones intercalated by conglomeratic sandstones up to 500 m thick (Morais et al. 1996). Morphologically, a series of NW tilted blocks marks this landform, with a SE-oriented "front" escarpment controlled by a sequence of parallel normal faults. The summit line at the top of the external front (known as the Urubu range) exceeds 900 m in elevation.

The dip-slopes are heavily dissected by consequent streams and exhibit tilting angles exceeding 15°, close to the threshold between cuesta-like homoclinal landforms and typical hog-backs. Drainage lines are deeply entrenched into canyons adjusted to the fracture zones. Further to the west, though, close to the village of Fatima, the sloping surface developed on the sandstones of the Tacaratu Formation merges gradually with rock pediments developed on crystalline rocks without any notable topographic unevenness separating both lithological units.

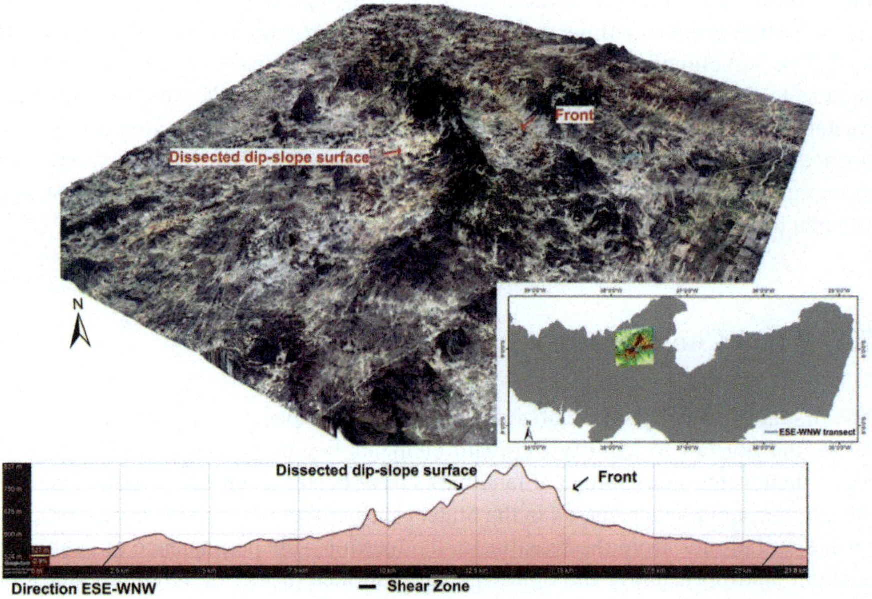

Fig. 7.17 DTM of the Fatima basin, indicating the homoclinal landforms, the front and the dissected dip-slope surface with canyons adapted to the fracture network. *Source* the authors

7.3.5.4 Depressions on Phanerozoic Sedimentary Basins

Several Proterozoic faults of the Borborema Province in Pernambuco have been reactivated during the Mesozoic rifting phase, leading to the opening of the Atlantic Ocean. Later, during the Cenozoic, repeated tectonic pulses along the newly formed passive margin caused the uplift of the region, including the Borborema Highlands. Finally, the establishment of an extensional regime and extensive crustal thinning, initially associated with the West Gondwana breakup, created rift basins on the passive margin and the continental interior, some more than 300 km inland (Schobbenhaus and Brito Neves 2003; Silva et al. 2003).

As in the case of the Araripe and Jatobá basins, the outer limits of these sediment accumulation areas follow the outline of the Brasiliano orogens, accompanying the network of reactivated shear zones. Hence, the Cenozoic reactivation of fault lines occurred primarily where the shear zones form sutures bounding different geological terrains or cutting through the sequences of metavolcanic-sedimentary supracrustal rocks. On many occasions, this structural configuration has promoted the topographic inversion of the rift basins, resulting in table-like plateaus like the Araripe or homoclinal, tilted tablelands like those of the Jatobá basin.

The circumdenudation processes of the Cenozoic resulted in the retreat of the sedimentary plateaus' escarpments, creating vast expanses of lowlands around them, or along the course of the main drainage lines, like the Moxotó River in the case of the Jatobá basin Plateau. These lowlands form large regional "erosive glacis"—planation surfaces structured on sedimentary rocks—at elevations ranging on average from 300 to 400 m. Sometimes they present sharp limits with the adjacent crystalline terrains, such as the mylonitic ridges associated with the Ibimirim fault, which delimits the western flank of the Jatobá basin and its Plateau. However, some depressions are formed on continental rift basins that have not been affected by the reactivation, and topographic inversion processes, like the São José do Belmonte and Cedro basins, situated to the north of the Sertanejo Highlands close to the Ceará state limit.

7.4 Final Considerations

The first academically systematized studies of Pernambuco's landforms in the mid-1950s did not fail to identify the main elements behind their spatial distribution. Since then, some morphological landmarks have been recognized as guidelines for the geomorphological research in the State, especially the " stepped" aspect of landform distribution—from the coastline to the interior—the permanence of exuberant residual landforms above the otherwise level surface of the topographic "steps"—massifs, ridges, inselbergs—and the fundamental role of the Borborema Highlands on the hierarchization and distribution of the Atlantic or inland (São Francisco River watershed) drainage networks.

The first explanations regarding the evolution of Pernambuco's landforms sought to tentatively establish a chain of erosive-depositional events that could provide a

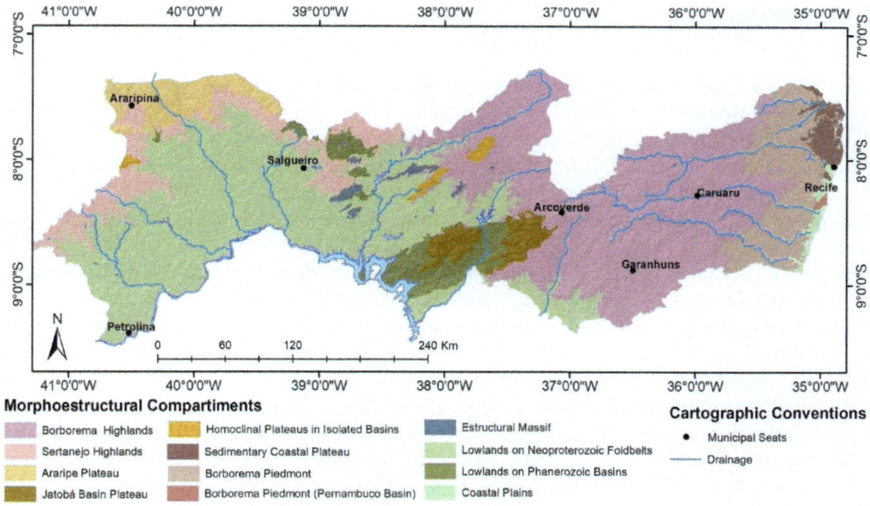

Fig. 7.18 Morphostructural compartments of the State of Pernambuco. *Source* the authors

chronological and evolutionary framework for the physical landscape. Unfortunately, the research rationale needed more detailed knowledge of the sedimentary basins' stratigraphy and geotectonic controls' role in the genesis and architecture of metamorphic complexes and intrusive plutons. In the lack of the above, the interpretive formulations tended to overemphasize the apparent correlation between certain sedimentary deposits and assumed "planation surfaces" upstream of them. However, by cleverly recognizing these propositions' practical and empirical limits, the early researchers pointed to alternative pathways to circumvent the lack of field evidence, especially the scarcity of continental Cenozoic deposits. One of such classical works (Cazjka 1959) went so far as to suggest the need to further the understanding of the role of Cenozoic deformation on the distribution of the inner regional escarpments that separate and individualize the geomorphic units of the State.

Empiricism, theory, and technique have been the guiding principles behind any exercise in geomorphology, be it a study on detailed aspects of surface processes or the regionalization of physiographic provinces. In all cases, the geomorphological problem permanently resides in the landscape. Not surprisingly, this very geographical arrangement of physical, natural, or even cultural attributes in a given space sets geomorphology in motion and inspires its practitioners. Thus, as presented above, the reading of Pernambuco's landscape (Fig. 7.18) will always be a narrative, a gaze in time.

References

Ab'sáber AN (1957) Significado geomorfológico da rede hidrográfica do Nordeste Oriental brasileiro. Boletim Geográfico 15(139):459–464
Allaby M (2008) Dictionary of earth sciences. Oxford University Press
Almeida JDM, Corrêa ACB (2020) Conectividade da paisagem e a distribuição de plainos aluviais em ambiente semiárido. Revista Brasileira De Geomorfologia 21:171–183
Andrade GO (1956) Quelques niveaux de terrasse eustatiques sur le littoral Nord-oriental et Nord du Brésil. Premier Rapports, 57p
Andrade GO (1958) A superfície de aplanamento pliocênica do nordeste do Brasil. Universidade do Recife, Diretorio Academico Faculdade de Filosofia. Recife, 44p
Andrade GO (1968) A gênese do relevo nordestino: estado atual dos conhecimentos. Estudos Universitários 2/3:1–13
Bates RL, Jackson JA (1983) Glossary of geology. United States: N. p., 1987. Web. BRASIL, Radam. Levantamento de recursos naturais. Rio de Janeiro, vol 31
Brito Neves BB, Santos EJ, Van Schmus WR (2000) Tectonic history of the Borborema Province, northeastern Brazil. In: Cordani UG, Milani EJ, Thomaz Filho A, Campos DA (eds) Tectonic evolution of South America. Rio de Janeiro: 31 international geological congress, pp 151–182
Carneiro CDR, Almeida FFM, Hasui Y, Zalán PV, Teixeira JBG (2012) Estágios evolutivos do Brasil no Fanerozoico. In: Hasui Y, Carneiro CDR, Almeida, FFM, Bartorelli A (eds) Geologia do Brasil. São Paulo: Beca, pp 131–136
Carvalho RR (2005) Origem e Proveniência da Sequência Siliciclástica Inferior da Bacia do Jatobá. 101 f. Dissertação (Mestrado em Geociências) – Universidade Federal de Pernambuco, Recife
Castellan P, Viegas G, Faleiros FM (2020) Brittle-ductile fabrics and P-T conditions of deformation in the East Pernambuco shear zone (Borborema Province, NE Brazil). J Geol Soc 178(1)
Caxito FA, Uhlein A (2013) Arcabouço tectônico e estratigráfico da Faixa Riacho do Pontal, divisa Pernambuco-Piauí-Bahia. Geonomos pp 19–3
Corrêa ACB (2001) Dinâmica geomorfológica dos compartimentos elevados do Planalto da Borborema, Nordeste do Brasil. 386 f. Tese – Instituto de Geociências e Ciências Exatas, UNESP, Rio Claro
Correa ACB, Cavalcanti LCS, Lira DR (2015) Stone and Sand Ruins in the Drylands of Brazil: the rustic landscapes of Catimbau National Park. In: Vieira BC, Salgado AAR, Santos LJC (eds) Landscapes and landforms of Brazil. Springer Netherlands, Dordrecht, pp 243–252
Corrêa ACB, Tavares BAC, Lira DR, Mutzenberg DS, Cavalcanti LCS (2019) The semi-arid domain of the Northeast of Brazil. In: Salgado A, Santos L, Paisani J (eds) Geography of the physical environment. Springer International Publishing, pp 119–150
Corrêa ACB, Tavares BAC, Monteiro KA, Cavalcanti LCS, Lira DR (2010) Megamorfologia e Morfoestrutura do Planalto de Borborema. Revista do Instituto Geológico 35–52
Cotton A (1968) Geomorphology. Witcombe e Tombs Limited, London, 187p
Czajka W (1959) Estudos geomorfológicos no nordeste do Brasil. Revista Brasileira de Geografia 20:135–180
Dresch J (1957) Les problèmes géomorphologiques du Nord-Est Brésilien. Bull Ass Géograp Français 263:48–59
Fonsêca DN, Corrêa ACB, Lira DR, Tavares BAC (2020) Chemical, physical and mineralogical attributes as markers in the identification of depositional events in the Eastern Northeast of Brazil. J S Am Earth Sci 103:102805–102825
Fonsêca DN, Corrêa ACB, Silva AC (2016) Compartimentação Geomorfológica da Região Metropolitana do Recife (RMR) a partir da análise morfoestrutural. Geo UERJ 29:201–219
Geramisov IP, Mescheirov JA (1968) Morphostructure. In: Fairbridge RW (ed) The encyclopedia of geomorphology—encyclopedia of earth sciences. Hutchinson e Koss Inc., Pennsylvania, Dowden pp 731–732
Gomes HA, Santos EJ, Polônia JAL, Dantas JRA, Coutinho PN, Manso VAV, Franco BA, Sobrinho ACPL, Santos CA, Medeiros VC, Oliveira RG, Lins CAC, Leão Neto R, Nóbrega M, Freire AG, Neves JACL (2001) Geologia e recursos minerais do Estado de Pernambuco. Brasília: CPRM, 198p

Gurgel SPP, Bezerra FHR, Corrêa ACB, Marques FO, Maia RP (2013) Cenozoic uplift and erosion of structural landforms in NE Brazil. Geomorphology 186:68–84

Hasui Y (2012) Compartimentação Geológica do Brasil. In: Hasui Y, Carneiro CDR, Almeida FFM, Bartorelli A (eds) Geologia do Brasil. São Paulo: Beca, pp 112–122

Klöcking M, Hoggard MJ, Rodríguez Tribaldos V, Richards FD, Guimarães AR, Maclennan J, White NJ (2020) A tale of two domes: Neogene to recent volcanism and dynamic upliftof northeast Brazil and southwest Africa. Earth Planet Sci Lett 547:1–13

Lima GG, Marçal MS, Corrêa ACB, Lima FJ (2021) Landscape evolution of the Salamanca watershed, Araripe Plateau: Insights from a river channel morphological classification. J Sou Am Earth Sci 107:103013

Lira DR (2014) Evolução geomorfológica e paleoambiental das Bacias do Riacho do Pontal e GI-8 no Sub- Médio São Francisco. 2014. 234 f. Tese (Doutorado em Geografia) - Universidade Federal de Pernambuco, Recife

Mabesoone JM (1984) Cenozóico nordestino: sistemas e seqüências deposicionais. SBG Simp Geol Nordeste 11(Atas):142–150

Mabesoone JM, Castro C (1975) Desenvolvimento geomorfológico do Nordeste Brasileiro. Boletim do Núcleo Nordeste da Sociedade Brasileira de Geologia 3:5–35

Maia RP, Bezerra FHR (2014) Condicionamento estrutural do relevo no nordeste setentrional brasileiro. Mercator 13(1):127–141

Maia RP, Nascimento MAL, Bezerra FHR, Castro HS, Meireles AJA, Rothis LM (2015) Geomorfologia do campo de inselbergues de Quixadá—NE do Brasil. Revista Brasileira de Geomorfologia 2(16)

Marques FO, Nogueira FCC, Bezerra FHR, de Castro DL (2014) The Araripe Basin in NE Brazil: an intracontinental graben inverted to a high-standing horst. Tectonophysics 630:251–264

Monteiro KA, Corrêa ACB (2020) Application of morphometric techniques for the delimitation of Borborema Highlands, northeast of Brazil, eastern escarpment from drainage knick-points. J Sou Am Earth Sci 103:102729–102741

Morais F, Rocha DEGA, Oliveira RG (1996) Bacia sedimentar de Fátima—Hidrogeologia. Projeto Bacia Interiores do Nordeste. Recife, CPRM

Morales N, Assine ML (2015) Chapada do Araripe: a highland oasis incrusted into the semi-arid region of northeastern Brazil. In: Vieira BC, Salgado AAR, Santos LJC (eds) Landscapes and landforms of Brazil, Springer, New York, pp 231–242

Murphy RE, Norman JW, Thrower (1968) annals map supplement number nine landforms of the world. Ann Assoc Am Geogr 58(1):198–200

Oliveira RG, Medeiros WE (2012) Evidences of buried loads in the base of the crust of Borborema Plateau (NE Brazil) from Bouguer admittance estimates. J Sou Am Earth Sci 37:60–76

Saadi A (1991) Ensaio sobre a morfotectônica de Minas Gerais: tensões intraplaca, descontinuidades crustais e morfogênese. 285 f. Tese - Instituto de Geociências, Universidade Federal de Minas Gerais, Belo Horizonte

Santos J, Galvincio J, Correa ACB, Silva DGDA, Machado CCC (2011) Classificação Geomorfológica dos Estuários do Estado de Pernambuco (Brasil) com Base em Imagens do LANDSAT 5/TM. Revista Brasileira de Geografia Física 01:118–133

Santos FG, Pinéo TRG, Medeiros VC, Santana JS, Morais DMF, Vale JAR, Wanderley AA (2021) Mapa geológico da província Borborema: projeto geologia e potencial mineral da província Borborema. CPRM, Recife

Schobbenhaus C, Brito Neves BBA (2003) Geologia do Brasil no Contexto da Plataforma Sul-Americana. In: Bizzi LA, Schobbenhaus C, Vidotti RM, Gonçalves JH (eds) Geologia, Tectônica e Recursos Minerais do Brasil. Brasília, CPRM, pp 5–25

Silva AJP, Lopes RC, Vasconcelos AM Bahia RBC (2003) Bacias Sedimentares Paleozóicas e Meso-Cenozóicas Interiores. In: Bizzi LA, Schobbenhaus C, Vidotti RM, Gonçalves JH (eds) Geologia, Tectônica e Recursos Minerais do Brasil. CPRM, Brasília

Silva DG, Corrêa ACB, Amorim RF (2017) Caracterização morfológica e dinâmica ambiental das marmitas de dissolução (weathering pit) no distrito de fazenda nova, Pernambuco—Nordeste do Brasil. Revista Brasileira de Geomorfologia 18(2):349–362

Suguio K, Bezerra FHR, Barreto AMF (2011) Luminescence dated Late Pleistocene wave-built terraces in Northeastern Brazil. Anais da Academia Brasileira de Ciências 83:907–920
Tavares BAC (2015) Evolução morfotectônica dos pedimentos embutidos do Planalto da Borborema. 251 f. Tese - (Doutorado em Geografia). Universidade Federal de Pernambuco, Recife
Van Schmus WR, Oliveira EP, Silva Filho AF, Toteu SF, Penaye J, Guimarães IP (2008) Proterozoic links between the Borborema Province, NE Brazil, and the Central African Fold Belt. Geol Soc London Spec Publ 294:69–99

Antônio Carlos Barros Corrêa He holds a degree in geography from the Federal University of Pernambuco (UFPE) (1994), an undergraduate exchange program at Radford University, Virginia, USA, a master's degree in geography (1997) from the Federal University of Pernambuco (UFPE), a Ph.D. in geography (2001) from the Paulista Júlio de Mesquita Filho State University with a Sandwich Internship at Durham University, UK, and a post-doctoral fellowship in geomorphology (2003) from the Júlio de Mesquita Filho Paulista State University. He is a professor and researcher at the Federal University of Pernambuco, a CNPq research productivity fellow (level 2), a permanent member of the graduate programs in geography and archeology, and the leader of the Quaternary study group of the Brazilian Northeast (GEQUA) and the Quaternary geomorphology laboratory at UFPE. He has experience in geoscience with an emphasis on geomorphology, acting mainly on Quaternary geomorphology, geomorphology of semi-arid environments, historical and structural geomorphology, application of morphometric indexes in geomorphological analysis, and geosystemic analysis.

Bruno de Azevedo Cavalcanti Tavares He holds a bachelor's degree (2007), a master's degree (2010), and a Ph.D. in geography (2015) from the Federal University of Pernambuco (UFPE), with all research focused on structural geomorphology and morphotectonics. He is a member of the Quaternary study group of the Brazilian Northeast (GEQUA). Since May 2016, he has been an adjunct professor at UFPE in the archaeology department, where he works with research focused on geoarchaeology and spatial archaeology. He is the vice-coordinator of the geoarchaeology laboratory (LABGEOARQ) of UFPE and vice-leader of the group for the study of archaeology, geoscience, and technology (GEAGT). He has experience in geoscience with an emphasis on geomorphology, working mainly on structural geomorphology, morphotectonics, morpho-stratigraphy, paleoenvironmental reconstruction, and platform relief evolution.

Drielly Naamma Fonsêca da Silva She holds a BSc. Degree in geography (2009) and a license degree in Geography (2022) from the Federal University of Pernambuco (UFPE), a master's degree in geography (2012) from the Federal University of Pernambuco (UFPE), and a Ph.D. in geography (2018) from the Federal University of Pernambuco (UFPE). She was a CAPES (Brazilian Federal Agency for Support and Evaluation of Graduate Education) post-doctoral fellow at the Graduate Program in Geography of the Federal University of Pernambuco (UFPE) from 2019 to 2022. She is a professor and researcher at the Department of Industry of Technical Metropolitan University of Santiago (UTEM), Chile, and an associate researcher of the Study Group of the Quaternary of the Northeast of Brazil (GEQUA). She has experience in geoscience with an emphasis on physical geography, structural geomorphology, morphotectonics, sedimentology, morphostratigraphy, and application of geochemical indices in geomorphology.

Daniel Rodrigues Lira He holds a degree (2008), a master's degree (2010), and a Ph.D. in geography (2014) from the Federal University of Pernambuco (UFPE), having experience in physical geography with an emphasis on geomorphology, pedology, and geo-technology, working mainly on geomorphology, paleoenvironmental reconstruction, pedology (soil × landscape), geochemistry, geoprocessing and remote sensing, and teaching of physical geography.

Chapter 8
Geomorphology of Alagoas State

Kleython de Araújo Monteiro and Antônio Carlos de Barros Corrêa

Abstract Alagoas lies in the southern or external sector of the Borborema Province. Its territory comprises Proterozoic lithologies with Archean and Paleoproterozoic nuclei bounded by Meso- and Neoproterozoic fold-belts. Phanerozoic sedimentary covers of the Tucano-Jatobá basins (Paleozoic) outcrop in the northwestern most sector of the State, along with other sedimentary units along the coastal strip, like the Cretaceous formations of the Alagoas marginal rift basin, the Neogene covers that overlie them, and Quaternary fluvio-marine plains. Since the early physiographical descriptions of its territory, the relief compartments of Alagoas have been arranged in four sets of landforms: the tablelands (tabuleiros), the crystalline highlands (Planalto da Borborema), the Peripheral Depression (situated between the two former units), and; the Sertanejo Pediplain (Pediplano Sertanejo). Beyond those, the residual landforms of Santana, Mata Grande, and Água Branca massifs, the fluvial and coastal plains, and the escarpments of the Borborema Highlands have also been identified as morphostructural sub-units. A new subdivision of the State's macro-relief units was proposed based on the relationship between the macro-morphological compartments and the lithostructural units. Namely, the morphostructural landforms are the crystalline highlands, the sedimentary plateaus, the semi-arid depressions and associated residual massifs, and fluvial and coastal plains.

Keywords Coastal landforms · Peripheral depressions · Residual massifs · Homoclinal plateaus

K. de A. Monteiro (✉)
Federal University of Alagoas, Maceio, Brazil
e-mail: kleython.monteiro@igdema.ufal.br

A. C. de B. Corrêa
Federal University of Pernambuco (UFPE), Recife, Brazil
e-mail: antonio.correa@ufpe.br

V. Claudino-Sales and J. F. Sobrinho (eds.), *Geomorphology of the Northeast Region of Brazil*, The Latin American Studies Book Series,
https://doi.org/10.1007/978-3-031-58874-7_8

8.1 Introduction: A Short Historical Background on the Study of Alagoas' Landforms

The study of processes that shape relief forms has long been the main focus for understanding their nature and the processes related to them. Sparse observations on the landforms made by European visitors to the Northeast of Brazil have produced accounts regarding harbor locations and their safety or the preferential routes to accessing the continental interior, especially those that followed the valleys of the main streams running to the coast.

With time, the focus has changed, analyzes have improved, and the understanding of landforms' generating processes have gained prominence as the central element in unveiling the dynamics of the natural components, as well as having become an essential tool for land use and occupation planning.

In the Northeast of Brazil, the first records concerning the description of landforms of what later became the territory of Alagoas date back to the seventeenth and early eighteenth centuries when the region's coast and the mouth of the main rivers were mapped, in an attempt to recognize adequate sites for establishing colonial settlements, harbors and defense structures.

In a more systematized fashion, the cartographic and geographic records regarding the topography and drainage networks gained strength in the nineteenth century. They were primarily associated with the separation of Alagoas's territory from the Captaincy of Pernambuco in 1817. This phase is followed by the need to erect a spatial identity for Alagoas, separated from Pernambuco's, both in the cultural and natural realms (Nunes Silva 2018).

In 1820, Sir Colonel Francisco Manuel Martins Ramos, commandant of the districts of Penedo and Poxim, commissioned the topographic map of the Captaincy of Alagoas to identify the most suitable areas for growing sugar cane. The areas along the coast and south of the territory were of great interest "since their geographical conditions were a facilitating factor for the implementation of sugar-cane plantations and, along them, human settlements" (Diegues Júnior 2012 apud Santos 2019).

Later, with the need to better characterize the main geographical aspects of Alagoas, three important works known as the Geography Reports (Herculano 2011) have been published:

- Booklet of Geographical, Topographical, Physical, Political and Historical description regarding the Province of Alagoas, Empire of Brazil (*Opúsculo da Descrição Geographica Topographica, Phizica, Política e Histórica do que unicamente respeita à Província das Alagoas no Imperio do Brazil—Antonio Joaquim Moura, 1844*);
- Elements of Geography and Cosmography offered to the youth of Alagoas (*Elementos da Geographia e Cosmographia oferecidos a mocidade alagoana—Thomaz do Bomfim Espíndola, 1860*);
- Geography and Cosmography (*Geographia e Cosmographia—Manoel Baltazar Pereira Diegues Junior, 1890*).

Such works proposed to contribute with an empirical fundament based on the description of natural elements (soil, climate, fauna, flora, landforms), which could account for the human character of social groups (Herculano 2011). For Almeida (2004, apud Nunes Silva 2018), these "booklets" also sought to construct an exemplary State of Alagoas.

During the twentieth century, several works and authors helped put together many texts, maps, and collections regarding Alagoas' landforms and natural aspects. However, one must highlight the contribution of the one considered the greatest geographer of that century in the State, Dr. Ivan Fernandes Lima. Among his various productions, two must be highlighted: Maceió, the sand spit town: contributions to the geomorphological study of Alagoas coastline (*Maceió a Cidade Restinga—Contribuições ao estudo geomorfológico do litoral alagoano, 1961*) and Geography of Alagoas (*Geografia de Alagoas, 1965*).

The first of them, "Maceió, the sand spit town," was a dissertation written by Dr. Lima as part of the requirements to be admitted as a lecturer at the Alagoas Lyceum in 1961. The work, with rich iconography and cartography, resulted from extensive fieldwork along the State's coastal region in 1959 (Lima 1990). It became a reference for his own future work as well as for other geographers'. In this essay, Dr. Lima shows a vast empirical knowledge of the subjects, which became his personal mark as a researcher, as later pointed out by the great Brazilian geographer Aziz Ab'Saber in the Prologue to the 1990's edition of the book.

In the "Geography of Alagoas," Dr. Lima maintained his descriptive qualities with powerful illustrations and graphics, aiming at presenting the geography of the State as a class book that conveyed the main aspects of its regionalization and landscapes. One must highlight the comprehensiveness of the geomorphological cartography offered by the book.

In this work (Lima 1965), the author identified four sets of relief morphologies in Alagoas: the tablelands (*tabuleiros*), the crystalline highlands (*Planalto da Borborema*), the Peripheral Depression (situated between the two former units), and; the Sertanejo Pediplain (*Pediplano Sertanejo*). Beyond those, he has also identified other essential morphological elements for proposing a basic geomorphic outline of the State: the residual landforms (Santana, Mata Grande, and Água Branca massifs), fluvial and coastal plains, as well as the eastern and western escarpments of the Borborema Highlands.

The geomorphic outline, coupled with block diagrams and descriptions of the morphological elements, constituted one of the most relevant chapters of Lima's book and an initiation for those willing to learn about the spatial organization of Alagoa's physical attributes. Moreover, the contents were presented alongside educational questionnaires to introduce the State's geomorphology to young students.

On a national level, the geomorphology of Alagoas would only come to the center stage of academic literature via Ab'Saber's (1956) contribution concerning the semi-arid peripheral depressions of the northeast. The author postulates that the residual landforms emerging from the level pediplain surfaces of Alagoas drylands had their summits truncated by a pre-Triassic planation surface, just like the inselbergs dotting Patos' depression in the State of Paraíba (p. 11). These lowlands of semi-arid Alagoas reached from "… Palmeira dos Índios to the foot-slopes of Tacaratu tablelands, endlessly extending as depressed, low-lying areas, corresponding to the peripheral depression elaborated around the southern end of the Borborema Highlands" (p. 5).

More recently and again not exclusively limited to Alagoas' territory, Correa et al. (2010) proposed a morphostructural and megageomorphological subdivision of the Borborema Highlands. Among the identified compartments, the "Pernambuco-Alagoas structural summit" and the "Remobilized Massifs of the Pernambuco-Alagoas Domain" are highlighted as some of the most significant regional landforms also comprising the State of Alagoas.

The "Pernambuco-Alagoas structural summit" corresponds to the elevated summit of the southern Borborema Highlands to the south of the Pernambuco shear zone, limited to the west by the Ipanema River watershed. This compartment is characterized by:

> … a more homogeneous topographic feature relative to the surrounding sectors, where ridges and residual landforms predominate (…). It is characterized by a topography composed of flat summits with thick pedogenic covers of acrisols and regosols at altitudes ranging from 600 to 700 m (Correa et al. 2010, p. 46).

To the authors, this differentiated compartment of the highlands resulted from a combination of controlling factors, such as the lithological homogeneity (the gneissic-migmatitic complex) and its relative distance from the axis of the regional up-warping (the Transversal Zone between the Patos and Pernambuco shear zones). As a result, some summit surfaces exceed 1000 m in elevation, like in the Garanhuns erosive "topographic step", structured in a band of quartzites. However, one cannot discard the vicinity of the uplifted margin of the highlands as another probable cause for the higher elevation of this sector (Correa et al. 2010).

In its turn, the compartment of the Remobilized Massifs of the Pernambuco-Alagoas Domain represents a small fraction of each State (Pernambuco and Alagoas). The Alagoas' portion of the compartment, the smallest one, separates the geomorphic units of the Jatobá basin plateau from the Ipanema River depression. In the State, this compartment corresponds to a series of lowered divides on the margin of the Recôncavo- Tucano-Jatobá basins tablelands. Consequently, this landform unit lies ahead of the residual massifs that emerge from the western semi-arid pediplain of the State, like the massifs of Mata Grande and Água Branca.

8.2 Physiographic Aspects that Contribute to the Understanding of the Geomorphology of Alagoas

Regarding its geological structure, Alagoas lies in the southern or external sector of the Borborema Province (Mendes 2017). It comprises Proterozoic lithologies with Archean and Paleoproterozoic nuclei bounded by Meso- and Neoproterozoic fold-belts. The exception to the crystalline rocks is the phanerozoic sedimentary covers of the Tucano-Jatobá basins (Paleozoic) that outcrop in the municipality of Olho

D'Água do Casado and the northwestern most sector of the municipality of Mata Grande. Other sedimentary units occur only along the coastal zone to the east, like the Cretaceous formations, the Neogene tablelands, and the Quaternary fluvio-marine plains.

The agroecological zoning of Alagoas (EMBRAPA 2012), based on the proposals of Nobre and Molion (1986), indicates that the State's rainfall regime is directly conditioned by the Intertropical Convergence Zone (ITCZ) behavior, combined with frontal systems and easterly waves (Eastern Undulatory disturbances). As a result of its low latitude, warm climates predominate in the State, with annual temperatures varying from 22 °C to 28 °C. However, precipitation totals differ significantly from wet coastal areas presenting annual averages above 2000 mm to semi-arid depressions further inland where rainfall values drop to less than 500 mm per year. There is a gradation in rainfall distribution from the coast to the western drylands, with a steady diminishment in totals further inland from the influence of the moist winds that blow from the Atlantic Ocean. Nonetheless, two areas of climatic exception occur amidst the drylands landscape, the massifs of Água Branca and Mata Grande. Due to the higher elevation, average annual precipitation in these two areas exceeds 1000 mm, contrasting sharply with the surrounding drought-prone lowlands.

This variation in rainfall also conditions the development of weathering mantles with varying degrees of mineralogic alteration. In general, the level of pedogenesis closely accompanies the modulation of precipitation totals. Thus, along the coastal zone to the east of the State, more developed soil profiles occur, with the prevalence of ferralsols, acrisols, and podzols, whereas to the west of Alagoas, fluvisols, planosols, and leptosols predominate (EMBRAPA op cit). Likewise, the vegetation cover varies from the evergreen Atlantic rainforest in the eastern sector, passing through a semi-decidual seasonal forest in areas of climatic transition, to the xerophytic formations, caatingas, in the western semi-arid lowlands (EMBRAPA op cit).

8.3 Landforms Macrocompartments of the State of Alagoas

Considering Alagoas's physical aspects and the early research discussed previously, a macro-compartmentalization of the State's landforms is proposed based on the interaction between the lithostructural elements and the surface's macro-morphologies. The mapping parameters proposed by Demek e Embleton (1978) will be applied in order to convey a representation of landforms at an intermediate cartographic scale.

We do not propose to offer an in-detail analysis of structural controls nor landscape surface processes accountable for shaping the morphologies of Alagoas' relief. Neither is our aim to establish new geomorphological mapping approaches. As a matter of fact, our pursuit was to provide a systematized grouping of the large landform units of the State, aiming at contributing to future in-depth studies.

For this purpose, initially, the lithological units of the State were identified and separated into sedimentary and crystalline terrains (Fig. 8.1). Phanerozoic sedimentary sequences occur predominantly along the eastern half of the Alagoas. They are

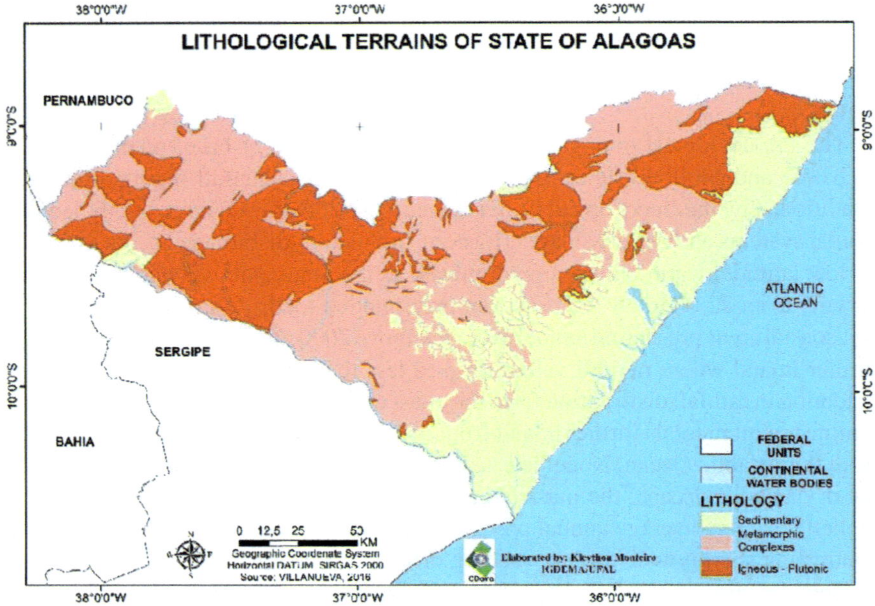

Fig. 8.1 Outline of the lithological terrains of State of Alagoas highlighting crystalline and sedimentary areas. *Source* Villanueva (2016)

represented mainly through the unconsolidated Neogene covers and Quaternary accumulations. The Barreiras Formation accounts for most of these young deposits, while the others are composed of fluvio-marine sediments of Pleistocene and Holocene age. In a smaller proportion, sedimentary rocks associated with the Tucano-Jatobá basins of Siluro-Devonian age occur in Olho D'Água do Casado and Mata Grande municipalities, further to the west (Mendes 2017).

Crystalline rocks present a variety typical of exhumed platform terrains. The main metamorphic complexes outcropping in the region are the Belém do São Francisco, Chorrochó, Cabrobó, and Arapiraca units. Furthermore, several Brasiliano plutons are found in the State, among which the Xingó, Serra do Catu, and Água Branca plutons deserve special attention due to their size and surface morphological characteristics (Mendes 2017). Morphologically, the metamorphic complexes are associated with the regional depressions, except when they have suffered the influence of the regional up-warping that elevated the Borborema Highlands (Correa and Monteiro 2021). In general, due to their compositional characteristics and relation to the surrounding host rocks, the plutons result in prominent residual landforms in response to differential erosion.

Altimetrically, the State of Alagoas can be subdivided according to the prevailing hypsometric intervals (Fig. 8.2). The northern sector of the State concentrates the areas with altitudes above 300 m, exceeding the 800 m threshold in the municipalities of Mata Grande, Maravilha, and Quebrangulo. Considering only the areas above

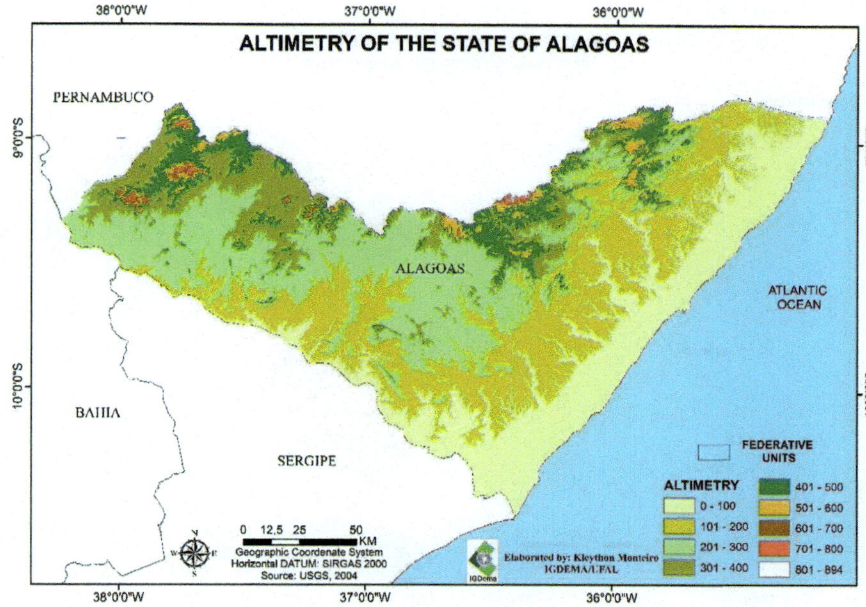

Fig. 8.2 Altimetry of the State of Alagoas in 100 m intervals. *Source* USGS (2004)

400 m, those are entirely encompassed by the Borborema Highlands, forming ridges and residual massifs, especially in the municipalities of Mata Grande and Água Branca, or associated with the Jatobá sedimentary plateau (Parafuso range), also in Mata Grande. As for the eastern, central, and southern sectors of the State, these are dominated by lowlands below 300 in elevation, associated with depressions, low coastal plateaus, and coastal plains. In the eastern sector, especially, the areas above the 100 m contour line seem structurally controlled, but this interpretation is still open to further discussion and studies.

It is essential to highlight that the altimetric classification enables us to define which areas are predominantly controlled by depositional or denudational processes. Consequently, general processual aspects can be anticipated, thus contributing to the delimitation of relief macro-compartments, such as highlands, depressions, and plains (Fig. 8.3).

Based on the subdivision of altimetric zones, morphological compartments, and lithological classes, we can draw an initial outline of morphostructural macro-units, initially differentiating between crystalline and sedimentary sets of landforms and their corresponding hypsometric level. This proposal applied the delimitation of the Borborema Highlands envisaged by Correa et al. (2010), later updated by Monteiro and Correa (2020) and Nascimento et al. (2016, 2017). Aiming at a better characterization of the residual massifs that emerge from the semi-arid lowlands, the contributions of Gois et al. (2019, 2021) were applied.

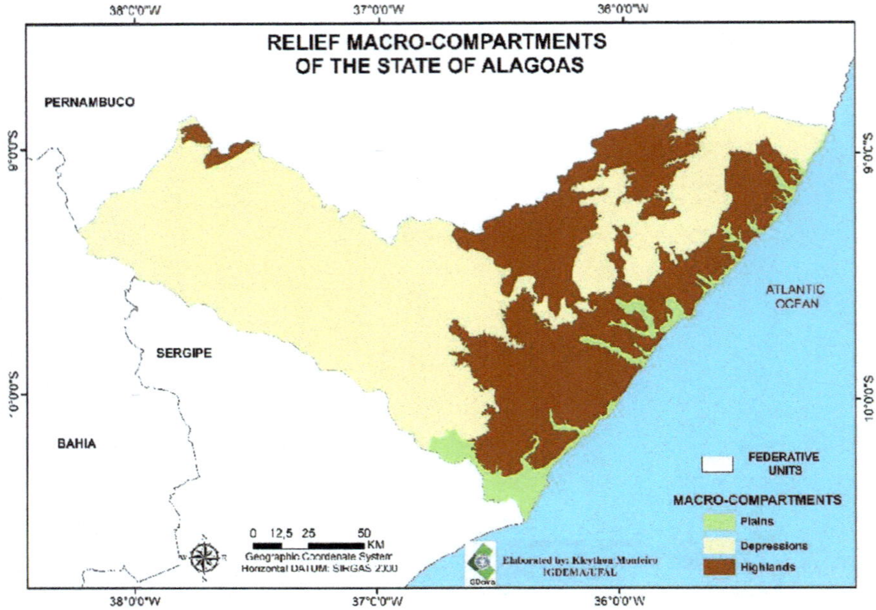

Fig. 8.3 Relief macro-compartments of the State of Alagoas, based on the relationship between the large sets of morphologies. *Source* The authors

Following the routine described above, four major landform macro-units have been defined (Fig. 8.4) based on the relationship between the macro-morphological compartments and the lithostructural units. Namely, the morphostructural landforms are the crystalline highlands, sedimentary plateaus, depressions, and associated residual massifs and plains. Each macro-morphology can be subdivided into several more detailed sub-divisions.

The crystalline highlands are, effectively, sub-compartments of the Borborema Highlands—the "Pernambuco-Alagoas structural summit" and the "Remobilized Massifs of the Pernambuco-Alagoas Domain". The sedimentary plateaus can be subdivided in two separate units. The "low coastal sedimentary tablelands" cover the largest area among the sedimentary plateaus. In the extreme northwest of State, covering a much smaller area, lies the homoclinal sedimentary plateau, presenting a cuesta-like morphology, locally known as the Parafuso range.

The depressions occur from the eastern to the western sector of the State, associated with crystalline lithologies. They are low-lying areas that correspond to the Piedmont of the Borborema Highlands in the vicinity of the coastline. Therefore, they behave as a true peripheral depression interposed between the eastern escarpment of the Borborema Highlands and the low coastal sedimentary tablelands. Within the depressions, the landscape is dominated by vast low-lying pediment ramps tilted toward the drainage of the São Francisco River. In order to differentiate between the depressions, their corresponding base levels were used. Consequently, there are

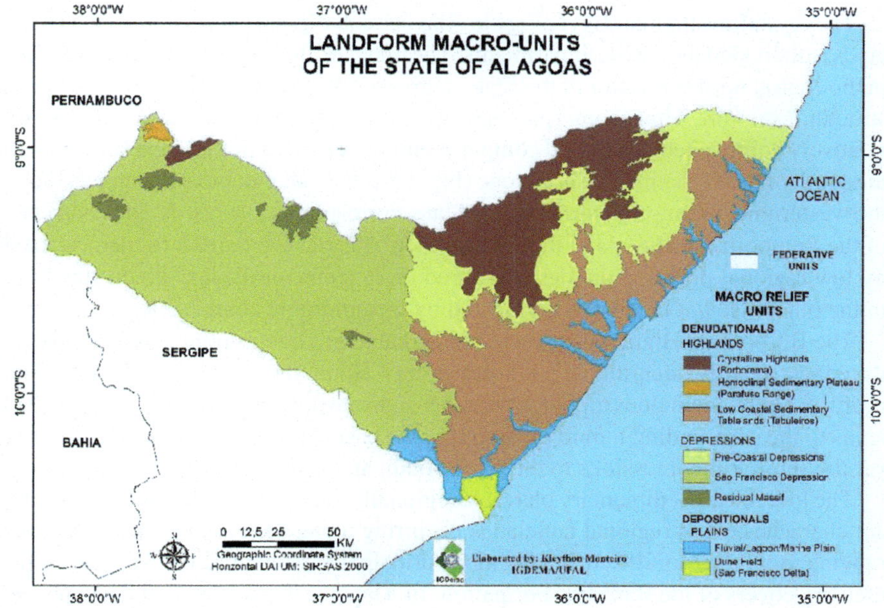

Fig. 8.4 Landform macro-units of the State of Alagoas. *Source* The authors

depressions whose drainages run directly to the Atlantic Ocean to the east, thus named Atlantic Depressions or Pre-coastal depressions. On the other hand, the extensive pediplains of the lower São Francisco River correspond to the São Franciscana Depression, above which lie quite a few residual massifs.

In the case of the low-lying surfaces that correspond to recent accumulation morphologies, the unconsolidated covers are related to sediments derived from the hillslope, fluvial, lagoonal, and marine dynamics, therefore grouped into fluvial, lacustrine, and marine plains. Among them, it is necessary to call attention to the plains that are directly influenced by the São Francisco River and the Atlantic Ocean alike, such as the dune fields of the Piaçabuçu region, at the mouth of the São Francisco.

The highlands units represent denudational morphologies of elevated landforms at several altimetric levels and are structured on various types of rocks. They display summit surfaces at different degrees of dissection commonly bounded by escarpments (Guerra 1993). In Alagoas, highlands correspond to ca. 35% of the State's area and are distributed from the coast to the semi-arid interior. Nonetheless, elevated morphologies tend to concentrate on the eastern sector of the State.

The contiguous elevated areas can be separated into two subtypes of large-scale landforms based on the dominant lithology: crystalline highlands and sedimentary plateaus. The latter, in Alagoas, is represented by the low coastal plateaus and the homoclinal sedimentary plateau associated with sediments of the Jatobá basin in the semi- arid domain.

The crystalline highlands typology in Alagoas correspond to the Borborema Highlands, subdivided into two compartments. The first one, the most extensive in area, lies in the central northern sector of the State, corresponding to the Pernambuco-Alagoas structural summit. This geomorphic unit was described by Correa et al. (2010) as a relatively homogeneous elevated compartment, with limited lithological variation, to the south of the Pernambuco shear zone (Fig. 8.5). The other compartment, located in the westernmost semi-arid corner of the State, corresponds to a restricted occurrence of the Remobilized Massifs of the Pernambuco-Alagoas Domain. It is characterized by low regional drainage divides structured on igneous intrusions, which stand out in the otherwise flat landscape dominated by metamorphic complexes.

The Borborema Highlands presents remarkable gradient ruptures mainly around its edges, readily identifiable as knick-points along the drainages' longitudinal profiles. These anomalous topographic zones indicate the transition from the elevated areas to the surrounding humid piedmont to the east (Monteiro and Correa 2020) or yet the semi-arid depressions to the west, south, and southwest.

The low coastal sedimentary plateaus, regionally known as tabuleiros (tablelands), are characterized as regional tablelands occurring along the coastline with summits ranging from 100 to 200 m in elevation, principally structured on Neogene sedimentary covers of the Barreiras Formation. In Alagoas, this macro-unit occupies the entire eastern façade, forming cliffs on the coastal zone and extensively flat divides that may extend up to 100 km inland (Fig. 8.6).

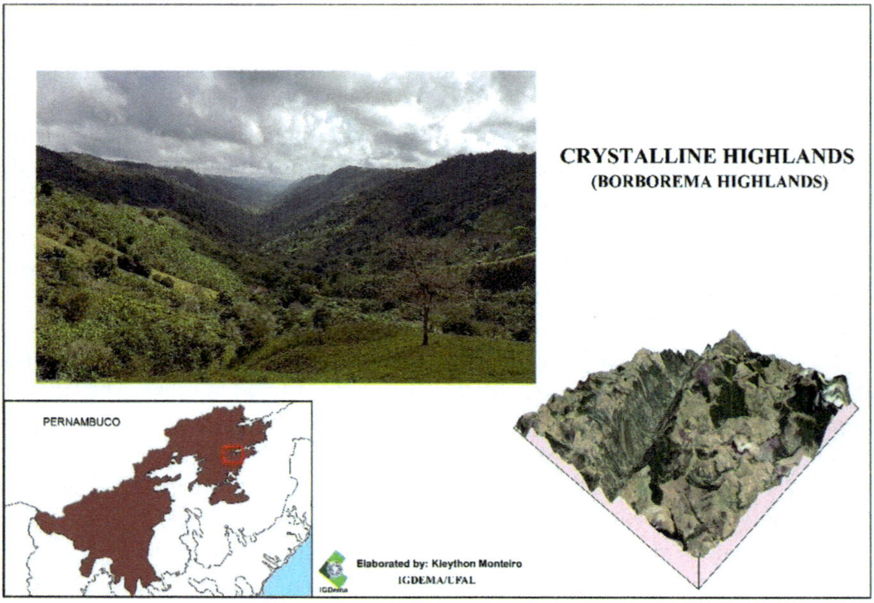

Fig. 8.5 Crystalline highlands (Borborema Highlands), exhibiting dissected landforms with steep escarpments and entrenched valleys. *Source* Map (Authors 2021); Block-diagram (USGS 2004); Photo (Authors, 2021)

Fig. 8.6 Low coastal plateaus (*Tabuleiros*) presenting an undissected morphology at the divide center and headward erosion by small stream along the cliffs. *Source* Map (Authors 2021); Block-diagram (USGS 2004); Photo (Authors 2021)

Whenever not cut through by the Atlantic drainage, these low plateaus present a rather undissected morphology with broad divides, only attacked by the headward erosion of the first-order channels along the steep cliff-lines that bound the geomorphic unit, especially to the east.

By its turn, the homoclinal plateau associated with the Jatobá basin occupies a smaller area on the northwest border of Alagoas with the State of Pernambuco. It is formed by Siluro-Devonian sedimentary rocks of the Tucano-Jatobá basins. This plateau presents a peculiar homoclinal morphology, with a dip-slip tilted to the northwest (Fig. 8.7). On its southeast margin, it presents steep escarpments (plateau's front) and buttes, characteristic of the scarp retreat process and exhumation of the sedimentary layers. The most prominent occurrence of this unit in Alagoas is the Parafuso range, which reaches more than 700 m in elevation, thus being one of the highest areas in the State.

The geomorphic units of depressions are also distributed from the coast to the State's interior. They represent low-lying denudational morphologies dominated by the surface transport of materials by channeled- and sheet-flows. Despite their several spatial expressions, in Alagoas, the depressions are large circumdenudation areas surrounding the higher landforms. Ahead of the Borborema Highlands, the piedmont is a type of peripheral depression interposed between the former and the low coastal plateaus. On the other hand, the Sertaneja depression presents itself as a sequence of vast rock pediments, gently tilted toward the drainage of the São Francisco River.

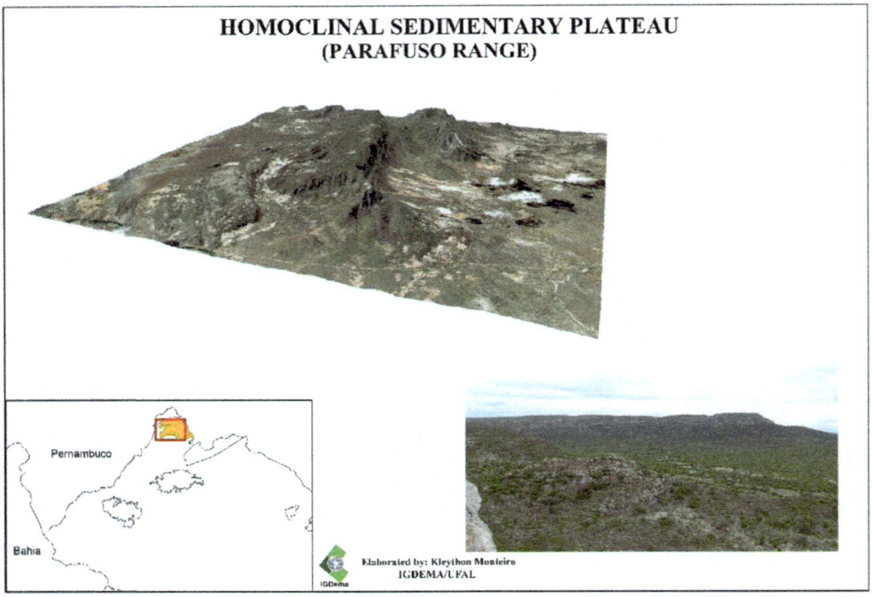

Fig. 8.7 Homoclinal sedimentary plateau associated with the Jatobá basin (Parafuso range), with a northwest tilted dip-slip and a southeast facing steep front. *Source* Map (Authors 2021); Block-diagram (Google Maps); photo (Authors 2021)

This extensive compartment is dotted with residual massifs and minor plutonic intrusions in the shape of inselbergs. The Sertaneja depression separates the elevated crystalline landforms to the east/northeast from the uplifted Paleozoic basins that form homoclinal plateaus to the west.

The Atlantic or pre-coastal depression (Fig. 8.8) is primarily characterized by dissected rolling landforms structured on crystalline rocks located on the piedmont of the Borborema Highlands, between the former steep escarpments and the dip-slope of the low coastal plateaus (*tabuleiros*). The intense dissection is due to this sector's high drainage hierarchization level due to the region's high rainfall totals. In some areas, the outcropping of plutonic rocks results in the appearance of sugar loaves and bornhardt- shaped residuals.

The São-Francisco depression occupies approximately 43% of the surface of Alagoas and, therefore, corresponds to the most expressive set of landforms in the State's territory. The area is poorly dissected, with a flat to gently rolling relief, receiving the lowest precipitation totals in the State. Nonetheless, this sector with vast pediments of the lower stretch of the São Francisco River is by no means a monotonous landscape; on the contrary, its flatness is interrupted by several residual landforms, of varied elevations, in the shape of massifs and smaller inselbergs.

The presence of these residuals also characterizes the morphology of this macro-compartment. In terms of area, the most expressive residuals are the massifs of Água

Fig. 8.8 Pre-coast or Atlantic depression presenting dissected, rolling landforms as a result of high drainage hierarchization under humid tropical conditions acting upon crystalline rocks. *Fonte* Mapa (Autores 2021); Bloco Diagrama (USGS 2004); Foto (Autores 2021)

Branca and Mata Grande, which surpass 800 m in altitude, thus adding significant morphological contrast to the region (Fig. 8.9).

The depositional (aggradational) compartments are represented primarily by the geomorphic unit of plains. These are characterized by flat surfaces dominated by depositional processes and can be formed by unconsolidated sediments of several origins, such as fluvial, lacustrine, and marine. Many times, they are classified as indiscriminate plains until further studies are conducted. In Alagoas, the plains are found along the lower reaches of the main rivers, around lagoons, and by the Atlantic coast (Fig. 8.10).

It is relevant to call attention to the fact that, besides the previously mentioned plains, another sector stands out at the southernmost corner of the State's coast. This area is characterized by a dune field (Fig. 8.11), displaying stable and mobile dunes close to the mouth of the São Francisco River. They form sand sheets that extend from the coastline to the foot-slope of the Neogene tablelands (*tabuleiros*) further to the west.

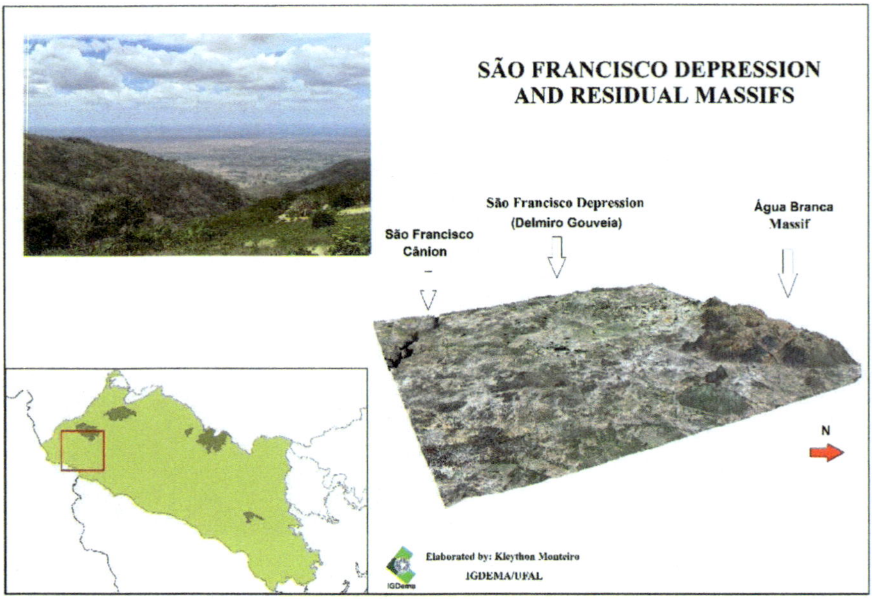

Fig. 8.9 São Francisco depression and residual massifs of western Alagoas. The sharp contrast between the flat pediments of the Delmiro Gouveia's region, that characterize the unit, and the lofty terrain of the Água Branca can be observed. *Source* Map (Authors 2021); Block-diagram (USGS 2004); photo (Authors 2021)

8.4 Conclusions

Despite not providing an in-depth look into structural analysis and surface geomorphological processes nor establishing a particular cartographic methodology for the landforms of the State, the present contribution focused on highlighting the importance of initial systematization to present the relief forms of Alagoas, seeking to establish a correlation between the overall lithological attributes and the sets of regional landforms.

We started from the understanding that the landforms are a product related to the disposition of certain types of materials at the crust's surface, where the general processes of denudation and accumulation may contribute to the grouping of landforms in separate compartments. In this way, denudational and depositional compartments may configure clusters of forms that are further subdivided based on the underlying materials at a larger spatial scale and the surface processes at a detailed level.

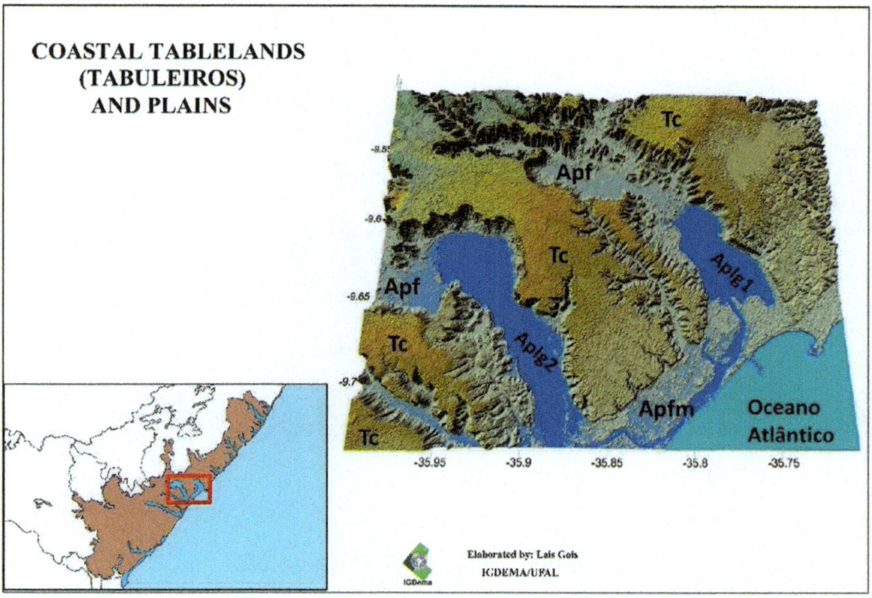

Fig. 8.10 Model illustrating the diversity of plains found in the lagoon area close to the city of Maceió. Apf—Fluvial plain; Aplg—lacustrine plain (1—Mundaú lagoon; 2—Manguaba lagoon); Apfm—Marine plain; TC—coastal tablelands. *Source* Map (Authors 2021), Block-diagram (USGS 2004)

Fig. 8.11 Depositional geomorphic compartment made up of sand dunes at the mouth of the São Francisco River. *Source* Map (Authors 2021), photos (Authors 2021)

References

Ab'sáber AN (1956) Depressões periféricas e depressões semi-áridas no Nordeste do Brasil. Boletim Paulista de Geografia, 22. São Paulo

Correa ACB, Monteiro KA (2021) Revisitando as superfícies de aplainamento: novos enfoques e implicações para a geomorfologia geográfica. Humboldt—Revista Geografia Física Meio Ambiente 1:1–26

de Corrêa, ACB et al (2010) Megageomorfologia e morfoestrutura do Planalto da Borborema. Rev Inst Geol 31(1):35–52

de Gois LSS, de Corrêa, ACB, de Monteiro KA (2019) Análise Integrada dos Brejos de Altitude do Nordeste do Brasil a partir de Atributos Fisiográficos (Integrated analysis of highland humid Brejos in Northeast Brazil based on physiographic attributes). Espaço Aberto, PPGG - UFRJ, vol 9, pp 77–98

de Gois LSS, de Corrêa ACB, de Falcão NAM, de Monteiro KA (2021) Caracterização sedimentológica dos colúvios do Maciço de Mata Grande - AL: uma comparação entre brejos de altitude do Nordeste do Brasil. Rev Geosci Nordeste 7:1–12

de Herculano EO (2011) Análise de compêndios de geografia (1844–1890). Trabalho de Conclusão de Curso em Educação (TCC), 2011. Universidade Federal de Alagoas, Centro de Educação. Maceió, 2011

de Monteiro KA, de Corrêa ACB (2020) Application of morphometric techniques for the delimitation of Borborema Highlands, northeast of Brazil, eastern escarpment from drainage knick-points. J South Am Earth Sci 103:102729

Demek J, Embleton C (1978) Guide to medium-scale. Geomorphological Mapping International Geographical Union. Commission on Geomorphological Survey and Mapping. Geomorphology, p 384

dos Santos RWD (2019) Escravidão, sociedade e economia da Villa Real de São José do Poxim – 1774 a 1854. Mestrado em História (Dissertação) 2019. Universidade Federal de Alagoas, Instituo de Ciências Humanas, Comunicação e Artes. Programa de Pós- Graduação em História. Maceió, 2019.

EMBRAPA (2012) Zaal—Zoneamento Agroecológico do Estado de Alagoas

Guerra AT (1993) Dicionário geológico-geomorfológico. Rio de Janeiro. IBGE, p 446

Lima IF (1965) Geografia de Alagoas. Editora do Brasil, São Paulo

Lima IF (1990) Maceió a cidade restinga – contribuição ao estudo geomorfológico do litoral alagoano. Imprensa Oficial Graciliano Ramos; Cepal, Maceió

Mendes VA (Org) (2017) Geologia e recursos minerais do estado de Alagoas: escala 1:250.000. CPRM – Serviço Geológico, Recife

Molion LCB, Bernardo SO (2002) Uma revisão da dinâmica das chuvas no Nordeste brasileiro. Rev Bras Meteorol 17(1):1–10

Moreno DC (1612) Livro que dá Razão do Estado do Brasil. Disponível em: http://arquivodigital.cm-porto.pt/Conteudos/Conteudos_BPMP/MS-126/MS-126_item1/index.html. Acesso em: 30 Nov 2021

Nascimento JPH, Leite PTB, Silva PEC, Monteiro KA (2017) Aplicação do índice de sinuosidade de frente de escarpa para identificação de setores sob influência tectônica em recorte da Borborema na bacia do rio Paraíba do Meio - AL. In: Filho AP, Amorim RR (Org.) Os Desafios da Geografia Física na Fronteira do Conhecimento, 1st edn. Campinas, vol 1, pp 6481–6485

Nascimento JPH, Leite PTB, Silva PEC, Lourenço da Silva R, Monteiro KA (2016) Identificação de anomalias de drenagem no Rio Paraiba do Meio (PE/AL) a partir da aplicação do índice de Hack. CLIO. Série Arqueológica (UFPE), vol 31, pp 157–173

Nobre CA, Molion LCB (1986) Boletim de Monitoramento e Análise Climática – Climanálise–Número Especial, Edição Comemorativa de 10 anos, 1986, INPE. São José dos Campos, SP. 125 p.

Nunes Silva PH (2018) Thomaz do Bomfim Espíndola e o pensamento geográfico na Alagoas da segunda metade do século XIX. In: XIX Encontro Nacional de Geográfos. 2018. João Pessoa. Anais [...] João Pessoa: ENANPEGE, 2018. Pág 1–7.

USGS (2004) Reprocessing by the GLCF. 2004. (1, 3, 30) Arc Second SRTM Elevation, Reprocessed to GeoTIFF. College Park, Maryland: The Global Land Cover Facility. Version 1.0.

Villanueva TCB (2016) Geodiversidade do estado de Alagoas. CPRM, Salvador, p 165

Kleython de Araújo Monteiro He holds a BSc. Degree in geography from the Federal University of Pernambuco (UFPE) (2007), a master's degree in geography (2010) from the Federal University of Pernambuco (UFPE), and a Ph.D. in geography (2015) from the Federal University of Pernambuco (UFPE). He is a professor and researcher at the Federal University of Alagoas, the leader of the Nucleus of Quaternary Studies of the Northeast of Brazil (NEQUA), and an associate researcher of the Study Group of the Quaternary of the Northeast of Brazil (GEQUA). He has experience in geoscience with an emphasis on structural geomorphology, mega-geomorphology, application of morphometric indices, morphogenesis, and applied physical geography.

Antonio Carlos de Barros Corrêa He holds a degree in geography from the Federal University of Pernambuco (UFPE) (1994), an undergraduate exchange program at Radford University, Virginia, USA, a master's degree in geography (1997) from the Federal University of Pernambuco (UFPE), a Ph.D. in geography (2001) from the Paulista Júlio de Mesquita Filho State University with a Sandwich Internship at Durham University, UK, and a post-doctoral fellowship in geomorphology (2003) from the Júlio de Mesquita Filho Paulista State University. He is a professor and researcher at the Federal University of Pernambuco, a CNPq research productivity fellow (level 2), a permanent member of the graduate programs in geography and archeology, and the leader of the Quaternary study group of the Brazilian Northeast (GEQUA) and the Quaternary geomorphology laboratory at UFPE. He has experience in geoscience with an emphasis on geomorphology, acting mainly on Quaternary geomorphology, geomorphology of semi-arid environments, historical and structural geomorphology, application of morphometric indexes in geomorphological analysis, and geosystemic analysis.

Chapter 9
Geomorphology of Sergipe State

Renata Nunes Azambuja, Ronaldo Missura, and Hélio Mário de Araújo

Abstract The state of Sergipe presents a varied mosaic of geomorphological landscapes, distributed in three different phytoclimatic regions known as semiarid, agreste and coast. In addition to the reworking of surfaces, caused by climate variability during the Pleistocene/Holocene, the action of neotectonic phenomena is remarkable in the rejuvenated structures of the mountains and in the morphotectonic control of the drainage network. In this way, it was decided to describe the morphology of the state of Sergipe by the subdivision of morphosculptures, individualized in the units of Depression Sertaneja, Domos de Itabaiana and Simão Dias and Planalto de Palmares in the western sector, coastal tablelands in the central/eastern range and river and coastal plains to the east. In the compartment unit of the Backwoods Depression, a flat relief predominates, punctuated by topographic caves of crystalline rocks. About landscapes of large rivers, there is evidence of neotectonic activity, exhibited by the Canyons of the São Francisco, Vaza Barris and also the Submarine Canyon of the mouth of the Japaratuba River. The coastal tablelands are found in several stretches under heavy dissection, when intersected by rivers of regional importance. On the short coastal strip are ubiquitous dynamic landscapes, dominated by wind and river action, associated with the dunes and lagoons, as in the Santa Isabel Biological Reserve, or by the great river plains. In the entire coastline it is possible to observe different sedimentary levels, corresponding to variability of marine levels in the Northeast Quaternary. Despite the state-of-the-art presented in this chapter, there are still many questions to be resolved that certainly future academic studies should complement.

R. N. Azambuja (✉) · R. Missura · H. M. de Araújo
Federal University of Sergipe, São Cristóvão, Brazil
e-mail: renatageo.ufs@gmail.com

R. Missura
e-mail: ronaldomissura@academico.ufs.br

H. M. de Araújo
e-mail: heliomarioaraujo@yahoo.com.br

V. Claudino-Sales and J. F. Sobrinho (eds.), *Geomorphology of the Northeast Region of Brazil*, The Latin American Studies Book Series,
https://doi.org/10.1007/978-3-031-58874-7_9

Keywords Geomorphology of Sergipe · Neotectonics · Morpho-sculptural compartmentation

9.1 Introduction

Although Sergipe is the smallest northeastern state in the Brazilian federation, it presents a varied mosaic of landscapes either conditioned, in specific cases, by endogenous mechanisms, given greater morpho-structural control, or superimposed by morphogenetic action concerning the Quaternary surface coverings in northeastern Brazil.

As occurs in all eastern northeast, the compartmentalization of the large morpho-structural units (Dantas and Shinzato 2017) follows a variation of the geological-geomorphological framework according to the accentuated reduction of rainfall in the east–west direction toward the semi-arid region. Thus, three phytoclimatic units are classified for Sergipe, subdivided into the coastal zone, countryside, and hinterland.

The coastal zone is affected by a humid tropical climate that extends over a strip that follows the coastline, traditionally described as the strip of Quaternary coverings of marine and fluvial-marine origin, as well as by the presence of the coastal tablelands. Average annual precipitation varies between 1300 and 1800 mm, with only a short dry season in summer and a rainfall concentration in winter.

The Sergipe countryside presents a narrow but elongated strip of sub-humid tropical climate, composed of flattened or collinear surfaces in the transition zone, structured both by the Sergipe Sedimentary Basin rocks, the Sergipe Meso/Neoproterozoic fold belt, and the Itabaiana Dome. The transitional-form vegetation comprises forest and caatinga in conditions of average annual precipitation between 900 and 1200 mm, with a dry spring–summer season and winter rains.

The hinterland displays a significant strip going toward Bahia to the west and Alagoas to the southwest and northwest. The phytoclimatic zone prevails over the Sergipe Meso/Neoproterozoic fold belt with an overlay of open caatinga vegetation (already quite uncharacterized due to pasture use) over the pediplain surfaces. In higher areas, distributed over the region as isolated mountain ranges, they present rupestrian fields.

Due to the semi-arid condition, the belt presents average annual precipitation levels between 500 and 800 mm, with autumn/winter rains (May to August) and a prolonged dry period.

9.2 Morpho-Structure and Geological Picture of Sergipe

The Sergipe geological composition presents a mosaic of ancient structures in the midwestern domain, with Paleoproterozoic and Neoproterozoic terrain fragments and Cretaceous sedimentary basins, a Cenozoic age. Traditionally, the area involves three significant structural provinces, the São Francisco Craton, Borborema and Coastal Provinces, and Continental Margin (Almeida et al. 1977). Detailedly, including tectonic and stratigraphic features covering Sergipe (Fig. 9.1), the morphostructural units are among the gneissic basement of the São Francisco Craton and the Itabaiana and Simão Dias Domes, the Sergipe fold belt, and the Tucano and Sergipe sedimentary basin units (CPRM 2005).

Corrêa et al. (2019) describe the São Francisco Craton as the sector of the Brazilian platform associated with Paleoproterozoic-aged orogens, in this case, represented by cratonic coverings to southern Sergipe and the Itabaiana and Simão Dias Domes.

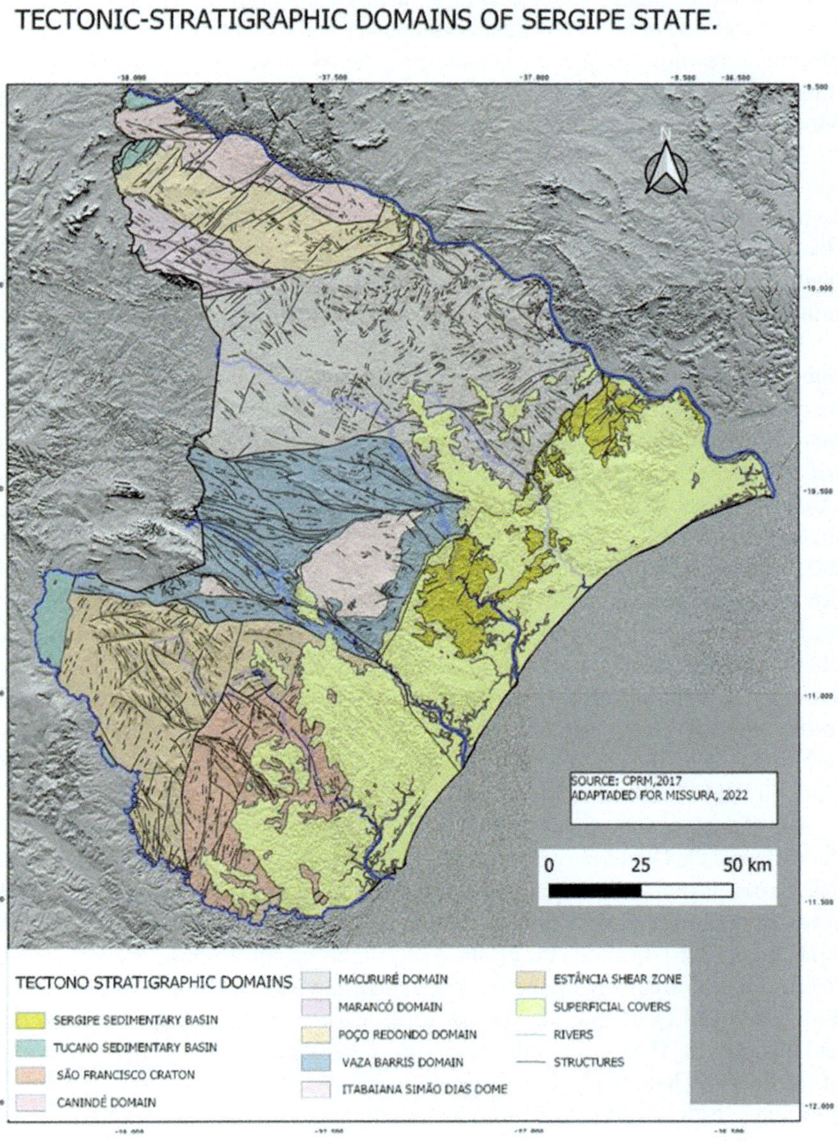

Fig. 9.1 Tectonic and stratigraphic domains in Sergipe

There is also the Sergipe Fold Belt, corresponding to the Borborema Province, a wide-ranging structure in semi-arid northeastern Brazil. This domain was strongly affected by the deformational regime resulting from the Neoproterozoic orogeny formed by the collision between the Congo-São Francisco and Pernambuco-Alagoas Cratons during the Brasiliano/Pan-African cycle (Brito Neves et al. 1977).

Extending from north to south, the belt presents six lithostratigraphic domains separated by rift zones, namely Canindé, Poço Redondo, Marancó, Macururé, Vaza Barris, and Estância (Santos and Souza 1988; Silva Filho 1998; Sial et al. 2010). The last three domains correspond to a passive-margin suture (D'el Rey Silva 1999) composed of low-grade metasedimentary and metamorphic stones. The others present rocks with a significant metamorphic degree besides plutonic bodies, outcropping to the northwest, the highest surface of the region, represented by the Black Mountain, an area of differential erosion (Corrêa et al. 2019).

After the platform stabilization in the Silurian (440 Ma), a denudational regime self-imposed throughout the belt, which only reactivated tectonically in the Gondwana rupture and the Atlantic Ocean opening in the Mesozoic (200 to 100 Ma) (Corrêa et al. op cit). The continuity of the denudation/erosion processes that occurred in this period gave rise to areas of syneclise filling, whose subsidence certainly influenced the action of the first stages of marine incursions (Pereira et al. 2012).

Thus, the third structural province in Sergipe is on this extensional regime separated into two members. The first one corresponds to syn-rift units to the west, the location of the Tucano/Jatobá Basin margin fragments. To the east, on the continental margin, the outstanding post-rift Cretaceous and Cenozoic sedimentary coverings belong to the Sergipe Basin. The Sergipe Sedimentary Basin goes from the São Francisco River to the Vaza Barris River, at the southern limit of the capital. Its end reveals the Itaporanga D'Ajuda Fault and has Tertiary and Quaternary sediments of the Barreiras Group. It penetrates Alagoas to the north and dives under the Holocene fluvial-marine coastal sediments to the east. The formation of the Sergipe Basin goes back to three great depositional cycles of the Cretaceous, related to different tectonic phases defined as continental, transitional, and marine. In the first, the sediments suffered deposition during the rifting phase between South America and Africa; in the second, there was the formation of sedimentary rocks; and in the third, the development of the Sergipe and Piaçabuçu Groups (composed by the Riachuelo, Cotinguiba, Calumbi, and Mosqueiro Formations).

The Cenozoic superficial formations include the Barrier Group Plio-Pleistocene deposits and the Quaternary coverings, predominantly Holocene (Santos et al. 1998). The extra-clastic sediments composing them vary in color and are little or non-consolidated, irregularly stratified, indistinct, and non-fossiliferous (Schaller 1969; Vilas Boas 1996). In a way, the clastic rocks of this group cover the Mesozoic terrains in erosive discordance with the well-delimited top and base. On the other hand, the Holocene-aged Quaternary coverings form marine terraces, coastal eolic deposits, and mangroves. The origin of these deposits relates to the paleogeographic evolution of the coastal Quaternary of Sergipe, from the oldest transgression maximum until nowadays. Bittencourt et al. (1983) lay out the most significant evolutionary events.

9.3 Geomorphology: Compartmentalization of the Landscape

Resembling other Brazilian northeastern states, Sergipe is in a passive-margin geotectonic and lithostratigraphic context. Its landscape reflects ancient regional geological morpho-structures, rejuvenated and reworked by the climatic conditions established in the Quaternary. In this work, the emphasis will be on the modeling in direct geotectonic action of greater expression to the following morpho-cultures: hinterland depression (Sertaneja depression), Itabaiana Dome, Palmares Plateau, coastal tablelands, and fluvial and coastal plains (Fig. 9.2).

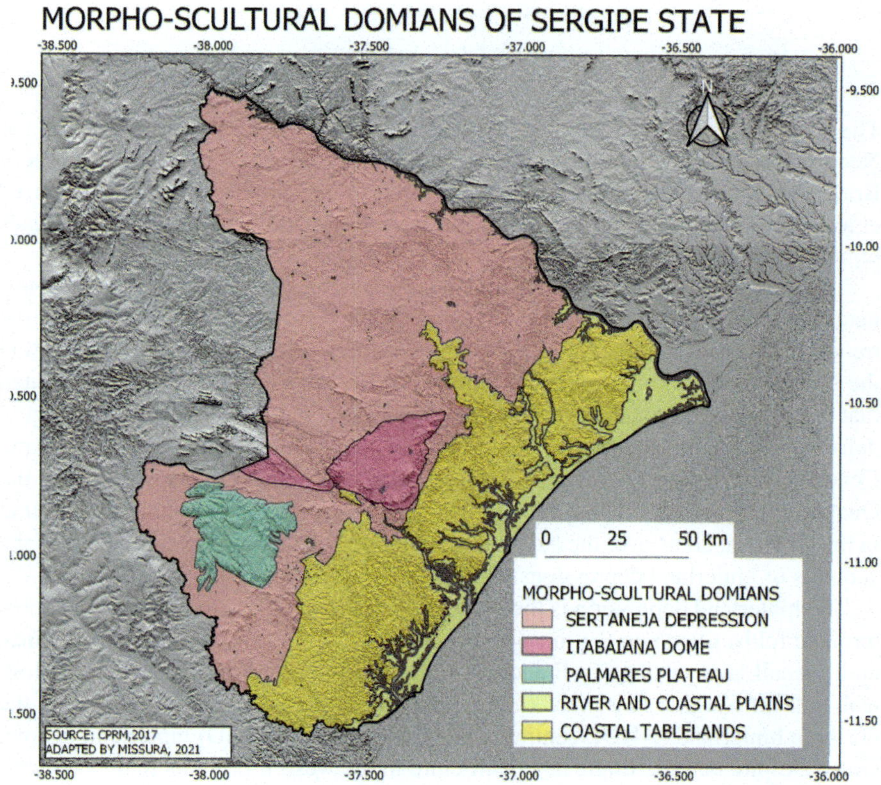

Fig. 9.2 Geomorphological domains in Sergipe

9.4 Neoproterozoic Fold Belt: The Sergipe Hinterland Depression

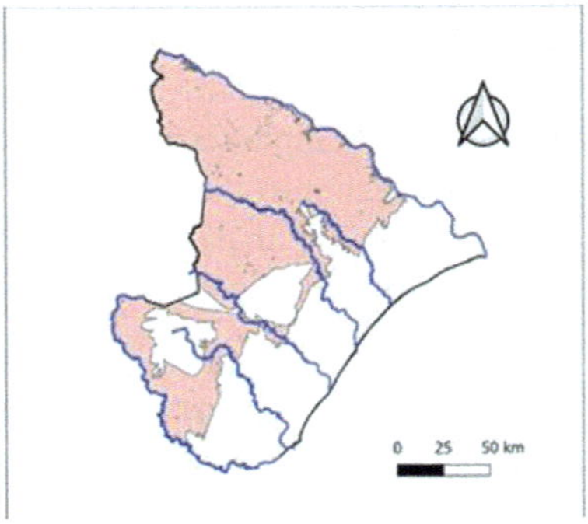

The orogenic belts—characterized by the terrain amalgamation that occurred in the Neoproterozoic with the collisions between Rodinia and Panotia—are in northeastern Brazil in the form of depressions, plateaus, and residual massifs. According to Corrêa et al. (2019), this sector hosts a set of intrusive suites and metamorphic complexes regionally affected by numerous shear zones.

The ductile deformation of the Brasiliano played a significant role in the continental structures of northeastern Brazil, producing structural massifs toward the trends of lineaments reworked by the action of differential erosion. When analyzing the structural relief conditions in northeast Brazil, Maia and Bezerra (2014) stated that the northeast-to-southwest and east-to-west lineaments, expressed by the ductile deformation, define the preferential zones of reactivation in the Cretaceous and Cenozoic and structurally control the drainage and dissection network over the Quaternary formations. Thus, the morpho-structural arrangements commonly found in the erosional areas over the Neoproterozoic fold belts in Sergipe individualize the subunits of hinterland depressions, mountains, and residual massifs.

The hinterland depression is the most frequent relief unit in the range of Neoproterozoic folds, seated on the morpho-structural domain of the Borborema Province and a small area of the São Francisco Craton. It is the prevailing geomorphological field in Sergipe (1005.65 km^2), subdivided into three portions, namely the northern hinterland (NHD), central (CHD), and southern (SHD) hinterland depressions. Despite being a single macrostructure in the west, it presents flattened forms and gently undulating relief associated with different morpho-structures in the three subunits. To the north, residual massifs and inselbergs are predominant, besides the valley fitted by the superimposition of the São Francisco River. In the center, the Itabaiana and Simão Dias Domes stand out, structures cut by the valley of the Vaza Barris River and Carstes morphologies. To the south, the Palmares Plateau and its surroundings are isolated and in contact with the Tucano Basin in Green Well to the west and São Francisco Craton to the east.

Although in most of the Neoproterozoic fold belt, the action of the erosive cycles has occurred homogeneously, establishing conserved or dissected peneplains, the erosion acted more vigorously in two compartments, submitting, at the end of its action, the crustal exposure in its initial thickness. It happens in the outcrop of the basement in the São Francisco Craton, in the southern fold belt, and in the emptied Itabaiana and Simão Dias Domes, with a Paleoproterozoic gneissic nucleus, in the center and west of Sergipe.

The northern compartment of the Neoproterozoic fold belt comprises four lithologic domains divided by Santos et al. (1998) in the southeast-to-northwest direction among the Macururé, Marancó, Poço Redondo, and Canindé Domains. What makes this sector to be unique in the relief morphology is the gradual sequence of transverse shear zones under a compressive regime of continental margin pouring lithofacies of meta-rhythmite, meta-graywacke, and schist (Macururé Complex) into zones of intrusive magmatic suites and volcanogenic, felsic, mafic, and ultramafic rocks (Marancó Complex). Poço Redondo and Canindé sequences have polyphasic and disharmonic pre-Brasiliano fold belts (Santos et al. 1998) besides volcanic arcs.

In general, the hinterland depression has an almost-flattened surface with preserved or dissected compartments at average altitudes between 180 and 250 m. The ubiquitous occurrence of intrusive granitoid suites in the northern depression gives the landscape an association of inselbergs, residual massifs, and ridgelines with summits above 300 m. The latter establishes the divisors of the main hydrographic basins of the region that flow toward the main local base level, the São Francisco River. Many drainages of these basins are oriented and fitted to the main tectonic structures and lithological contacts in this compartment.

The Black Mountain stands out (Fig. 9.3) among the innumerable smaller massifs in the area. Composed by Black Mountain granitoid (Santos et al. op cit.), it is in the northwestern limit with Bahia, between the municipalities of Poço Redondo (SE) and Pedro Alexandre (BA) and is the highest point in Sergipe (742 m). The Black Mountain is also a significant water divisor of the basins draining into the São Francisco River to the north and those into the Sergipe and Vaza Barris River basins to the east and south. Because of the south-to-north transversal position, the seasonal semi-arid signs controlling the distribution of precipitation determine differentiated processes between the south and north of the massif, originating slopes covered with colluvial deposits to the south (already in Bahia) and detrital pediments to the north toward the São Francisco River.

The northern flank, face to Sergipe, functions as a leeward region reflected in the progressive precipitation decrease in the São Francisco hinterland. The drainage system developed in the zone presents a dendritic character with low dissection over detrital pediments (Fig. 9.4). The mostly-ephemeral rivers have an intermittent flow also on the larger channels because of the short rainy season. Mabesoone (1981) characterizes these channels as deposition areas of sandy and coarse materials. Some inselbergs, originating from granitic plutons, are frequently spotted in this portion, mainly between Monte Alegre, Poço Redondo, and Canindé de São Francisco, with peaks reaching between 400 and 600 m, according to Dantas and Shinzato (2017).

Fig. 9.3 Black Mountain massif and the Sergipe hinterland depression. *Photograph* Daniel Rodrigues Lira (2021)

Fig. 9.4 Inselbergs in the background of the image and detrital pediments in the foreground, northern Sertaneja depression, Poço Redondo. *Photograph* Paulo Vitor (2020)

In the extreme north toward the São Francisco River, there is one of the most scenic reliefs derived from the drainage incision by tectonic readjustment, the São Francisco Canyon. The incision limiting Alagoas, Sergipe, and Bahia among Paulo Afonso (BA), Canindé de São Francisco (SE), and Piranhas (AL) has an altimetric difference around 100 m between the tops and bed of the São Francisco River, shaping the tributaries coming from the depression as valleys with pronounced dissection in the mouths (Ferreira et al. 2018).

Considering the geomorphological analysis of the northeast by Ab'Saber (1969) and Mabesoone (1978), the development of the flattened surfaces of the hinterland depression in the lower São Francisco River occurred in the Cesozoic without significant variations of the base level, denoting relative tectonic stability. However, reaffirming the classic assumption of Almeida et al. (2000), Ab'Saber (1997) identified two to three intermittent excavation phases of the São Francisco Canyon.

According to Ab'Saber (1997), during the first phase, at the end of the Pliocene (three million years), a discrete inlaid valley floor occurs a few tens of meters below the base level of the modern hinterland depression, following the consolidation of the river in the region. A new uplift episode establishes an intermediate erosional plateau on top of the previous valley floor, creating alternating stretches of a relatively open and narrow valley. From the steepened valley floor, the São Francisco River has definitively deepened the vertical canyon walls.

The São Francisco River inlet exposes a weakly dissected hilly relief in its margins. The portion encompassing such a feature is on the downstream bank of the Xingó Dam. Besides, canyons—composed of sandstone walls of the Tacaratu Formation and belonging to the Jatobá Group of Silurio-Devonian age (Fig. 9.5) with a length of 30 m—are found on the right margin of the Xingó Dam. The intermittent anastomosed default—resulting from the action of the old Xingó River, currently drowned—presents escarpments over crossly-and-parallelly sedimentary structures, labyrinths, and valleys of singular beauty (Claudino-Sales 2020).

The next subunit in Sergipe has features of great prominence in the regional relief. According to Martins (2017), the origin dates back to the constitution of passive margin, characterized by carbonate-pelitic sedimentation, generated and deformed in the Neoproterozoic, belonging to the Vaza Barris Group. The gradual contact to the north with the Macururé Domain—characterized by turbidite, pelite, and psammite—belongs to the northern hinterland depression. The faulted central sector of the hinterland depression is limited to the north by the Great Mount shear zone, to the east covered, discordantly, by the coastal tablelands, to the south by the Itaporanga shear zone, and to the west by the Tucano/Jatobá Basin in Bahia.

Tectonically, the region corresponds to the outer domain of the Borborema Province, known as the embedded Vaza Barris Valley (Radambrasil 1983; Ferreira et al. 2018). As a result of the ductile deformation of the Proterozoic, the various east-to-west oriented faults play a significant role in the structures rejuvenated by structural factors.

Generally, Dantas and Shinzato (2017) classify the Sergipe hinterland depression in the central and western southwestern portions as a vast flattened surface. Characterized by an incipient dissection, the relief is in the form of broad and gentle

Fig. 9.5 Master Valley in Canindé de São Francisco is a canyon structured on the sandstones of the Tacaratu Formation on the shores of the Xingó Lake/SE. It is a specific transition area between the hinterland depression and the Paleozoic Sedimentary Basin of Jatobá. *Photo* Renata Azambuja (2015)

hills, sometimes more dissected, when the drainage tends to present an adjustment to the regional base level. The extensive flattened areas have shallow soils with a low water storage capacity and vegetation typical from the countryside and caatinga. In a higher proportion, the drainages in the Sergipe hinterland depression have an intermittent character, with alluvial plains clogged with sediments and interrupted cut and fill structure due to the seasonal signs of the semi-arid climate, as described by Mabessone (1981), Azambuja (2012), and Corrêa and Azambuja (2020).

Moreover, in the portions with greater control of structural factors, topographic rises of residual aspects appear in the landscape in the form of alignments of quartzite ridges, such as hogbacks, structured on old thrust faults between the municipalities of Frei Paulo and Ribeirópolisa. Although the regional altimetry varies between 100 to 300 m, which corresponds to the interplanaltic depression sector, Corrêa et al. (2019) classify it in the semi-arid northeast as the region associated with the rejuvenated Neoproterozoic massifs.

Given the significant number of drainages with rectangular patterns fitted in a fault network of the Itaporanga shear zone (D'el Rey Silva 1995) to the south of the Itabaiana Dome, the superimposition of the Vaza Barris River occurs consecutively in the area. Such a feature extends from Pinhão, the western boundary with Bahia, until the Itaporanga D'Ajuda, where it meets the coastal tablelands domain with the river reaching its lower course toward its mouth in São Cristóvão. Curiously, flattened reliefs occur in some stretches of the Vaza Barris River Basin in the central compartment of the hinterland depression, sometimes equivocally classified as coastal and sub-coastal tablelands. With a composition somewhat different from the Barreiras Group sequence, these surfaces, classified as Acrisol zones, in reality, present an overlay of quartz pebbles, phyllites, and quartzite, corresponding to the proximal remobilization of the lithology frequently found in this region. Thus, the term dissected tablelands of Vaza Barris, previously described by Radambrasil (1983), was chosen for this description.

In the height of Macambira, southwest of the Itabaiana Dome, the Parrot Rock is a relevant geological site and potential morpho-pathological heritage of Sergipe. Considered a relief structured in an inverted fold over intercalations of metapelites, the valley stretch, fitted in the Vaza Barris River (Fig. 9.6), shows an impressive rejuvenated Meso/Neoproterozoic massif (orogen), reaching a height of about 30 scarps. As well as in the São Francisco River, the area shows adjustments resulting from the last tectonic reactivation, which occurred in the Miocene between 50 and 20 Ma. In the Captain Farm section, it is possible to see small canyons carved in the contact among the quartzite, diamictites, and metapelite rocks in the Parrot Rock (Fig. 9.7). Other relevant features in the external sector of the Itabaiana Dome, home to the Miaba and Ribeira Mountains, are the waterfalls, such as the Macambira, Saboeiro, and Ribeira Big Well Waterfalls, demonstrating a steady neotectonic control.

In Simão Dias, due to the borderline presence between the high-angle ductile–brittle contractional shear zone and the cratonic surface of the ranch domain (meridional hinterland depression), the other dissident portion of the Itabaiana Dome occurs. The surroundings have lithologies of slightly-deformed carbonate rocks of the Eye Water Formation and phyllites rhythmically interbedded with meta-arenite, impure

Fig. 9.6 Superimposition of the Vaza Barris River between Macambira and Lagarto, Sergipe. *Photograph* Renata Azambuja (2021)

Fig. 9.7 Parrot Rock at Captain Farm is an example of a rejuvenated Neoproterozoic massif at the margins of the Vaza Barris River, which establishes the boundary among Macambira, Lagarto, and São Domingos. *Photo* Renata Azambuja (2021)

meta-carbonates, and meta-siltstone (Macedo 2019) of the Frei Paulo and Jacaré Formations, respectively.

The Fox Cave in the vicinity of Simão Dias is the largest one in Sergipe, which, according to Martins and Carvalho (2017), is 250 m long. The cave has a variety of speleothems and stalactites besides Pleistocene-aged mammal fossils. Another karst feature that stands out in the area is the Dorinha Cave, also known as Simão Dias Cliff. It is a cavity with a 43-m depth until it meets the water surface, hard to access, and highly dangerous. The genesis and conservation of the features happen with the rock of incipient primary porosity, characterized by dolomitic limestones, which, after Paleozoic tectonic reactivations, led to a secondary porosity, enabling the corrosion and dissolution of the material (Macedo 2019).

In a survey on the Neoproterozoic karst features in Sergipe, Macedo (op cit.) cataloged thirty-two sinkholes divided between abating, collapsing, and subsidence features on the Eye Water and Frei Paulo units. The region presents a poorly organized drainage network of low density and predominance of poorly developed soils, such as the eutrophic Litholic Nesol and the Natric Planosol. The presence of Proterozoic-aged endo-karsts and exo-karsts in Sergipe occurs among Simão Dias, Macambira, Pinhão, Lagarto, São Domingos, and Poço Verde (Fig. 9.8).

The third compartment is one of the smallest within the Neoproterozoic fold belt in Sergipe. In addition to the characteristics described previously regarding the hinterland depression, this domain differs from the other compartments due to the

Fig. 9.8 Exo-karst and endo-karst features. **a** Polje Jaci in Saco do Camisa Village, Poço Verde. *Photograph* Luana Pereira Lima (2019); **b** Lapiás Field in Caríbas Village, Poço Verde; and **c** Fox Hole in Raposa Village, Simão Dias. Photographs (**a**) and (**b**) by Heleno dos Santos Macedo (2016)

lithology, which, according to Martins (2017), has mixed platform deposits (siliciclastic and carbonate) and is a little deformed. Its development occurred at the northeastern edge of the São Francisco Craton, after the distensive phase marked by the Arauá volcanism, in adjacent contact to the Neoproterozoic Sergipe fold belt. This adjacent transition is noticeable with the abrupt change in landform dissection, where the craton outcrops giving rise to a surface of gentle hills.

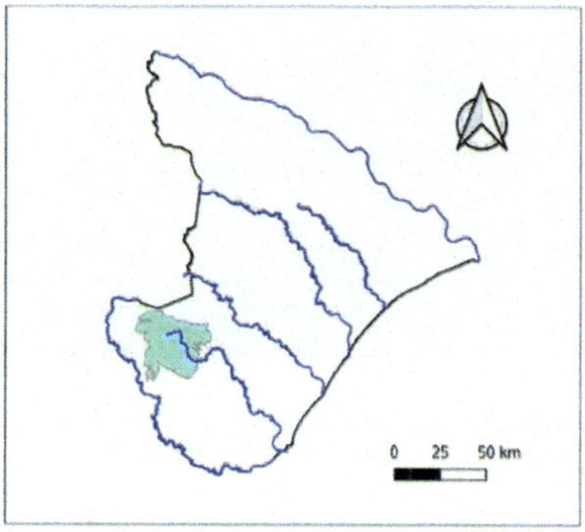

9.5 Palmares Plateau

The Palmares Plateau is part of the morpho-structural compartment of the São Francisco Craton (Fig. 9.9), which stands out when observing the relief compartments of Sergipe because of its higher geomorphologic region, around 350 and 500 m, while the surroundings are 200 and 300 m high. It is notorious the landscape map and observation because its escarpments vary between 150 and 200 m of unevenness with the entrance depending on the sector of the slope, whose maximum variation is in its northwestern sector, presenting 300 m between the base and the top of the escarpment.

The plateau is on the basement of ancient rocks and sedimentary coverings of the São Francisco Craton. The Palmares formation lithology and tectonic structures provided more resistance to erosion than the surrounding areas. Its irregular shape, surrounded by the hinterland depression, is similar to the rectangle in the southwest. The limits to the east, south, and north are the retreat scarps between the depression and the plateau. Besides the plateau compartment, there are faults in compressional zones and tectonic structures that manifest inland, making it a principal but not unique plateau structure surrounded by smaller plateau surfaces and ridges. The landscape reveals a previous higher plateau that, during its evolution, had a reduction caused by the recurrent relief dissection in the hinterland depression development.

Some examples of mountain ranges composing the plateau are Aguilhada, Big, Pururuca, Cumbi, Palmares, Evangelista, Boqueirão, Caveira, Madeiro, Arruda, Jabiberi, Canina, Embira, and Macota. Such mountains, whose names usually refer to the distinguished, elongated, and scarped altimetry, show the well-dissected plateau, highlighting the relief compared to the surrounding lower areas.

Fig. 9.9 Southern hinterland depression with views of the Palmares Plateau in the background. *Photograph* Ronaldo Missura (2019)

Even though its lithology is a cratonic sedimentary cover, its relief does not present flattened tops typical of sedimentary plateaus. However, it is an almost mountainous relief due to the altimetric difference between the depression and the plateau tops, with well-dissected hills, faults, invoices, deeply embedded valleys, and waterfalls. It deserves studies to explain the origin and evolution of this unique compartment in the Sergipe geomorphology.

Thus, the Palmares Plateau and Craton transition happens discontinuously to the east with a landscape characterized by pediplain surfaces or reliefs composed of hills and spikes that range between 150 and 250 m (Dantas and Shinzato 2017). The variability is subordinated to the change of the substrate of a low-grade metamorphic constitution (Neoproterozoic) and by outcrops of granulitic and gneiss-migmatite complexes of Paleoproterozoic age, as well as the transition zone between the semi-arid climate regime for the southern Sergipe countryside.

9.6 Simão Dias and Itabaiana Domes

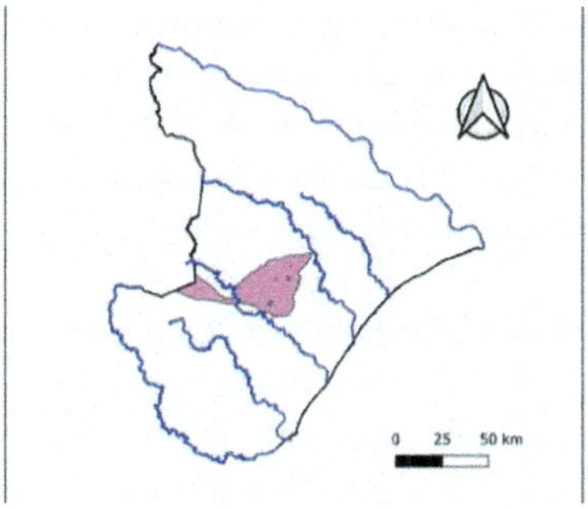

Recognized as a hollow core of the Paleoproterozoic gneissic basement, the Itabaiana Dome is a morpho-structure composed of mountain ranges structured in metasedimentary rocks. However, D'el Rey Silva and McClay (1995) more precisely defined the dome as a structural high formed by an antiformal structure. The deep weathering probably happened with the Cretaceous tectonic reactivations with the opening of the South Atlantic Ocean associated with the edge effects of the underplating magma identified by Moraes Neto et al. (2009) as responsible for the reactivations of the shear zones in the Borborema Province, described here as the central hinterland depression and the dome. The only National Park of Sergipe—the PARNA Itabaiana Mountain—is in the Itabaiana Dome.

Although the external terrain of the Borborema Province suffers little impact in its extension, remaining undulated, the Itabaiana and Simão Dias Domes represent, in reduced scale, a kind of old plateau structure compared to the Borborema Plateau, however, eroded inside. In its surroundings, there is much evidence of tectonic reactivation, probably Neogene-aged, as seen to the east of the Itabaiana Mountain in Areia

Fig. 9.10 Hanging valleys in the background over quartzitic rocks in the outer portion of Itabaiana Mountain, east of the Dome with the central hinterland depression. *Photo* Renata Azambuja (2018)

Branca. Azambuja et al. (2018) identified in detail the geomorphological mapping of the Itabaiana Mountain National Park, numerous normal-type faulting features, erosional amphitheaters in the form of hanging valleys and correlative deposits on the unit underlying the quartzitic composition scarps (Fig. 9.10). In the stretch, the pediment units are very dissected, exposing sequences of frank sandy sediments, indicating direct relation with the residues deposited by gravitational action at the slope foot.

As a geomorphological unit, the Itabaiana Dome presents residual mountains in its contour with altitudes varying between 400 and 559 m, structured in quartzite rocks (Itabaiana Formation), meta-arenites, and basal conglomerates belonging to the Miaba Group and ages of Meso/Neoproterozoic. The front, which faces the inner dome, experiences dissection on the former top, which along with tectonic reactivations, favored the accelerated removal of the cover, promoting an emptying and exposing a pediplain surface composed of a rocky pediment structured between 250 and 150 m (Fig. 9.11). The drainage network obeys structural control only at the western edge of the Itabaiana Mountains through a parallel pattern in a transition from the litholic neosol at the foot of the mountain to acrisol due to the local exposure of the gneiss/migmatite basement identified by Azambuja and Lima (2018). In the rest of the dome, a sub-dendritic pattern of an incipient incision tends to occur over Natric Planosol. The eastern flank is home to the Itabaiana, Ribeira, Comprida, Quizungo, and Capunga Mountains, while the western margin shelters the Miaba (Fig. 9.12), Tauá, and Cágado Mountains. The only not-dissected unit is the Capunga Mountains (Araujo and Mendonça 2003).

The Itabaiana and Simão Dias Domos lithology is banded mylonite orthogneisses of intrusive origin with granite composition and intercalations of amphibolites. The first formation phase relates to the tangential Brasiliana tectonics. However, later reactivations of extensional faults to contractional faults of the basement (Mocambo,

Fig. 9.11 Panoramic view of the emptied interior of the Itabaiana Dome showing a pediplain evolution modeled by the Itabaiana, Comprida, and Cajueiro Mountains. *Photograph* Renata Azambuja (2021)

Fig. 9.12 Structured ridges in quartzitic rocks of the Itabaiana Formation, Miaba Mountain. Highest elevation point (630 m) of the Itabaiana Dome. *Photo* Renata Azambuja (2022)

Simão Dias, and Itaporanga faults) promoted the shearing of the southern part of the Itabaiana Dome where it meets the Simão Dias Dome (Araujo and Mendonça op cit). This displacement redefined the Simão Dias portion as an apophysis of the stock, derived from the orthogneiss basement similar to the larger body of the Itabaiana Dome.

9.7 Sergipe Tableland

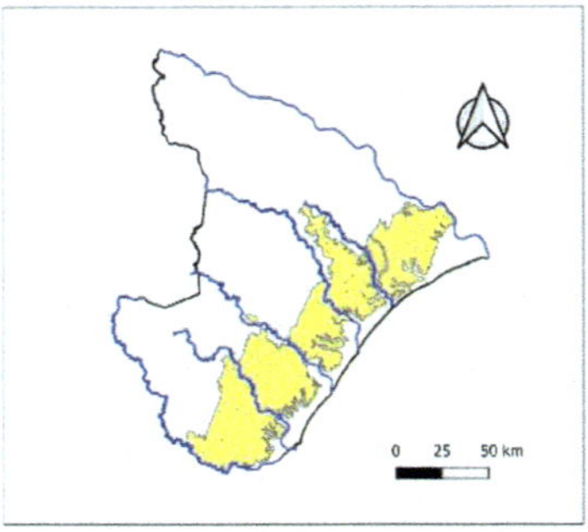

The tablelands are part of the morpho-structure of the Phanerozoic basin and sedimentary coverings. The 5910.57 km^2 makes it the second-largest geomorphological compartment in Sergipe. It occurs in a strip of approximately 188 km long from northeast to southwest and an average of 35 km width from east to west—a maximum of 66 km and a minimum of 13 km. The division into three parts is (1) between the São Francisco and Japaratuba Rivers (northern Sergipe tablelands—NST); (2) between the Japaratuba and Vaza Barris Rivers (central Sergipe tablelands—CST); and (3) between the Vaza Barris and the southern limit of Sergipe with Bahia (southern Sergipe tablelands—SST).

The coastal plain limits the compartment to the east in the form of paleo-fallen escarpments between the tableland and the plain, sometimes advancing inland, mainly associated with the largest watersheds of Sergipe. It causes the boundary between the compartments to be irregular in the form of festoons of the escarpment of the tablelands concerning the plain. The hinterland depression limits it to the west, also in the form of scarps caused by the dissection of the depression relief toward the tableland surfaces.

The coastal tablelands are structured in the Barreiras Group sediments, covering in some localities the sedimentary formations belonging to the Sergipe Alagoas Sedimentary Basin, namely the Cotinguiba, Riachuelo, Calumbi, and Bananeira Formations. The relief, predominantly composed of flattened tops, evolves into gentle hills in some sectors because of the dissection of the tops. The most preserved areas are in the NST, the northernmost compartment that covers Japoatã, Pacatuba, and Neópolis. In this sector, the tabular tops are extensive, with low drainage density and relief dissection. This region is popularly known as the Neopolis Plateau and is widely used for a great diversity of irrigated crops thanks to the flat relief and easy mechanization.

The following sector, the CST, presents greater diversity of relief forms, such as the most inland areas with extensively preserved tabular tops among Nossa Senhora das Dores, Capela, and Cumbe, being widely used for the cultivation of sugarcane. In its more coastal sector—Japaratuba (Fig. 9.13), Carmópolis, Rosário do Catete, Siriri, Maruim, and General Maynard—predominate gentle hills structured on rocks

Fig. 9.13 Sergipe tablelands covered by eolic sediments in the central portion of Japaratuba, Sergipe. *Photograph* Jean Conceição de Sousa (2019)

of the Sergipe Basin and Barreiras Group, in which there is an extensive field of oil extraction on land in the Sergipe Sedimentary Basin.

The most preserved SST sectors of tabular relief are in Lagarto, Cristinápolis, and Itabaianinha in an advanced dissection, in comparison with the abovementioned sectors, due to the drainage toward the coastal front and inland related to the hinterland depression. The SST also presents more restricted sectors of gentle hills structured in the Barrier Group sediments and São Francisco Craton lithologies. Other regions have tabular relief dissected as gentle hills, where it is still perceptible sedimentary packages that discontinuously structure the relief compartment, mainly toward the hinterland depression.

In the tablelands, there is a general surface inclined from west to east, with the highest altitudes of around 280 m further inland and the lowest ones at the base of the scarps that border the coastal plain, around 10 m (Fig. 9.14). The exception is in the Japoatã highland, where geological structures in the Sergipe Sedimentary Basin, despite 23 km far from the coastline, present reliefs structured in the Barreiras Group sediments that reach 280 m high.

Furthermore, altitudes below 150 m predominate in this compartment, making up more than 90% of the tablelands. The frequent presence of karstic relief in places where the Sergipe Sedimentary Basin outcrop is interesting for having a significant limestone concentration besides the pronounced local rainfall. Thus, they provide an accelerated dissolution of the soil, usually localized and without much expression and development. Besides that, the micro depressions—resulting from the doline

Fig. 9.14 Southern sector of the Sergipe tablelands in Itaporanga D'Ajuda, Sergipe. *Photograph* Ronaldo Missura (2019)

formation in tabular reliefs, more expressive in Nossa Senhora das Dores, Capela, and Cumbe—occur in dozens and require further studies on the origin and evolution.

9.8 Coastal Plain

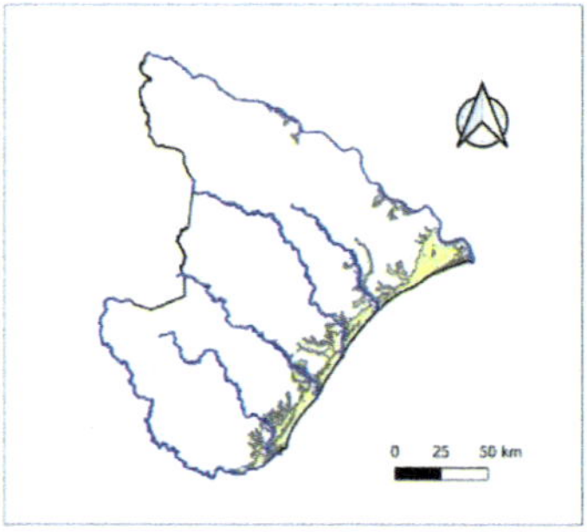

The coastal plain in Sergipe has 1851.79 km^2. The division of the compartment into three sectors—Northern and São Francisco Coastal Plain (NSFCP), Central Coastal Plain (CCP), and Southern Coastal Plain (SCP)—helps to understand it.

The São Francisco River Plain dominates the northern NSFCP due to depositional dynamics in the vicinity of the mouth. The São Francisco River has in this sector fan-shaped Quaternary deposits that, together with its margin at the Alagoas coast, form a rhombus cut in half by the São Francisco River, having its axis 64-km long, oriented northeast to southwest. In the minor axis, the maximum extension is 23 km between the mouth and the beginning of the deposit strands at the limit with the coastal tableland. Its width tends to decrease toward the south and north, where the strands coalesce with the coastal plain between the northern Sergipe and southern Alagoas coastlines.

The plain has strands of sandy and clayey sediments, oriented northeast to southeast, in the form of gentle hills with tops of 10 m, besides low slope and relief dissection. In some sectors, they reach more than twenty meters, as in the zones of wind accumulation in the form of stabilized or semi-vegetated dunes. In the lower areas, the drainage does not form channels, but intermittent and perennial lagoons instead, some of them in the same direction as the cordons. Sometimes these lagoons with elongated forms develop a complex of small lagoons disconnected by gentle interfluves, composed of sedimentary deposits of at most a few tens of meters.

The formation of these deposits exerts a barrier effect for the drainages that flow from the tableland compartments toward the coastal plain. Thus, channels such as the Betume River, which presents stretches of its bed well spread out, develop wetlands that flow toward the São Francisco River to the northeast. The conjunction of swamps and lagoons between the coastal cords form the Pacatuba Wetland (Fig. 9.15), a region rich in scenic beauty, fauna, and flora in the northeastern Sergipe. According to Fontes (2015), the NSFCT has five morphosedimentary units.

Morphosedimentary unit 1 occupies the innermost strip of the coastal plain in contact with the Barrier Group sediments, having deposits of coalescing alluvial fans and marine terraces (coastal ridges).

Morphosedimentary unit 2 is in contact with the first and occupies the broad elongated depressions on both sides of the deltaic plain border, north and south of the São Francisco River. The low and flat areas are home to the floodplains of the Bitumen River in Sergipe. The unit has a little expressive coastal bar and paleo-laguna deposits, in which the current marsh and fluvial floodplain deposits overlay it.

The morphosedimentary unit 3 has marine terraces occupying an intermediate position in the coastal plain constituted by coastal ridges. The outermost portion of the two sets of coast has eolic dunes inactive cover. The limits of the cordon sets are erosive truncations evidenced by changes in the orientation of their alignments. The boundaries mark episodes of changes in the coastal dynamics related to the sea level and variations in the relationships among marine, coastal, and fluvial processes.

Morphosedimentary unit 4 involves two sets of coastal ridges: (1) the innermost has the barrier island deposits, spurs, and sub-actual pontoons with eolian dunes cover; (2) and the outermost includes barrier island deposits and a current island covered by mangroves, inactive and active eolian dune fields, and the beach zone.

Fig. 9.15 Coalescence of lagoons limited by eolic sediments that originated the Pacatuba Wetland. *Photograph* Ronaldo Missura (2021)

Morphosedimentary unit 5 has fluvial deposits of the São Francisco River cutting across the sets. The deposits constitute the lower fluvial terrace and the flood surface formed by the current fluvial channels with fixed islands and the overflow plain with marginal lagoons that overlie past deposits.

There are several rice fields in lowland areas, as in Ilha das Flores and Brejo Grande. The use is mainly in the fluvio-marine deposits, but clayey sediments are present near the São Francisco River mouth forming islands covered with mangrove vegetation, and many of these areas have carciniculture, with the removal of the original vegetation and the construction of tide flood tanks. In the sector closest to the ocean occurs the formation of a 20 km-long sandy restinga and, consequently, in its internal part, there is a large saltwater lagoon channel under the action of the ocean tides. In the southernmost sector, there are stable dune deposits that, according to Bittencourt (1983), are Pleistocene-aged, coalescing with the beginning of the central sector of the Sergipe plains.

The CCP covers the coastline of Pacatuba, Pirambú, Barras dos Coqueiros and Aracaju. In the central compartment are the mouths of the Japaratuba, Sergipe, and Vaza Barris Rivers. At the end of the vertex of the southern sector of the NSFCP is the beginning of a dune field that coalesces with the initial sector of the CCP. The sector is the smallest on the coast of Sergipe, averaging 2 km from the beach line to the limit with the coastal tablelands. Stable to unstable dune fields predominate in the landscape. Their formation is due to the availability of sandy sediments coming from the beach, the action of the oceanic winds (especially southeast and northeast trade winds depending on the season), and the low density of vegetation cover.

According to Bittencourt (1983), the eolic deposits present on the Sergipe coastal plain are of two generations, one from the Pleistocene and the other more recent, Holocene. The deposits found in the northern stretch present variations between stretches covered by vegetation and others with exposed eolic sediments. Oliveira (2008), in his work about the eco-dynamics of the coastal dunes of Sergipe, attributes to this sector of the Pacatuba and Pirambu coast the occurrence of a northern sector of stable dunes, and a southern sector of intergraded and unstable dunes, where, in general, pedogenesis processes predominate in the eolic deposits. However, in certain sectors, it is still possible to observe the advance of a linear cord of dunes toward the interior of the plain. The altitudes of the dunes in this sector rarely exceed 10 m. The region of the coast of Sergipe is delimited by the Santa Isabel Reserve (Fig. 9.16), classified as a coastal marine biome, occupying 4,109.88 hectares. Besides the absence of any kind of human occupation, this protected area protects the wind formations described here and is home to the endangered hawksbill turtle (Eretmochelys imbricata).

The end of the aeolian deposits to the south is marked by the encounter with the Japaratuba River marine lowland-fluvial plain, where the urban site of Pirambu is developed. The plain at the mouth of the Japaratuba River has sandy deposits of the Quaternary age (Pleistocene and Holocene). These deposits can be presented on the morphologies of marine terraces, fluvial-lagoon deposits, estuarine deposits, and clayey deposits in the estuary areas of the Japaratuba River, covered by mangrove vegetation.

Fig. 9.16 Aeolian dunes in ReBio Santa Isabel in the background and river plains. *Photograph* Ronaldo Missura (2021)

The lowland is cusp-shaped, with its greatest extensions of 5 km in width in the northeast-to-southeast direction, and the fluvial-marine plain extends from the coastline to 6 km inland, where it is still possible to observe mangrove vegetation. However, the river plain of the Japaratuba River continues continentally, averaging 2 km in width. The Japaratuba River Plain has a development near the town with the same name as the river, where it has a width of 1 km, 18 km from the coast. The Japaratuba River, despite being a river of little extension compared to neighboring basins, presents some characteristics at its mouth that denote having been a river of greater extension and energy in its past erosive processes. Besides the great extension of its fluvial-marine plain, the existence of a submarine canyon is 21 km wide and 1400 m deep. Figueiredo Jr. et al. (2011) denote the existence of a more vigorous river in its geological past.

The altitudes in the plain present maximum levels of 5 m, but the average is around 2 m. Another particular characteristic is the presence of great meandering in this sector, presenting several abandoned channels in the form of oxbow-type lagoons. Near the coast, lagoons develop in its northern sector, covered by hydrophilous and halophytic vegetation, forming marshy fields that sometimes extend to the foot of the coastal tablelands.

After the fluvio-marine plain at the mouth of the Japaratuba River, there are several generations of marine terraces that develop parallel to the coastline in the northeast-to-southwest direction, interrupted by the mouths of the Sergipe and Vaza Barris Rivers. In this sector also occur the largest fluvial-marine plains of the Sergipe coast, which are the plains of the Japaratuba, Sergipe, and Vaza Barris Rivers.

The third and last sector of the SCP, after the Vaza Barris river plain, presents the continuity of the marine terraces of the previous sector, finishing at the mouth of the Piauí-Real Rivers, forming fluvio-marine and fluvial-lagoon plains. In this part of the plain, according to Bittencourt (1983), one can find marine terraces of Pleistocene and Holocene age, Pleistocene alluvial fans, three generations of Pleistocene and Holocene dunes, lagoon deposits, and mangrove deposits (Fig. 9.17).

The most recent terraces formed after the last marine regression at 5,000 AP. The terraces are composed of Pleistocene sediments near the coastal tablelands and gradient to Holocene deposits toward the coastlines. They present altitudes varying from centimeters to five meters, interspersed with coastal ridges, lagoons, and the occurrence of stabilized and active dune fields. The relief of this sector was partially mapped by Ramos et al. (2021), where they found from the limit with the coastal terraces, relief models of internal marine terraces, coastal ridges, paleo-dunes, fluvio-marine plains, external marine terraces, to tidal plains/mangroves and current dunes/interdunes. In this sense, one can see that these are pleasant relief forms whose altitude variations are mostly below 10 m, being higher than this in the paleo-dune and dune sectors.

Fig. 9.17 Estuarine area of the Vaza Barris River. A place of intense tourist activity in, Aracaju, São Cristóvão, Sergipe. *Photograph* Ronaldo Missura (2014)

9.9 Sedimentary Deposits

This geomorphological domain is dispersed along several other compartments. It is because the recent sedimentary deposits are established in areas with relief of marine, fluvial-marine, or lagoon plains, resulting from the processes of weathering and erosion. The sediments carried in great part by hydric and sometimes eolic transport have as deposit surfaces the flat sectors along the hydric courses, littoral regions, or dune cordons.

In Sergipe, these compartments can be found mainly on the coastal fronts as a result of the accumulation of marine deposits and also in the form of fluvial deposits along the course of the Sergipe Rivers and fluvial-marine deposits at the mouths of the São Francisco, Japaratuba, Sergipe, Vaza Barris, Paiuí Rivers. The fluvial-marine and marine plains are present along the 163 km Sergipe coast. They are the most extensive compared to the neighboring states, with an average of 4 km from the beach line to the limit with the tablelands, the stretches with lesser extension have around 2 km, and those with greater extension reach around 20 km at the mouth of the main rivers with the ocean.

9.10 Considerations on the Sergipe Geomorphology

Despite being the smallest state, Sergipe presents a great geomorphologic diversity, the fruit of the interaction between its diverse geological structures because the limits of the São Francisco Craton terrain, Borborema Province, and Basin and Phanerozoic Sedimentary coverings are found there. The geological diversity, therefore, derives distinct geomorphological compartments that with the passing of the geological time and mainly during the Quaternary, interacted with the diverse climatic regimes and remodelings of its landscapes during its evolutionary process until today, allowing the state to possess this great diversity of compartments in such a small territory. However, few works seek to understand how the evolution of the relief of Sergipe happened, there are more unknowns than answers to its evolution both in a regional context and in localized contexts that are evidenced in some of the items of this chapter that it is necessary to list them for future needs of understanding its evolution and geomorphological context.

- Evolutionary context of the coastal plain of Sergipe is different from those of the other northeastern states.
- Understand the evolution of the Japaratuba River Canyon, which has almost the same depth as the São Francisco River Canyon, and the evolution of the geomorphological context in its watershed.
- To understand the geomorphologic contexts of the coastal upland terraces of Japoatã and the preservation of the Capela and Nossa Senhora das Dores terraces, the most inland in the state, and their dozens of shallow micro depressions.
- To understand the geomorphology and the evolutionary context of the São Francisco canyons and their tributaries in the evolutionary context of the Sergipe hinterland depression.
- To understand the evolution of the superimposition process of the Vaza Barris River to the Simão Dias and Itabaiana Dome and the lands of the Sergipe hinterland depression.
- To understand the geomorphology and evolution of the Palmares Plateau, its tectonic structures, and analogies with other terrains of the São Francisco Craton.

References

Ab'sáber AN (1969) Participação das superfícies aplainadas nas paisagens do Nordeste Brasileiro. Universidade de São Paulo, Instituto de Geografia

Ab'sáber AN (1997) O Homem dos Terraços de Xingó. Relatório de visita e pesquisas na área de Xingó (nov.de 1997). Projeto financiado pela CHESF. Doc. n. 6. Projeto Arqueológico Xingó. Universidade Federal de Sergipe

Almeida FFM (1977) O cráton do São Francisco. Rev Brasil Geociênc 7(4):349–364

Araujo HM, Mendonça SMO (2003) Feições morfológicas da geotectônica regional em Sergipe: Domos de Itabaiana e Simão Dias. In: Anais do IX Congresso da Associação Brasileira de Estudos do Quaternário—ABEQUA

Azambuja RN (2012) Dinâmica superficial em uma bacia de drenagem semiárida: um estudo de caso da bacia do riacho Salgado-PE. Recife: Tese de Doutorado-PPGEO/UFPE

Azambuja RN, Lima AHS (2018) Geoprocessamento Aplicado Análise Morfométrica E Caracterização Da Paisagem Da Serra De Itabaiana—SE. Anais do XI SINAGEO: Paisagem e Geodiversidade—Crato/CE

Azambuja RN, Lima AHS, Missura R (2018) Mapeamento Geomorfológico de detalhe do Parque Nacional da Serra de Itabaiana-SE. Anais do XI SINAGEO: Paisagem e Geodiversidade—Crato/CE

Bittencourt ACSP et al (1983) Evolução paleogeográfica quaternária da costa do estado de Sergipe e da costa sul do estado de Alagoas. Rev Brasil Geociênc 13(2):93–97

Brito Neves BB, Sial AN, Albuquerque JPT (1977) Vergência centrífiga residual no Sistema de Dobramentos Sergipano. Rev Brasil Geociênc 7:102–114

Claudino-Sales V (2020) Como de formaram os impactantes cânions do Rio São Francisco. In: Eco Nordeste: Agência de Conteúdos endereço eletrônico: https://agenciaeconordeste.com.br/como-se-formaram-os-impactantesos-canions-do-rio-sao-francisco/. Acessado em 3 Oct 2021

Corrêa ACB, Tavares BDAC, Lira DR, Mutzemberg DS, de Souza Cavalcanti LC (2019) The semi-arid domain of the Northeast of Brazil. In: The physical geography of Brazil. Springer, Cham, pp 119–150

Correa ACB, Azambuja RN (2020) Dinâmica superficial em uma bacia de drenagem semiárida mediante o uso do césio-137: estudo de caso da bacia do riacho Salgado-PE. GEOUSP Espaço e Tempo (online) 24(2):381–406

CPRM (2017) (Serviço Geológico do Brasil). Geodiversidade do Estado de Sergipe. CRPM, Salvador

Dantas ME, Shinzato E (2017) Origem das paisagens do Estado de Sergipe. In: Carvalho LM, Martins VS (eds) Geodiversidade do estado de Sergipe. CPRM, Salvador, pp 35–50

D'el Rey Silva LJH (1995) The evolution of basement. Gneiss domes of the Sergipano fold belt (NE Brazil) and its importance for the analysis of Proterozoic basins. J South Am Earth Sci 8(3/4):325–340

D'el Rey Silva LJH (1999) Basin in-filling in the southern-central part of the Sergipano Belt (NE Brazil) and implications for the evolution of Pan-African/Brasiliano cratons and Neoproterozoic sedimentary cover. J South Am Earth Sci 4–5:1–18

D'el Rey Silva LJH, McCLAY KR (1995) Stratigraphy of the southern part of the Sergipano Belt, NE Brazil: Tectonic implications. Rev Brasil Geociênc 25(3):185–202

Ferreira RV, Mariano G, Lima RA, Guimarães TO, Santos EM (2018) Projeto Geoparque Cânion do São Francisco. CPRM, Brasília

Figueiredo Jr AG et al (2011) Geomorfologia da plataforma continental da bacia Sergipe-Alagoas. In: Anais do XIII Congresso da Associação Brasileira de Estudos do Quaternário

Fontes LCS (2015) Da fonte à bacia: interação continente-oceano no sistema sedimentar Rio São Francisco, Brasil

Mabessone JM (1978) Panorama Geomorfológico do Nordeste Brasileiro. São Paulo. Geomorfologia 56

Mabesoone JM et al (1981) Ambiente semiárido do nordeste do Brasil: Os Rios Efêmeros. Série B: Estudos e Pesquisas 4:83–91

Macedo HS Ambientes cársticos em Sergipe: vulnerabilidade e instrumentos para sua gestão. São Cristóvão, Tese de doutorado em Geografia—Universidade Federal de Sergipe, p 403

Maia RP, Bezerra FHR (2014) Condicionamento estrutural do relevo no Nordeste setentrional brasileiro. Mercator (fortaleza) 13:127–141

Martins VS (2017) Geologia do Estado de Sergipe. In: Carvalho LM, Martins VS (eds) Geodiversidade do estado de Sergipe. CPRM, Salvador, pp 19–34

Martins VS, Carvalho LM (2017) Geodiversidade: Adequabilidade/Potencialidades e Limitações Frente ao Uso e Ocupação. In: Carvalho LM, Martins VS (eds) Geodiversidade do estado de Sergipe. CPRM, Salvador, pp 67–116

Morais Neto JM, Hegarty KA, Karner GD, Alkmim FF (2009) Timing and mechanisms for the generation and modification of the anomalous topography of the Borborema Province, northeastern Brazil. Marine Petrol Geol 26(7):1070–1086

Oliveira ACA et al (2008) Ecodinâmica das dunas costeiras de Sergipe

Pereira MF, Solá AR, Chichorro M, Lopes L, Guedes A, Silva JB (2012) North-Gondwana assembly, break-up and paleogeography: U-Pb isotope evidence from detrital and igneous zircons of Ediacaran and Cambrian rocks of SW Iberia. Gondwana Res 22:865–881

Radambrasil (1983) Projeto. Folha SC 24/25—Aracaju-Recife. Departamento Nacional da Producao Mineral 30

Ramos LP et al (2021) Mapeamento geomorfológico da zona costeira sul de Sergipe. Pesquisas em Geociências 48(3)

Schaller H et al (1969) Revisão estratigráfica da bacia de Sergipe/Alagoas. Boletim técnico da Petrobrás 12(1):21–86

Santos RA, Souza JD (1998) Folha Piranhas SC-24-X-C-VI-Estado de Sergipe, Alagoas e Bahia. DNPM/CPRM, Brasília

Santos RA et al (1998) Geologia e recursos minerais do estado de Sergipe. CPRM

Sial AN et al (2010) C-, Sr-isotope e Hg quimiotratigraphy de neoproterozoic cap carbonates do Sergipano Belt, Nordeste do Brasil. Precambr Res 182(4):351–372

Silva Filho MA (1998) Arco vulcânico Canindé-Marancó e a Faixa Sul-Alagoana: sequências orogênicas mesoproterozoicas. IX Congresso Brasileiro de Geologia, Belo Horizonte, Minas Gerais, p 16

Vilas Boas GS (1996) As coberturas Paleozóicas e Mesozóicas. In: Barbosa JSF, Dominguez JML (eds) Geologia da Bahia: Texto Explicativo. SGM, Salvador, p 382

Renata Nunes Azambuja She has a degree (2004) and a bachelor's degree (2006) in geography, a master's degree (2007), and a Ph.D. (2012) from the Federal University of Pernambuco (UFPE). She has been a member of the Quaternary Study Group of the Northeast (GEQUA) since 2004 and has links to the UFS Department of Geography in the adjunct 4 class since 2014. She coordinates the Laboratory of Environmental Analysis and the Research Group of environmental dynamics and geomorphology from the UFS. She has experience in processual geomorphology and pedology, acting mainly on space–time environmental analysis—morphodynamics, soil erosion in urban and rural areas, semi-arid environments of the Brazilian Northeast, geomorphological dynamics of small basins, detailed geomorphological mapping and Quaternary geomorphology.

Ronaldo Missura He holds a degree (2003) and a bachelor's degree (2002) in geography and a master's degree (2005) in physical geography from the Júlio de Mesquita Filho Paulista State University and a doctorate in geography (2013) from the Federal University of Pernambuco (UFPE). He is a professor at the Federal University of Sergipe (UFS), having experience in geoscience with an emphasis on geo-cartography, working mainly on phisical geography, watershed, geomorphology, remote sensing, and geography teaching.

Hélio Mário de Araújo He has a post-doctorate in physical geography from the Federal University of Ceará (UFC) with a specialty in coastal geomorphology. He is currently a full professor at the Department of Geography and a permanent member of the undergraduate program in geography at the Federal University of Sergipe (UFS)—São Cristóvão Campus, where he teaches and supervises undergraduate, graduate , master's, and doctoral courses. He develops research in geosciences with emphasis on geomorphology and geographic cartography/geomorphology, working on the Quaternary geomorphology and environmental change, geodiversity and geoconservation of geomorphological heritage, beach morphodynamics, coastal and rural/urban erosion, natural risks (geomorphological/hydrological), dynamics and management of watersheds, urban climate and health geography, land/environmental planning, and GIS and cartographic representations of spatial data.

Part V
Southern Northeast

Chapter 10
Geomorphology of Bahia State

Jémison Mattos dos Santos

Abstract The text written about the geomorphology of Bahia, from a regional point of view, aims to contribute to relief studies, presenting some of the main scientific works in evidence, with the intention of clarifying readers' understanding, through consistent data and information. The central objective is to provide a concise and unifying document, to expand and popularize the understanding of the geomorphology of the State of Bahia. We begin with a brief characterization of geology and geomorphology, in order to present the general aspects that make up the Bahian landscape. We adopted as the main reference to reveal the Bahian landscape two official public documents, both from the 70s, the works of the RADAM BRASIL Project, Folha Salvador—SD.24 and the Atlas of the State of Bahia, as well as numerous classic and contemporary scientific documents, produced by specialists in geography and geosciences. We highlight that these two fundamental works and other studies are extensive, sometimes difficult to understand, and inaccessible to the majority of the public. Therefore, a careful and synthetic geomorphological discussion is outlined. The reflections produced in this scientific work confirm the existence of eight (08) main units of the Bahian Relief: Coastal Plain, Plateaus, Sedimentary Basins, Sertaneja Planation Surface, Diamantina Plateau, Espinhaço Residual Mountain System, São Franciscana Planation Surface and Great Western Plateau ("chapadão").

Keywords Bahian relief · Plateaus · Planation surfaces · Sedimentary basin · Mountain system

J. M. dos Santos (✉)
State University of Feira de Santana, Feira de Santana, Brazil
e-mail: jemisons@uefs.br

V. Claudino-Sales and J. F. Sobrinho (eds.), *Geomorphology of the Northeast Region of Brazil*, The Latin American Studies Book Series,
https://doi.org/10.1007/978-3-031-58874-7_10

10.1 Introduction

Our main objective is to provide a concise, unifying text, aiming to broaden and popularize the understanding of the geomorphology of the State of Bahia, through the historical context of its formation, the brief geological characterization and the relief; we adopted as the main reference to present the landscape Bahia two official public documents, both from the 70s, the works of the RADAM BRASIL Project, Salvador Sheet—SD.24 and the Atlas of the State of Bahia (produced by the Secretariat of Planning, Science and Technology—SEPLANTEC and the Bahia Planning Center—CEPLAB), as well as through numerous classical and contemporary scientific documents, produced by experts in geography and geosciences.

These two fundamental works and other studies are extensive, sometimes difficult to understand and inaccessible to most of the public. A careful and synthetic geomorphological discussion is then weaved. The chapter is added to the most recent documents (scientific articles, theses and dissertations) and, also, information, data based on the professional experience gained over more than two decades of research on the geography and geomorphology of Bahia.

The reflections of this document are closed presenting eight main units of the Bahian Relief: coastal plain, plateaus, sedimentary basins, lowered surface Sertaneja, Tableland Diamantina, residual mountainous system Espinhaço, lowered surface San Franciscana and the western large tableland ("chapadão").

We began through the succinct characterization of geology and geomorphology, in order to present the general aspects that make up the landscape of Bahia.

10.2 Brief Characterization of the Geological and Geomorphological Context of Bahian Territory

The state of Bahia is in the northeast region of Brazil, in the southern context of the region, between the parallels of 08 31′58″ S and 18 20′55″ and meridians 46 37′02″ W and 37 20′28″ W. It is limited to the states of Alagoas, Sergipe, Pernambuco and Piaui (to the north), Espirito Santo and Minas Gerais (to the south), Goiás and Tocantins (to the West) and with the Atlantic Ocean (to the east).

Bahia is a state of great territorial extension, with a total area of 564,760,427 km^2 (IBGE 2021), marked by a hydro climatological, geological, geomorphological, and vegetational diversity (for e.g., domain of the xerophyte "Caatingas" and the Atlantic Forest). It is, the landscape, the Bahian relief was configured over billions of years (Archean), from conditions of temperature and rain that acted on the litho-structural properties associated with the topographic layout of the terrain (current and past climatic dynamics as the paleoclimates and quaternary climatic oscillations).

From the geological point of view Bahia is formed by rocks dating from the Archean and the Pre-Cambrian Inferior to the Quaternary. They form ancient rocks of the Inferior-Archean Proterozoic that are the granulitic lands; granulitic and

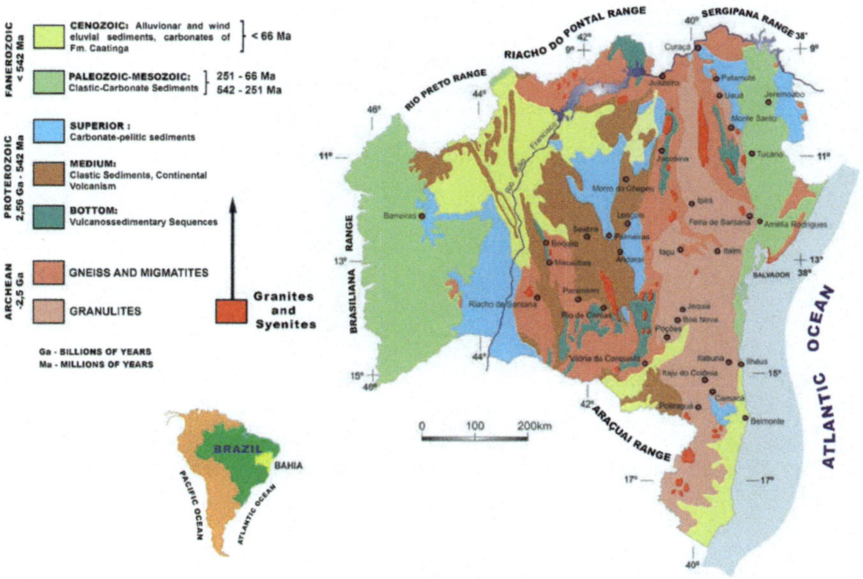

Fig. 10.1 Geological outline of the state of Bahia. Figure adapted by Santos, J. M. dos. *Source* Carvalho, L. M. CPRM—2010

migmatitic terrain; sedimentary volcanic complexes (greenstone belts); rocks of the Middle Proterozoic (Espinhaço Super Group)–siliciclastics, in general felsic volcanics; Upper Proterozoic rocks (Supergroup San Francisco) of carbonate origin; Phanerozoic sedimentary basins; Tertiary-Quaternary sedimentary coverings (Bahia 1990) (Fig. 10.1).

The Bahian relief is composed of a set of different forms; dynamized prominently by the semi-arid tropical climate, hot and humid with dry season in the winter and displays also singular topographic variations, mainly in the central region of the state. A significant part of the landscape is between 200 and 900 m above sea level, covering a total area equal to 466,846,5636 km^2 of extension, which represents a percentage of 82.3% in the territory of Bahia (Figs. 10.2 and 10.3).

Among the most commonly known compartments there are: the coastline (highlighted in Fig. 10.3, in dark blue) with a striking scenic beauty (plains, estuaries, mangroves and dunes); the São Francisco River Valley (presence of depressions); the Diamantina Tableland (plateaus and extensive mountain ranges of flattened tops); Vitória da Conquista Plateau, Jaguaquara and Maracás (planed or wavy reliefs, mountains, and landings (low hills), intermediate surfaces or steps between higher sectors and lower areas.

To this end, before starting the path, it is crucial to understand the historical context of the evolution of geomorphological knowledge of Bahia, which is presented below.

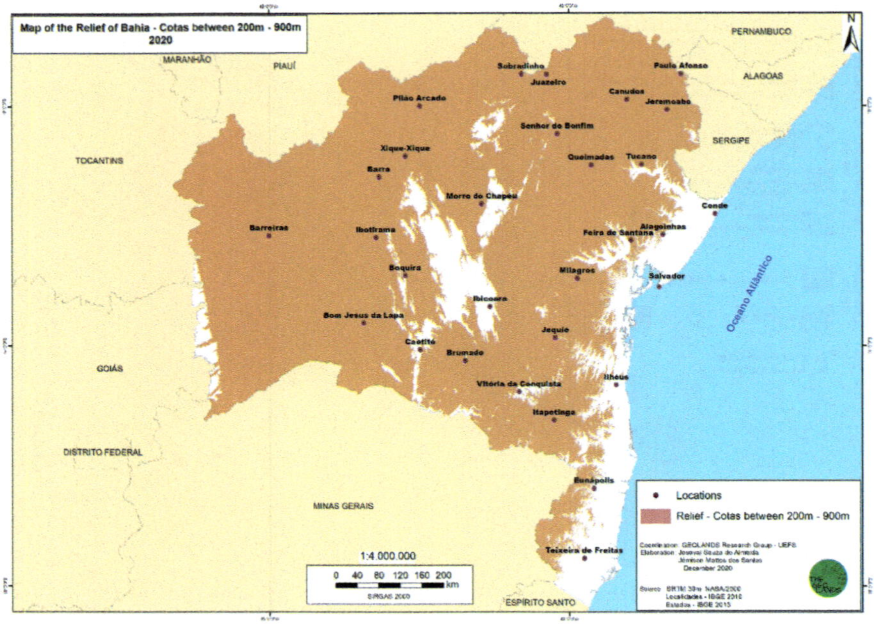

Fig. 10.2 Map of the relief of Bahia—elevation lines between 200 and 900 m. *Source* Santos, J. M., dos & Almeida, J. S.

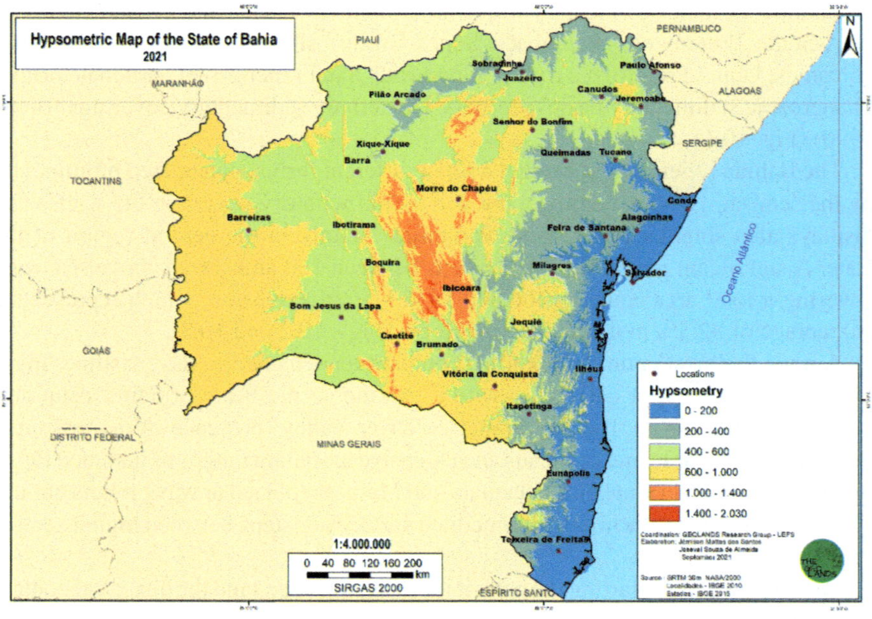

Fig. 10.3 Hipsometric map of Bahia—2021. *Author* Santos, J. M. & Almeida, J. S.

10.3 Conditional Variables of Regional Geomorphology: Large Climatic and Structural Sets

The information worked seeks to reveal the structural panel, as well as the fundamental aspects that mark the morphoclimatic evolution of Bahia and contribute to the understanding of the geomorphological framework.

10.3.1 Paleo Morphoclimatic Frame (1)

In the lower and middle tertiary, the conditions were favorable to the sculpture of extensive planation surfaces. The Paleo Quaternary is distinguished as a period of greater effectiveness for the changes of relief, due to the occurrence of successive climatic oscillations observed in it. These contrasting paleoclimatic conditions are characterized by the alternation of moisture and dryness, which prepare the rocks and then erode them.

The drier climates, with rains concentrated on rocks decomposed during the previous phase, influence more effectively in the formation of the landscape. The typical floods of semi-arid climates elaborate pediments and terraces from the massifs of inselbergs and the escarpments of the plateaus.

Climates with regular moisture retouched the inherited forms without erasing them completely, because the chemical actions act slower. Inheritances of the various types of ecological conditions are thus found in all areas of the state.

The current climatic zonation, relatively recent, characterized by the decrease of moisture from the coast to the interior, with differences introduced by the orographic arrangement, prints its features in the inherited forms modifying or preserving them.

The semi aridity characterizes the current climate of a vast area of the interior of Bahia, where the caatinga occupies the inter-highlands dissection levels. It is preferably located on semi-outcrop rocks of the crystalline basement, carved in the form of terraces and pediments dominated by residual reliefs. Climate agents collaborate to improve and maintain the old shapes.

In the reverses, in the plateaus facing dryness and rainfall irregularity there is an ecology favorable to the appearance of dry forest and Savannah. In these places, as in the tops of the tablelands and plateaus, morphogenesis is slow, tending the conservation of inherited features.

In the area of plateaus and coastal tableland and in the piedmonts of the east-facing mountains (E), humidity favors the presence of humid tropical forest. In these areas the decomposition of the rocks is more intense and the slow retouching of the slopes prints convex shapes to the landscape. On the coastal tableland, which preserve a vegetation of a dry forest based on paleogeographic heritages, some nuances are introduced after deforestation, but the general forms are preserved, depending on the excavation of rivers in the proximity of the coast.

In the coastal plain the forms found reveal influences of paleoclimatic variations combined with fluctuations in sea level during the Quaternary. The current humid climate impairs the formation of cliffs, prevents the accumulation of coarse material on the beaches, and favors the appearance of vegetation, fixing the dunes and sandy plains.

In an attempt to contribute to the updating of data and information on paleoclimatology, some classic and contemporary titles are listed below. However, it is noteworthy that Branner (1910a, b) inaugurated the discussion and contributed to the understanding of the relief of Bahia (limestone plains) associating with the climate.

The researcher Lima (2017) revealed the existence of 05 depositional cycles in a sector of the Itapicuruna Bahia River (Tucano-Jatobá plateau), where 04 are in the Upper Holocene, indicating ages between (2200 and 460 years A.P.) and the remaining cycle was confirmed in the Lower Holocene–Middle Holocene (– 8500 years A.P.).

Góes (2019), also, while working in the Sertaneja Depression, in another sector of the Itapicuru River basin, he revealed ages ranging from the Upper Holocene to the Middle Holocene (150–6.15 ka A.P.), as well as the Upper Pleistocene (12.6 ka A.P.).

In another research Lima and Lupinacci (2019) investigate the depression (in the North Center) and, using a set of modern techniques and LOE dating (optically stimulated thermoluminescence) defined "isolated events of deposition by sporadic flows (– 4100 to 1115 years A.P.)", where the main processes that occur in the valleys and canals, as well as the "destruction of depositional features", were conditioned "to extreme rainfall events associated with dry climate".

Regarding paleoclimatic reconstruction of South America and the Brazilian Northeast (NEB), Landerlei Santos says that:

> Paleoclimatic reconstructions of NEB have been considered as a key area for the advancement of knowledge about the functioning of continental atmospheric circulation (Stríkis, 2018; Stríkis et al., 2015). In recent years, isotopic records of speleothems have been the main proxy used for paleoclimatic reconstructions on a regional scale and even postulated for climatic circulation on a continental scale. However, as clearly revealed by Wang et al. (2004), in the Caatinga biome of the NEB semiarid, low precipitation and high evapotranspiration oppose the development of speleothems and travertines. The authors also concluded that in a period of 210ka AP, more than 90% of the time the NEB region was subjected to arid conditions (Santos 2020, p. 23).

Mescolotti et al. (2021) pointed out that two events occurred between 23 and 18 thousand years and, between 15 and 10 thousand years past were responsible for the process of formation of the dunes of Xique-Xique. However, the consolidation stage of this dune environment occurred in wetter periods and with more vegetation cover in this region of the middle of São Francisco. And the stabilization period of these began 5000 years earlier and continues to this day.

10.3.2 Summary of the Regional Structural Framework

The northeast of Brazil compartment, especially the eastern sector of the state of Bahia, displays varied types of models and associated topographic expressions (for example, the processes of rifting and rupture of the continent), called: eroded surfaces (depressions); tablelands residual mountainous system and escarpments. The list of geological events in the context of the region in focus, which may be tectonic and/or erosive, defined a singular geomorphology, with expressive scenic beauty was dynamicized by tectonic movements linked to the Brasiliano-Pan-African Cycle (approx. 650–550 MA), the evolution of the continental margin, as well as the episodes of neotectonics (1.0 MA—5000–6000 years, more modern civilization), without disregarding present-day tectonic, that is, recent events for example earthquakes.

> It is increasingly evident, from morphotectonic indicators, that the geomorphological evolution of the Brazilian Northeast was carried out in a much more complex way than the pediplanation model proposes, given that this model is quite limited in relation to the recent conceptions related to intraplate tectonics. This is because the pediplanation model does not incorporate rift mechanisms, history of basins and Cenozoic reactivation, a limitation that is derived from the idea of stability of the Brazilian territory. It also does not incorporate post-rift reactivation data, summarizing an uplifting and flattening model that describes the eastern equatorial passive margin of South America and West Africa as successive staggered surfaces, developed from a uplifting and subsequent erosion (Maia et al. 2010, p. 12).

The Borborema Province (PB) is noteworthy in the northeastern regional context, which "consists of land or strips of folding associated with Meso and Neoproterozoic orogenesis and related granite genesis, including ancient fragments of the Archean/Paleoproterozoic, Mesozoic sedimentary basins and (…) recent surface coverings" (Angelim and Kosin 2001).

The PB or Borborema Orogenic System refers to a region of folding of the type *branching system of orogens*, extending for about 380 km^2, inserted in a semi-arid climate (Almeida et al. 1977). It is limited to the west with the Parnaíba basin, to the south with Cráton San Franciscano and north and east with the Coastal Province and reveals a structural compartmentalization associated with the orogenic event Brasiliano/Pan-African (650–530 Ma) that conditioned the addition and amalgamation of crustal blocks during the Neoproterozoic in the NE of Brazil (Angelim and Kosin 2001; Almeida et al. 1981, 1996; Santos and Medeiros 1999; Neves et al. 1995; Brito Neves et al. 2000).

> Possibly, BP is still suffering from epirogenesis post-depositional deformations found in the Barreiras Formation strata, reactivations and faults at the end of the Quaternary and thermochronological analysis of FTA suggesting the existence of a cooling stage between 20 and 0 Ma may be geological evidence of the continued action of epeirogenesis to date (De Oliveira et al. 2012).

In the broadest sense, it is apprehended that the relief of the Northeast and Bahia are conditioned to geological structures, regional lithological variations and pedoclimatic factors, aspects that denote the diversity of landscapes mostly settled in pre-Cambrian lithologies, respectively:

Synthetically, the residual ridges aligned according to the main tectonic directions, edges of sedimentary plateaus affected by uplifting crystalline shaped by differential erosion, residual mass individualized by flattened depressions where erosive processes supplanted the pleasant supplanted the agradational and the coastal plains landscape by eustasy, and coastal plains landscape by eustasy, make up the complex mosaic landscape that documents important episodes of its morphotectonic and paleoclimatic evolution. Among these units, the coastal plains, river valleys and pre-coastal tableland show important indicators of neotectonic events from various indicators. In view of this situation, recent studies have shown that seismic activity intraplate is an important mechanism of morphotectonic interpretation (Bezerra et al. 2007). For sedimentary areas, the effects of these paleo tensions are expressed in different indicators (Maia et al. 2010, p. 16).

In depressions, these lithologies are generally morphologically devastated by successive cycles of erosion; remains of folded chains subjected to ancient orogenetic cycles include pseudo-Appalachian features and large structurally oriented crystalline massifs disturbed by tectonism (domain of the Crystalline Plateaus). In turn, in the domains of the Plateaus in folded structures, metasedimentary platform coverings, intensely tectonized, result in deflated anticline-type inverted reliefs and suspended slopes, characterizing high mountainous massifs and pediplaned highlands of the Diamantina Tableland (Brazil 1981, p. 197).

Dissected Coastal Tableland with "hill seas" (convex slopes) features extend into sub-coastal strips, accusing the effect of paleoclimatic fluctuations of wetter tendencies on the lithologies of the basement. The tectonic abatement pits of the Recôncavo-Tucano, Camamu and Almada contain morphologies, eventually structural, carved over thick packages of Jura cretaceous deposits. To the epirogenetically uplifting plateaus below unconsolidated detritic sediments Cenozoic, worked on vast summit areas, (...), which in the coastal tableland are recognized with members of the (...) Barreiras at the edges of the Diamantina Tableland, Paleozoic depositions of carbonate lithologies constitute plateaus and buried karstic depressions (...) (ibidem).

10.3.3 Structural Framework of Bahia

In general, the Bahian territory consists of rocks dating from the Archean and the Precambrian Inferior to the Quaternary, presenting a wide structural and lithological variety (Halla et al. 2020; Bahia 1990; BAHIA-SEPLANTEC 1976) (Fig. 10.4).

The Archean and Paleoproterozoic lands that constitute the foundation of the São Francisco Craton in the State of Bahia can be grouped into four important Archean crust segments, called Block Gavião (BG), Block Serrinha (BS), Block Jequié (BJ) and Block Itabuna-Salvador-Curaçá (BISC) (...). These blocks collided in the Paleoproterozoic, becoming an important mountain range that is currently at the level of deep erosion, exposing poly-deformed and grainy rocks from its roots (BARBOSA & SABATÉ 2002, 2004). During this collision the BISC was extremely deformed and largely grained and as presented by the vertical foliations/bandages in the N10° direction for almost 100 km, it was called the Itabuna-Salvador-Curaçá Belt (CISC) (Barbosa et al. 2012) (Souza 2013, p. 11).

It is important to note that part of the State of Bahia is inserted in the São Francisco Structural Province or as called Craton São Francisco, which was stabilized at the end of the Transamazonic geotectonic cycle (1800 a 2000 m.a), that is, it constitutes the macro geotectonic unit of the South American continent, in which the evolution

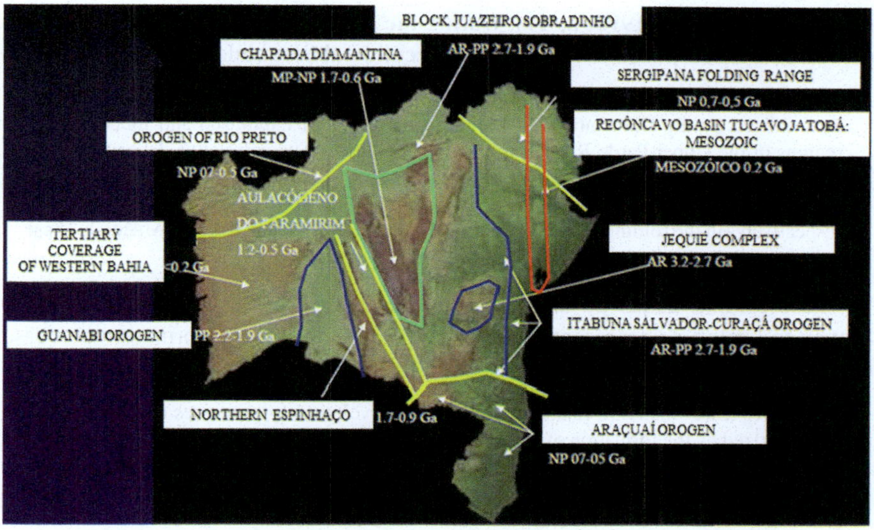

Fig. 10.4 Sketch of the tectonic structure of the State of Bahia—2021. Figure adapted by Santos, J. M. dos. *Source* Didactic material of the discipline Tectonics and Global Geophysics—Luís Cesar Correa. IGEO: UFBA

of the basal rocks stopped at the end of the Paleoproterozoic (Bahia 1990, p. 11; Almeida 1969, 1977).

> The oldest rocks have undergone successive tectonic cycles resulting in intensely metamorphic series, sheared and penetrated by intrusive rocks of various ages. They form the crystalline basement and are distributed in a large area that occupies the eastern center of the state. The actions of humid climates make erosion resistant to these rocks weakened by tectonics and containing more alterable minerals than other younger quartz rocks. The extensive hydrographic basins currently cover large areas consisting of rocks of this type (Bahia 1990; BAHIA-SEPLANTEC 1976; Tricart and Silva 1969).

This Craton is delimited by bands or folding systems associated with the development process of the Brasiliano Cycle (1100–450 m.a), consisting of metasedimentary and metavolcanic-sedimentary rocks with consisting of metasedimentary rocks and metavolcano-sedimentary with convergence to the cratonic direction in the cratonic sense (Ibidem, p. 11).

> The basement of the Craton is formed by rocks of high, medium, and low degree of metamorphism, granulites, charnockites, enderbites, which were structured with mobile belts which were structured with mobile belts that were structured with movable belts: gneisses, granites and migmatites derivatives comprising remains of Vulcano-sedimentary sequences and greenstone belt structures, with ages ranging from 3000 to 2000 m.a. (Ibidem, p. 11).

The craton's cover is composed of metavolcanic-sedimentary and metasedimentary sequences with medium and low degree of metamorphism (1800–450 m.a) (Ibidem, p. 11). From the Jurassic to the Early Cretaceous, the rift valley and inland basins were implanted as responses to the processes of continental separation of

Brazil-Africa (Bahia 1990, p. 11). During the Tertiary-Quaternary, morphogenetic processes were responsible for alluvial and colluvial accumulations (Ibidem, p. 11).

> The correlative values from 1800 to 1,100 m.a. correspond to the so-called Espinhaço Cycle, materialized by the Diamantina Tableland, in the northern Espinhaço and its correlative regions of the Upper Pardo River and other smaller areas. The correlative deposits located between 1,100 and 450 m.a. refer to Super-Group São Francisco, locally known as Group Bambuí and one (cratonic area) by Group Vaza-Barris, Rio Preto, Rio Pardo and Macaúbas, respectively (in the folding ranges of Sergipe, Riacho do Pontal and Ribeira-Araçuaí). These values are the same in the Brasilia Range (Idem, p. 13).

These folding bands go around and delimit what Almeida (1977) called São Francisco Craton, to represent the area that is not subject to deformation processes and geossinclinal magmatism at the end of the Transamazonic cycle.

> The values lower than 450 m.a. are related to the evolutionary processes of Phanerozoic, characterized by the formation of the Juro-Cretaceous sedimentary basins, known as Recôncavo Tucano, Urucuia and the marginal basins (Almada and Espírito Santo), in addition to Tertiary, Quaternary and Recent deposits. These also represent the geomorphological processes that elaborated the current relief, in which the carstic reliefs are included (Idem, p. 14).
>
> Two metasedimentary series originated in the middle pre-Cambrian overlap the crystalline base and present distinct tectonic styles: the Minas series, represented by quartzites, phyllites and conglomerates of the Jacobin Group was very bent and faulted; the Lavras Diamantina series composed of quartz sandstones, conglomerates and metavolcanics suffered large foldings, often displaced by faults (BAHIA-SEPLANTEC 1976).

These metasediments behave like resistant rocks and appear on projections. They form a large mountain massif in the central part of Bahia, in the extension of the Espinhaço mountain range.

> The State also has other sets of Paleozoic-age rocks belonging to the Bambuí Group. The siltstones, slates and limestones of this group form highlands located between the Espinhaço mountains or at the foot of them: they are the plateaus of Chapéu Uphill, Irecê, Salitre River and Andaraí. The Mesozoic sedimentary rocks form two large sets: the W of Bahia, horizontal layers of siltstones, sandstone and shales, deposited in an extensive basin, gave rise to a plateau whose edges turn to the valley of the São Francisco River; in the tectonic trench of the Recôncavo-Tucano, sandstone, siltites, limestone shales and conglomerates were very broken by faults and form a complex relief of monocline and elevations of various forms, limited by the escarpments of the edges of the basin, of approximate direction NE-SW (Ibidem).

Cenozoic rocks occur throughout the State of Bahia, with greater thickness and extension in the coastal strip, being represented by the Barreiras Group (friable arkoses, with conglomeratic or clayey levels). However, they are concentrated in three main areas: central and northwest, Diamantina tableland and Vitória da Conquista (composed of laterites and associated with crustal uplifting Also, originate cliffs on the south coast, at Ilhéus (Japsen et al. 2012; BAHIA-SEPLANTEC 1976).

Quaternary sediments were deposited in the coastal zone and in the valleys of the main rivers. They form models of fluvial-marine accumulation, marines, beaches, dunes, sandy plains with sandbanks, floodplains and alluvial terraces.

In the Cryogenic period of the Neoproterozoic Era (700–650 million years ago):

Fig. 10.5 Bambuí Sea. *Source* Silvafilho, R.B. CBPM, 2010

> The Bambuí Sea has progressed by taking advantage of the sinking of this part of the crust, which extended to about 670 million years ago, configuring the greatest advance of water ever to happen on our land. Typically, continental interior, it was a shallow sea, with the areas of greatest drawdown having average depth between eighty and one hundred meters, with the deepest locations reaching probably about two hundred meters. Despite this great advance of waters over the continent, nothing indicates a tendency of continental breakage, representing this sinking a local reflection of the Rodinian raters in its long crisis. It had lost the full coherence of its old set around 750 million years ago (Silva Filho 2010, p. 109) (Fig. 10.5).

The various reactions and complex structures, lithologies impelled to the dynamism of paleoclimates during the evolution phases of the geomorphological framework, especially in the Quaternary period, produced a unique set, different forms of relief of Bahia (landscape).

Large-scale **structural lineaments**, translated into alignments of ridges, large mountains, mountains (e.g., Residual Mountain System of Espinhaço, Fig. 10.6), shorelines and valleys of some rivers are the result of great mechanical discontinuities of the crust, forming zones of constant ruptures.

10.4 Large Geomorphological Compartments: Existing Proposals and Specific Aspects

The expressive spatial dimension of the landscapes of Bahia associated with geological-geomorphological substrates, presents a lush, complex and diverse scenario, marked by large surfaces or relief compartments, called: planation surfaces

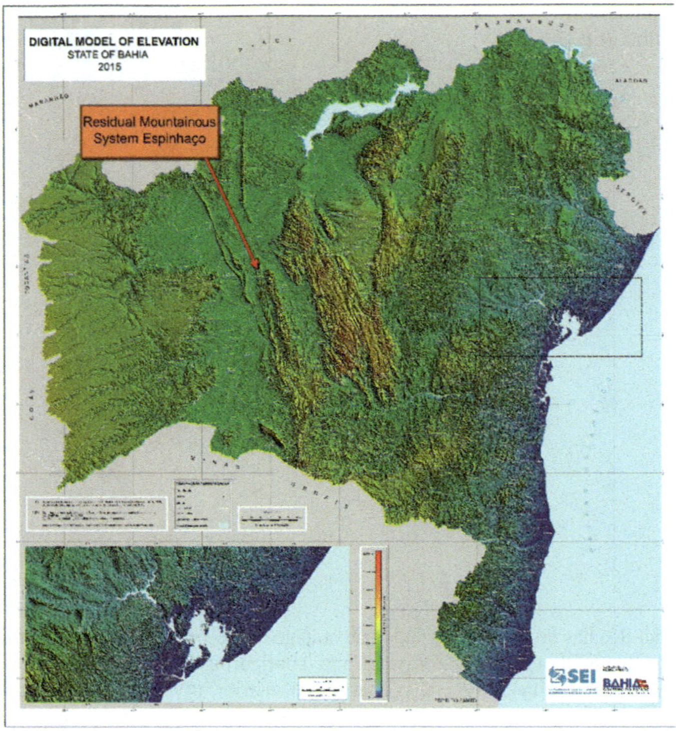

Fig. 10.6 Digital model of elevation of Bahia. Figure adapted by Santos, J. M. dos. *Source* SEI—Bahia, 2015

(depressions), plains, coastal tableland, plateaus, tablelands/plateaus, mountain and residual mountain systems (King 1956; Silva 1959; Tricart and Silva 1969; BAHIA/SEPLANTEC 1976; Ab'sáber 1977; Radambrasil 1983; Lage 1986; Valadão 1988; Seabra 1991, 2017; Sei 1991; Oliveira 2020; Santos 2016). The official maps and other representations consulted on the relief of the Bahia, as in Fig. 10.7, do not show the occurrence of the landscape: residual mountainous systems and lowered surfaces.

The considerable set of lowered areas of Bahia are generally known as depressions (Tricart 1958; Tricart and Silva 1959; Silva 1960; Falk 1977; Mendes 1978), that is, these lowered surfaces constitute the largest areas of the state of Bahia (occupies 230,183,333 km^2). From the regional point of view, these were mapped preliminary, using 02 guide criteria (slope and altimetry), formed by a flat or smoothly wavy topography, with low altitudes and slightly sloping areas, which vary predominantly between 200 and 500 m (Fig. 10.8).

The planation surface Sertaneja (east), which covers most of the central-north of Bahia and San Franciscana area (west) are worth mentioning, also called inter-mountains and inter-highlands depressions, that is, altitude areas lower than the plateaus surrounding them and peripheral depressions, that occur in the contact zone

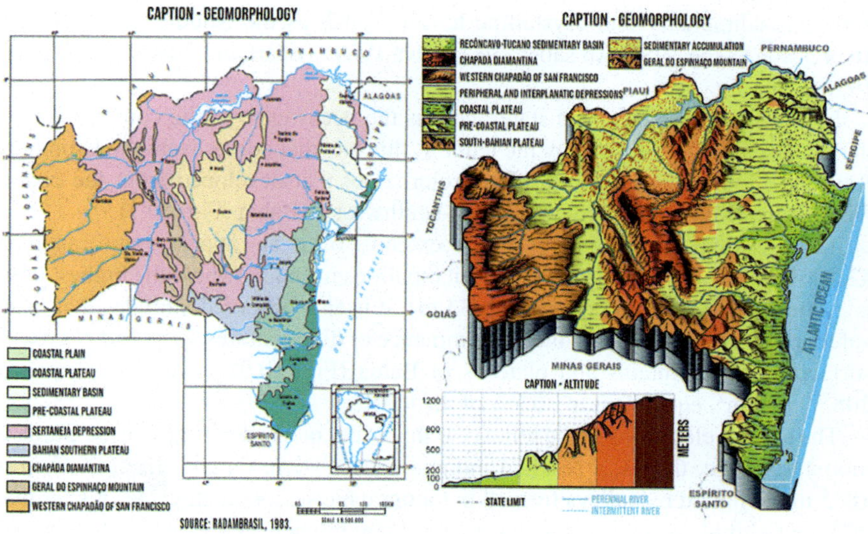

Fig. 10.7 Map of the relief of Bahia (left) and diagram block of geomorphology of the State of Bahia (right). Figure adapted by Santos, J. M. dos. *Source* SEI, 1991 e FIORI, 2012

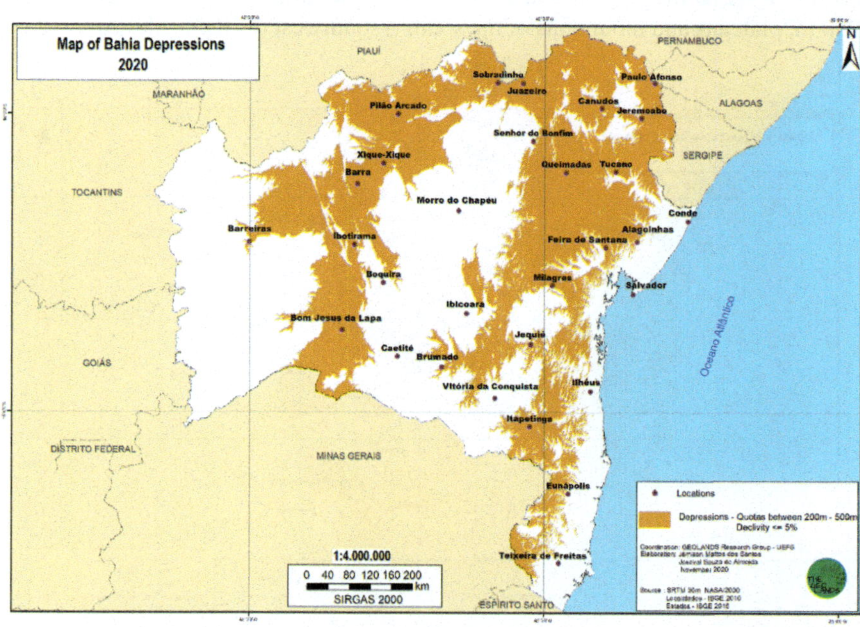

Fig. 10.8 Map of the depressions of Bahia—2020. *Author* Santos, J. M. dos & Almeida, J. S.

between sedimentary and crystalline terrains, such as the Middle San Francisco Depression (King 1956; Ab'sáber 1977; Silva 1959; Tricart and Silva 1961; Barreto 1996).

In the central sector of the state, there is the occurrence of the compartment of the highest areas, delineated by the letter V, this one that comes from the Espinhaço Residual Mountain System (Middle Proterozoic) that developed from the city of Diamantina (MG) and entered the south-southwest of Bahia (Fig. 10.7).

This residual mountainous sector extends over an approximate range of 42,000 km^2, covering more than 30 municipalities in Bahia, where the highest point is in the Barbado mountain range (2033 m). It is also observed altitudes that reach values close to 2000 m. Note the predominance of hilly features, displaying rounded surfaces. The mountains and plateaus of Bahia (Fig. 10.9), cover a total area of 208,396.0 km^2, equivalent to 36.9% of the territory.

The highest elevations of the relief of Bahia occur in Rio de Contas, in the Barbado mountain range, that is, a residual mountainous system, ranging from 2033 m to 1845 m, delimited by valleys formed in topographic sectors with values higher than 200 m of altitude and levels that have values ranging between 1000 and 540 m.

We established a classification (based only on altitude), defining 04 (four) large compartments, that are: plateau and mountains of the north center of Bahia: elevations above 600–1216 m; plateau and Bahia central mountains: elevations above 600–1930 m; plateaus and mountains of the south—southwest of Bahia: elevations above

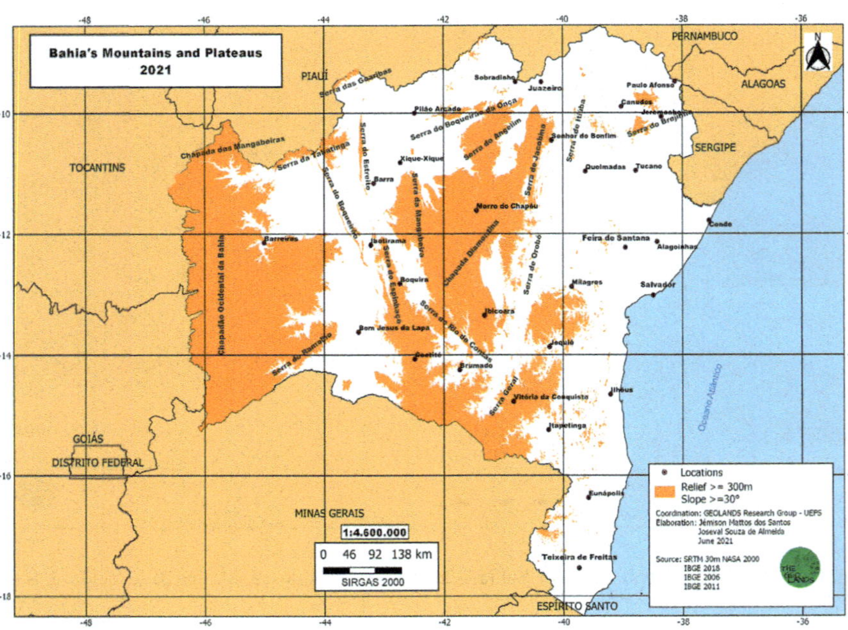

Fig. 10.9 Map of the mountains and plateaus of Bahia—2021. *Author* Santos, J. M. dos & Almeida, J. S.

700 m and the plateaus and western large tableland—elevations above 800–1000 m of altitude.

The municipality of Piatã serves as the main landmark, as it has the highest city in Bahia and the northeast region. It is located on a plateau between the Tromba and Santana mountains (Fig. 10.9), at an altitude of 1280 m in relation to sea level. In addition, it also has the lowest temperatures recorded in Bahia territory (values lower than 10 °C in winter), that is, Piatã is the highest and coldest municipality in the state of Bahia, where one of the most important rivers of this region is born—known as Rio de Contas.

The highest peaks of Bahia are traditionally known as Barbado (2033 m), Peak Almas (1958 m) and Itobira (1930 m). But, according to Barros (2020), in the work entitled Summits of Bahia, it updates the list of the "Mountain of Bahia" putting in second place the peak of the Elephant (1.980 m); third the Lapa Grande (1.970 m), fourth the peak of Prateleira (1965 m) and fifth the Peak of Almas (1.958 m) (Fig. 10.10).

According to Alcoforado (2003), approximately 70% of the territory of the State of Bahia is between 300 and 900 m and 23% below 300 m altitude. The morphological picture comprises three units: the coastal lowland, the edge of the plateau and the plateau. The coastal lowland constitutes the set of lands located below 200 m of altitude. They stand there, dominating the beaches and sands of the shoreline coast, tabular feature lands and the so-called sandstone coastal tableland.

The coastal plain refers to the compartment of lowlands flattened or gently steep (up to 200 m altitude), which is connected with the flat-rate sectors (more expressive coastal plateau to the south and the pre-coastal plateau that extends from the central range to the south of Bahia).

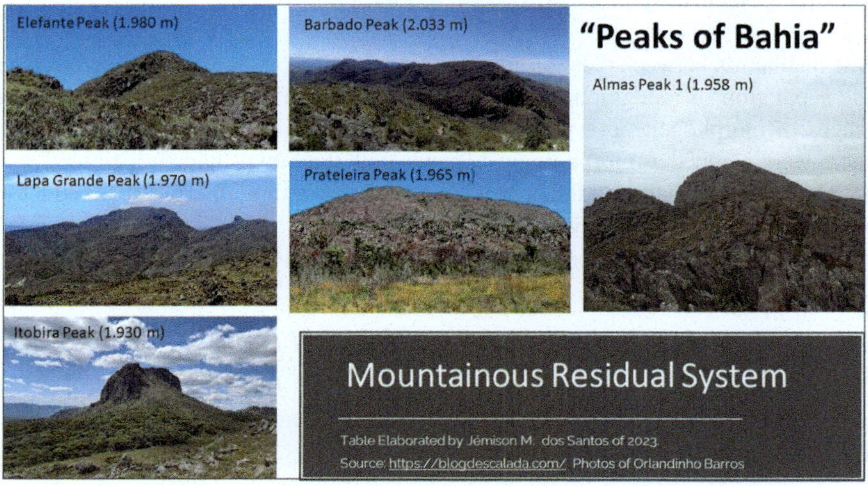

Fig. 10.10 Peaks of Bahia. Residual mountainous system (2023)

The coastal plateau is an area consisting of coastal tableland (tertiary deposits) and "hill seas" (convex hills) (Proterozoic crystalline massifs), demarcating a sloping surface, starting from the interior to the coast, usually formed by flat tops, and some surfaces lowered in crystalline rocks (in the Pardo River Basin). The pre-coastal plateau represents the elevated crystalline sectors.

For the interior of the state, these lands give way to a strip of hills and clayey hills, of thick soil, relatively fertile, especially in the Recôncavo, where the famous massapê soil of Bahia is located. Along them extend large alluvial plains (floodplains) subject to floods that periodically renew the soils with the deposition of new alluvials (Alcoforado 2003). The St. Franciscan depression, in turn, extends to the west of Espinhaço, with similar disposition, forming a north–south direction range. The lands have low altitude (400 m on average) and relatively flat that, with gentle inclination, fall to the São Francisco River, hence the name of this geomorphological region.

The western plateau, consisting of sedimentary rocks, rises to the west of the San Franciscan depression, with an approximate height of 850 m. Its regular top prints it tabular feature and the character of extensive large tableland to which applies the generic name of Espigão Mestre. Finally, the pediplain comprises the entire northeastern part of the Bahian plateau. There develop large surfaces that lean gently toward the coast, to the east, and to the San Francisco trough to the north, with altitudes between 200 and 500 m (Alcoforado 2003).

10.4.1 Geomorphological Compartmentalization of the State of Bahia (2)

10.4.1.1 According to SEPLANTEC-CEPLAB

The distinction between these groups was established from hypsometry, that is, from the separation of topographic compartments related to the structure and the great stages of morphogenetic evolution in Cenozoic. The identifiable units in each of them were mapped as morphological frames with their own characteristics.

These aspects make out the predominance of one or more morphogenetic factors in each unit, described below.

10.4.2 Set of Crowning Morpho-Structural Units

This first set consists of five units that include the predominance of structural litholic factors in the formation of the landscape.

Highlands and Espinhaço Chain Mountains—Located in the center of the state, they form its "dorsal spine". It contains the highest elevations, located between 900 and 1300 m, and the culmination of Bahia that reaches about 1850 m in the

Almas Mountain range (Contas River Mountain). In addition to these, the most important reliefs are: the monoclinal-shaped Sincorá mountain range facing W, with an extension of several hundred kilometers; the serra da Mangabeira, bordered by escarpments in two directions and ending in the plain of the Paramirim River, a tributary of São Francisco, with a dislocation of more than 300 m; the Piatã mountain range, in the south-central part of Diamantina Tableland, reaching 1300 m of altitude; the Tombador mountain range, above which is the Morro do Chapéu, whose steep front in the shape of cuestas facing E.

Also, part of the mountain massif of Espinhaço has its extensions to NE and NW, composed of the Jacobina mountain range and the narrow and Boqueirão mountains. These mountains have a linear arrangement, in the form of rocky bars, forming protrusions of quartzite rocks and presenting phyllites in the valleys.

Sedimentary Large Tableland—The tops of the various morphological units are related by plains that form tablelands. These plains resulted from the pediplanation of different lithological types during the oldest phases of the evolution of the landscape, in the lower tertiary. The indications of these planations are found in the form of ferrified and silicified deposits in the remains of high erosion surfaces located between Barra da Estiva and Mucugê, between Ibitiara and Santa Quitéria, in the vicinity of Piatã and in several other places known from the Gerais to W de Caetité, on the Jacobina mountain and in the Western large tableland. These remains were correlated and interpreted as parts of one or more pediplanation that occurred from the lower tertiary to the middle tertiary, which was called the South American Surface (L. King) or Eogen surface (Tricart e Cardoso Da Silva), corresponding to Pd2 in northeastern Brazil (Bigarella, Osorio, Buerlen, Mabesoone, Campos e Silva). They are topographically called ridges of residual elevations, testimonies of higher erosion levels.

The elevations of the Diamantina Tableland are supported by anticline folds that have been emptied and by slopes suspended by flank erosion, facilitated by tectonic accidents. The structure of the Jacobina, Boqueirão and Estreito mountains is widely discussed. The ideas most accepted in this regard are favorable to the hypothesis of very intense folding that occurred during a tectonic cycle prior to the deposition of the sediments that formed the Diamantina Tableland. The folding was cut by faults in several directions. The tectonics and the geological constitution contributed decisively to the current disposition of the relief of the Espinhaço chain. It is assumed that this chain also suffered the effects of block movements in the vertical direction after the Paleozoic, which caused the excavation of the folds and shifted the assembly upward in relation to the crystalline base.

South-Bahian Plateau—The litho-structural influences are also notable in the morphology of this plateau, divided into two sectors: the Maracás-Jaguaquara plateau and the Vitória da Conquista plateau, separated by the Contas River Basin.

The Plateau of Vitória da Conquista is separated from the eastern compartment by a continuous, festooned escarpment, represented by the Marçal mountain range, with the direction approximately N-S, whose origin is controverted. The tops, with average altitudes located between 800 and 900 m, present remains of ferrification phenomena, developed during the evolution of the flattened surface that can be correlated with that found in the Diamantina Tableland and over the Jacobina mountain range.

The plateau has a thick sandstone ferralytic cover, inherited from the various phases of its morphogenetic evolution. This cover favored, under the actions of more recent climates, the excavation of depressions on the plateau and headwaters of suspended valleys, in the form of alveolus, and on its edges.

Crests and Residual Bars—Scattered over several groups of units are crests and residual bars. They are aligned according to the predominant structures in the state, with an approximate direction N-S; some, however, are distributed in several other directions. Because of the resistance imposed on erosion by acidic intrusive rocks, the ridges sometimes turn into bands of residual hills that follow a preferential orientation: the Itiúba mountain range, the Quijingue mountain range, Orobó mountain and Juremal mountain resulted from enhancement made by erosion of cores of syenites, granodiorite, quartz dikes and others.

Escarpments and Shoulders—The units whose lithological resistances allowed the differential erosion of other neighbors are separated by escarpments and shoulders. They become notable on the western and eastern edges in the Diamantina Tableland, where the most frequent uneven ranges from 100 to 300 m, and on the eastern margin of the conquest plateau.

Intermediate Set of Morphoclimatic Units—Formed by features that resulted from the predominance of paleoclimatic actions that occurred after the tertiary superior, the units of this set are in a lower topographic position than the previous ones (between 100 and 600 m), and often constitute inter-mountains and inter-highlands sub-compartments. Lithological differences and relative positions generated particular characteristics.

Piedmonts, Inter-fluvial Baselines and Remains of Terraces—Piedmonts located near the mountain reliefs are often covered by materials derived from them; coincide with the foot of the escarpments of Diamantina Tableland, the Jacobina mountain range and the large plateaus and large tablelands. Its extent depends on the area affected by the accumulation of debris. Between Capim Grosso and Lord of The Bonfim, the deposits of the piedmonts of the Jacobina mountain range, dissipated by the hydrographic network, extend for hundreds of kilometers to E.

Most inter-fluvial baselines levels and the remains of terraces (glacís) located in the river basins attest to the existence of one or more phases of pediplanation. The sandy ferralytic coverings that are found on them are correlative deposits of the planation called by L. King of Cycle Velhas, considered as a trace of Neogene-forming massif hills (Tricart; Cardoso Da Silva) correlated with Pd1 (Bigarella, G. Osório; Ab Saber; Mabesoone e Campos e Silva).

Coastal Tableland and Plateaus—The Neogenic planation affected the sedimentary and crystalline rocks in the southern part of Bahia and allowed the conservation of the plan of the coastal plains and plateaus. On these are also located detritic top levels, sometimes ferrified which support the flat shapes and originate from small cornices on their edges.

Massifs of Hills—In areas of more abundant rainfall, coinciding with rugged sectors, the crystalline basement rocks released from the mantle of alteration gave structural shapes, forming massif hills. The traces of the previous evolution were erased by the aspects related to the decomposition processes, resulting in the elaboration of the convex forms of Espinhaço. When the rocks are more alterable, hammocks zones appear. These aspects are most notable in the areas of Mairi, Mundo Novo, Miguel Calmon, in the Tombador piedmont and in parts of southern Bahia, among the tablelands.

Slopes–Terraces—The open areas-shaped strands form a lower unit topographically converging on the current valleys. They are glacis carved from residual reliefs by surface flow processes, coincident with phases of hydrographic network installation, whose climatic conditions would be more contrasting than the current ones. In some basins, such as Paraguaçu, Rio de Contas, Itapicuru, Cachoeira, the inselbergs are dotted. They show signs of erosion and tendency to damping. In the terminal parts of the slopes located in the middle courses of the current rivers there are accumulations of quartz and quartz rocks debris, sometimes carved and forming very high terraces in relation to the current riverbed. The mechanisms of formation of these systems of inclined planes combine the actions of the non-concentrated flow of rainwater, the floods, with the facilities created by successive changes in climates and with the relations with regional base levels. They are distributed, and more or less conserved, in almost the entire area of rocks of the crystalline basement, in the semi-arid part of the state.

Recent River and Marine Erosion and Accumulation Levels—The levels result from the most recent geomorphological evolution present usually embedded in the other units and dominated topographically by them. They denounce the latest climatic variations that occurred in the Quaternary, correlated with changes in sea levels.

Floodplains—After finding its current riverbed in the middle course, San Francisco, cutting its own alluvials, divided into arms around numerous islands. Its frequent floods cause flooding in marginal terrain, forming the floodplains. Several other Bahian rivers such as Vaza-Barris, Itapicuru, also form lowland land, mainly in their medium and low courses. Other rivers such as Paraguaçu, Cachoeira, Rio de Contas, attesting to a different evolution of its lower courses, end in deep "rias".

Sandy Plains and Dune Fields—The most recent episodes of geomorphological evolution caused the accumulation of sandy, silt and gravel alluvial that were carved by the rivers themselves, forming several levels of terraces. On the left margin of São Francisco, between the Boqueirão Mountain, the border with Piauí and Ocidental large tableland, the material accumulated by this river during a phase that proceeded to its most regular flow formed a huge sandy sheet. These sands, later turned by the winds, gave rise to a dune field that extended through several municipalities: Barra, Pilão Arcado, etc.

Low Coastal Tableland and Recôncavo Hills—Correspond to a sector of sedimentary rocks, with small elevations formed as a result of the excavation of the Bay de Todos os Santos, under humid conditions.

Coastal Plain—It can reach several kilometers wide in the vicinity of the mouths of the Rivers Jequitinhonha, Pardo, up to only a few meters to S of Ilhéus, in Salvador, Porto Seguro and several other points of the coast. Consisting of sand and rare pebbles, this plain includes topographic and morphological aspects that attest to various stages of formation, depending on the clashes between transgressions, marine regressions and climatic conditions that controlled the river regimes that have ended there.

The main features found there are sandy terraces, strips of barriers, dune fields, beaches, littoral bars and, separating it from the ocean, barriers reefs and organic reefs and, by sea, sandstone Barreiras and organic reefs.

The dynamics of the current processes, specific to each of the morpho-climatic zones, introduces continuous modifications in the landscape, adding the pre-existing forms to the traces of older forms. To this end, it has also contributed, since a few hundred years, human beings.

10.4.3 According to Radam Brazil Project

10.4.3.1 Domain of Inter-Highlands Depressions

Geraizinhos Plateaus—The Geraizinhos Plateaus have a total of about 17,563 km^2 in map SD 24 Salvador.

The altimetry is more than 500 m, being, in general, between 600 and 1000 m, as the Periperi mountain range, immediately north of Vitória da Conquista. There have been no high-level actions. The plateaus of Geraizinhos are mainly subjected to semi-humid climate, which occurs throughout the northern sector and in most of the sector located southwest of map SD 24 Salvador, while the semi-arid tropical climate predominates in narrower stretches, such as around the plateau towns in Poções.

The unit of the Geraizinhos Plateaus is geomorphologically characterized by extensive areas of tabular topography, consisting of detritic deposits of the Tertiary and Quaternary.

The most widespread relief forms are inclined planation that follow constituting pediplanes, more or less conserved, mapped respectively as Pgi and Pri.

Locally form wide spines that sometimes reach up to 4 km long.

The topography is completely flat, only weakly carved by many wide valleys, covered by colluviums ramps or widening forming depressions, and closed, shallow depressions, usually containing water and associated or not with the outcrop of slabs, and block chaos, more common in the northern part of the area under study.

There are residual elevations that emerge from the surface of the Geraizinhos Plateaus. The Periperi mountain range, north of Vitória da Conquista, is one of these quartzite ridges, which stand out above the general level of the plateau, mapped as ridge lines.

Colluviums ramps are at the foot of this residual relief, sculpted by lower inclined planation that converge to the depressions or wide valleys. At the northern end of the unit the pediments join the basis of characteristic residual reliefs, bare hills that rise, on the planation surface, mapped as inselbergs.

Geraizinhos plateaus present topographic inclination in various directions, as well as valleys with asymmetric and uneven edges. North of the city of Vitória da Conquista, the Plateau is clearly ramped up, merging with the unity of the levels of the Middle Rio de Contas.

Flattened tops of the Geraizinhos Plateaus are festooned and often in rectilinear clippings that accompany structural directions, with very fractured rocks nearby, in relation to the adjacent planation depressions or intense dissection areas, with high inclined escarpments dominated by cornices.

Region of the low plateaus—tabular reliefs of tracks located northeast and southeast of map SD 24 Salvador. These are discontinuous and lowered areas with altimetry ranging from a few meters and up to about 400 m of altitude.

They are covered by detritic and lateritic materials of the Tertiary and Quaternary. The performance of tectonics in the area can be inferred only by the direction of some ridge lines or drainage channels, which sometimes form angles or have rectilinear segments, such as interior coastal tableland and coastal tablelands.

Interior Coastal Tableland—A track located northeast of map SD 24 Salvador, reliefs arranged in the Direction SSO-NNE, encompassing areas northeast of the V-D grid. With a total of about 3590 km^2 it has varied width, usually between 15 and 20 km.

They have more frequent altitudes around 200 m, only exceptionally approaching 300 m, by reliefs of very high concordant tops, characterized by low reliefs and concordant tops. It is a pediplain that is preserved in the area closest to Feira de Santana, mainly south of Cruz das Almas and Santo Antônio de Jesus. It is observed the common occurrence of closed, circular or oval depressions, many of which form true lagoons with more than 1 km in diameter, sometimes covered by low vegetation, especially in the vicinity of Cruz das Almas and Feira de Santana (Est.2 VI B).

Residual reliefs of inselbergs type are also integral parts of the landscape in the northern area of the unit, as "whale backs" and chaos of blocks, the latter not mappable to the scale of 1:1,000,000. The coastal tableland has well-marked edges mainly in the dissected areas of flat tops. In other places, however, contact with neighboring unit is made without topographic rebound, through ramping, as on the descent into the Paraguaçu River.

Surface erosion acts through diffuse and concentrated flow by carving ravines and grooves, while mass movements are evidenced by the existence of solifluction and reptation marks. They are also observed on the slopes alveolus of bedside, which often develop as amphitheaters suspended almost to the top of the reliefs. The area of coastal tableland is drained by rivers of the Jequiriçá and Paraguaçu Basins.

As occurs in stretches of the Jequiriçá River, it appears quite embedded and meander between the dissected reliefs of tabular or slightly bulging tops and convex slopes, with flat valley and asymmetric margins in some stretches.

Coastal Tablelands—Totaling about 3100 km^2, these coastal tablelands are in a very low altimeter position, not exceeding 100 m of altitude, and are inclined to the sea.

The relief bands almost always coincident with the sediments of the Barreiras Formation, which is subjected to intense and uniform dissection, mapped as landscape of homogeneous dissection. The interfluves have tabular tops of small extension, generally deeply integrated with a high density of drainage channels. The coastal tableland without points locally submitted to podzolization processes which cause the appearance of sandy soils on the surface. The valleys usually have a flat bottom and sometimes are rectilinear, cases in which they have been mapped as adapted to the structure, with preferential meaning SO-NE and NO-SE.

Dominion of plateaus in folded structures—is represented in sheet SD 24 Salvador, by the geomorphological regions Diamantina Tableland and Espinhaço Mountain Range, which comprise elevations and flattened surfaces remaining of stilt and anticline bend structures in failed metasedimentary rocks.

The regions highlight the peculiarities mainly related to lithology, tectonics and morphology, which individualize them. Region of Diamantina Tableland located in the central part of the State of Bahia, the plateaus of this region extend through almost the entire northwest portion of the aforementioned sheet.

The BR 242 crosses them, to the north, toward O-E, facilitating access to some of the most important cities in the area, such as Seabra, Lençóis, Palmeiras, Andaraí and Mucugê. Rio de contas and Barra da Estiva, which stand out in the southern part of the region, and the city of Livramento do Brumado, located at the foot of the Serra in the Contas River, in the Region of the Sertaneja Depression. It is a topographically elevated set, with altimetry almost always exceeding 800 m, in the flattest areas and passing quickly at more than 1500 m altitude. It includes a culmination of the State of Bahia, the Barbado peak, in the Mesa Mountain, with 2033 m, with more moisture stains in the Sincorá mountain range, around Lençóis and in the Barra da Estiva area.

The reliefs that make up this region reflect to some extent a remnant structure of foldings, comprising mainly residual elevations that correspond to fragments of excavated anticline and suspended slopes and vast intermediate areas flattened. There are also formations of the Middle Precambrian, included as coverage of the Craton of São Francisco and corresponding to the Diamantina Tableland Group, belonging to the Espinhaço Supergroup. They mainly cover the double and faulted metasedimentary rocks of the Caboclo Formation and the Tombador-Lavras Formation.

The arrangement and affection of the folds, almost always of wide radius of curvature, represent traces that particularize the region of Diamantina Tableland in the face of other regions such as the Espinhaço Mountain, where the foldings are narrower and elongated. For the extensive longitudinal faults have a clear preferential alignment approximately with NNO-SSE steering. Minor transverse faults overlap with folding, characterizing above all the eastern part of the region.

Western Edge Mountain—This unit encompasses the mountainous reliefs of the western part of the geomorphological region of Diamantina Tableland. It is an area that is not occupied, where the city of Piatã and Contas River represent the most important urban centers.

The Western Edge Mountains are in an altimetric position almost always above 1000 m and often with the highest dimensions of the SD 24 Salvador sheet greater than 1400 m. It includes the highest points of the State of Bahia such as Barbado Peak with 2033 m and Almas Peak with 1958 m of altitude located, respectively, in the Mesa mountain and the Almas mountain.

The relief was identified in the mapping as landscape of differential dissection, governed by tectonics and lithology, where the interfluve present, in general, steep slopes and sharp tops, in the case of hills of irregular or possibly convex-concave slopes, almost always containing rocky outcrops and fallen blocks in which fragmentation marks are noted.

The remaining forms of the ancient folded structures almost always configure asymmetric ridges, mapped as hogbacks. To the west of the city of Piatã, where they are most conserved, they were marked on the map as edges of suspended slopes, corresponding to the elevations known locally by Atalho Mountain and Tromba Mountain. It consists of a rocky cornice at the top, followed by a detritic talus that is amended to glacis already dissected at the foothills.

The Western Edge Mountains represent an area subjected to folding, recognized by the bars and ridges aligned in a characteristic way, and to faults, identified through deep rectilinear grooves and escarpments that rise parallel to these valleys adapted to the structure and the fault escarpments.

To the north of the Contas River, in the interior of Piatã, and to the west of it are recognized planed stretches considered as pediplanes elaborated in successive phases and partially dissected, in the place where a dam was built on the Brumado River, in the vicinity of the city of Contas River. The drainage network of these Western Mountains is headed for the Contas River, which has its springs at the southern end of the Piatã slope. It is composed of small rivers, with intermittent regime.

Eastern Slopes—It is a geomorphological unit representing the relief printed on the extensive eastern flank of the Diamantina Tableland. In Sheet SD.24 Salvador extends from north to south for almost the entire central part of the V-A and V-C grids, except for the north-central and the center-south, respectively. With about 3180 km^2, it has an elongated shape and approximately constant width of 20–25 km, narrowing only at its southern end. This unit stands out for being an area of mountainous relief where altitudes are almost always above 800 m, although there are areas, usually in river valleys, where altimetry can drop to 600 and up to 400 m. The highest altitudes exceed 1200 m at the tops of the Sincorá Mountain.

It is composed of lithologies of the Middle Cambrian Pre-Cambrian belonging to the Diamantina Tableand Group, treating quartzites, metarenites, metassiltites and conglomerates of the Tombador and Caboclo Formations. Throughout its extension it has very thin coverings, which correspond to litholic soils.

As a wide structural surface with east-to-east trim and sectioned across numerous valleys controlled by the structure, this unit is constituted, as to the geomorphological aspect, mainly, by interfluves represented by inclined and bare structural planes, where the rock surfaced or is close to the surface. Often the back of these reliefs is carved into approximately triangular facets (chevrons) that are formed due to the selective work of erosion, accompanying the diving planes of superimposed layers and different resistances.

Mostly interfluves are separated by valleys of tectonic origin, which deeply cut the structure of the great anticline in various directions, cloned by almost entirely bare walls, steep escarpments dominated by very pronounced cornices.

The reverse of this extensive plateau formed by the domed slope corresponding to the structure of the anticline ends in ramps quite inclined and sectioned, descending from the mountain to the valleys of the tributaries of Paraguaçu that cut the structural planes. Another feature of common occurrence in this unit are monocline front and steep rocky cornice.

In some cases, it is observed that the topographic trim does not coincide with the dip of the layers, evidencing the presence of erosion plains. Flat top residuals in layers with sub-horizontal diving, such as The Hill of Pai Inácio, contrast with asymmetric ridges in very inclined layers. The more conserved parts of the relief are supported by the resistant layers of quartzite, as attested by the rocky cornices that top the summit. The dissected stretches correspond to intercalations of more tender rocks, where erosion sculpts ravines and convexized forms such as bumps and hills, often in areas embedded in the plateau and lined by escarpments.

In the landscape of the Eastern Slopes stand out typical features of inverted relief as the emptied anticlinal where the hill of Pai Inácio is located, presenting residual ridges (crêtes) and hills testimonies of the anterior arrangement of the layers. The western boundary of this unit in its central-southern sector is marked by escarpments adapted to the faults that line up in the NNO-SSE direction and are disposed of in an almost continuous wall, interrupted only by the passage of rivers that accompany areas of weakness resulting from fractures transverse to the escarpment.

The performance of tectonics on the geological formations that constitute the unit in question is expressed in the numerous grooves and valleys that have adapted to fractures and faults by cutting the entire area of the anticline flap, in a dissection entirely governed by structural elements.

In this unit can be mentioned the existence of a depression formed as a result of faults, bordered by the escarpments of hogbacks tilted to east, southeast and west. Some of these reliefs rise as residual within the depressed area, very sandy, covered by Savanna vegetation and already in the process of dissection. It is the place where the locality of Afrânio Peixoto is located, in the northern part of the eastern slopes.

This unit is drained by rivers belonging to the Paraguaçu Basin, some with springs in the mountain itself, while others come from the western part of Diamantina Tableland, superimposing the anticline of the Sincorá mountain range on its descent to the Paraguaçu trainers. The Paraguaçu River penetrates the area of the mountainous reliefs through an open canyon approximately west of the city of Mucugê and is little carved while crossing the back of the anticline.

Bonito River Plateau—corresponds to a segment of the eastern buttresses of the Diamantina Tableland, placing immediately north of the unit representing the eastern slopes. It occupies the extreme north in the central part of the V-A grid, exceeding the limits of Sheet SD.24 Salvador to Sheet SC.24 Aracaju. It totals about 1490 km^2 in the highlighted map, with altitudes between 600 and 1200 m, the highest corresponding to the most conserved stretches of flattened relief.

It is an area of occurrence of quartzites and conglomerates of the Tombador-Lavras Formation and siltites with sandstone intercalations of the Caboclo Formation, both Diamantina Tableland Group, of the Middle Precambrian, in addition to stretches of detritic coverage of the Tertiary-Quaternary. It constitutes a plateau with more conserved sectors of tabular reliefs, and others already dissected at different intensities, mapped as facies of differential dissection. In the most dissected areas the relief configures large convexized shapes, knolls and hills separated by valleys usually very deep, with slopes of strong slope, covered by forest vegetation that reaches the margins of the watercourses.

The drainage of the area goes to the Utinga River, obeying a certain structural control. Some rivers exhibit rectiline stretches or anomalous curves and have been mapped as adapted to the structure. The Bonito River stands out, entirely rectiline and with uneven and asymmetric margins due to faults that raised the corresponding block to its left margin.

Central Pediplane—This unit is represented mainly by the highlands and residual reliefs of the central part of the Diamantina Tableland, extending diagonally toward NNO-SSE between the western part of the V-A grid and the south-central part of V-C and extending to Sheets SD.23 Brasília, SC.23 São Francisco and SD.24 Salvador totals an area of 8800 km^2.

It comprises large expanses of high plateaus, with altimetry always exceeding 1000 m and residual strains of more than 1200 and 1400 m in altitude. Dissected reliefs occurring on the southern and northern edges correspond to the lower areas. The Central Pediplain is characterized by almost completely flat topography in the set, with residual reliefs that stand out on the flattened surface. It is a pediplain that presents different altimetric levels, related to different stages of preservation. The most conserved is in a higher position, exceeding 1200 m of altitude, located mainly to the east of the Bastião Mountain and west of the locality of Cascavel.

The less conserved interfluves constitute most of these plateaus. They correspond to lower areas and landscape in wide slopes, very small slope planes, one to two kilometers long, elaborated during successive phases of planation and dissection. These are very regular ramps, usually colluviums, that perfectly mend with the detritus talus, always present, or sometimes form an abrupt (knick) at the foot of the escarpments. Residual reliefs appear almost always isolated or forming extensive ridge lines, usually in the form of asymmetric ridges, mapped as hogbacks or monocline escarpments, dominated by quartzite cornices and, in certain cases, as inselbergs.

The tops of the residual reliefs correspond to exhumated structural plains and often constitute the fragmented edges of excavated anticline or suspended slopes, such as the elevations of the Bastão Mountain, representing crumble sections of the shoulders of a large anticline that is now emptied and flattened, constituting the most extensive part of the Central Pediplain.

The Central Pediplain has a marked border at its western boundary, where it comes into contact with the reliefs of the mountains of occidental border. At the southern end of the unit there is an extensive depression, molded into a synclinal carved in limestone rocks of the Bambuí Group, where it was mapped patterned of dissolution, in covered karst (Kc) or karst patches in exhumation (Ke). This depression has raised edges that end in escarpments facing outward, dominated by rocky cornice followed by detritus talus. Long inclined planes accompanying the diving of the layers of the synclinal border the relief of the interior of the depression, where there are also forms such as limestone hills (one of which has the cave of Ituaçu, three kilometers long), forming "islands" in the alluvial planes of the Várzea stream, a tributary of the Ourives River.

Most watercourses are intermittent. Many streams and rivers flow perpendicularly to the alignment of residual reliefs forming superimposition gorges, such as those that the Cochó River dig stake in the Bastião mountain range taking advantage of lines of weakness. Valleys are usually shaped like "U". In its high course, in the southern part of the pediplain, the Paraguaçu River flows slowly, fed by sources from the Sincorá mountain range, where the resumption of recent erosion has not yet had any sensitive effects. The trend presented during some phases of the basin formation continues, which is characterized by the occurrence of temporary lagoons with diatoms deposits.

Espinhaço Mountain Region—includes reliefs belonging to the domain of the plateaus folded structures, common aspects that differentiate them from the Diamantina Tableland region, and represented in Sheet SD.24 Salvador, by the geomorphological unit Pediplain of Alto do Rio Pardo.

Alto do Rio PardoPediplain—Narrow range of high and mountainous reliefs that penetrates the area of Sheet SD.24 Salvador to the northwest of the southern sector of the Geraizinhos Plateau, isolating the most lowered area where the city of Condeúba is located and separating these plateaus from the Levels of the Middle Contas River. [...], approximately 790 km^2 in the study sheet.

Altimetrically it is always above 800 m, with stretches between 1200 and 1400 m in altitude. These are small planation sectors and landscape with differential dissection. The relief forms are arranged on the folded structures of the mica schist and conglomerates of the Inhaúmas Mountain Group.

The flattened areas are composed of planes drawn up by successive phases of planation, where the interfluves constitute spines only slightly inclined, in an incipient way by the grooves and streams of diffuse surface runoff and elemental concentrate. It is drained by small tributaries of the Muquém River, belonging to the Pardo River Basin, and the Galvão River, from the Contas River Basin.

10.5 Domain of the Crystalline Plateaus

The denomination given to these plateaus intends to show the opposition between them and the other plateaus, sedimentary and metasedimentary. This domain is represented in Sheet SD.24 Salvador by three geomorphological regions: Raised Plateau, Lowered Plateau and Rejuvenated Plateau. It is characterized by a geological history dating back to very ancient orogenetic cycles, Guriense and Transamazonic, with rocks already in high degree of metamorphism and clear alignment of the relief, following preferential directions, and areas of granite rocks removed later. It includes granitized marginal bands incorporated into the Craton of The San Francisco, which are traversed by extensive faults directed at the SO-NE and NO-SE directions, where fault-underlined folds are noted in the form of alignments in the relief. It also covers areas of the Ribeira orogenic range, which underwent a reactivation in the Brasiliano cycle, with folding and inclusion of acid and basic plutonism.

Raised Plateau Region—This region is in the center-east of Sheet SD.24 Salvador. It occupies almost the entire western part of the V-D grid and portions northwest of Y-B, as well as fragmented sectors southwest of V-B and southeast of V-A, as well as a small spot southeast of the V-C grid. It comprises elongated relief bands in the north–south direction and which follow gradually decreasing altitudes from west to east. Residual elevations reaching 800–1000 m of altitude are indicated, related to the conservation reliefs of the top plateaus, to the west and southwest.

The region consists of lithologies of the Lower Precambrian, mainly, charnockites and garnet-biotite-gneisses. It is subdivided into two geomorphological units: Massif Central and Marginal Mountains.

Central Massif—a higher nucleus, located in the mid-west of the raised plateau region, from which the relief strips are available, in the domain of the crystalline plateaus, in a roughly concentric way, from the interior to the coast. It occupies parts to the southwest and west of the V-D grid, with a total of about 3810 km^2. Altitudes ranging from 300 to 700 m predominate, but residual tops of up to 1200 m occur and in the valley bottoms you can reach quotas around 120 m.

The performance of tectonics left its most remarkable impression in the strict direction of the valleys, preferably in the directions NO-SE, SSO-NNE, ONO-ESSE and OSO-ENE, composing an intricate network where no direction stands out, but the block of crisscrossed valleys itself imposes itself overall. The interludes constitute elongated hills, of rounded slopes and tops sometimes resembling slightly flattened residual bars. Some alignments or groupings of these hills are locally named "mountain", such as the Gentio Mountain, the Sucupira Mountain and the Capim Grosso mountain range, to the west and southwest of the city of Manuel Vitorino. These are structural forms almost always with differences of more than 60 m high, often reaching about 80, 100 and up to 150 m difference between the top and bottom of the valley.

It was found that the structural hills have irregular slopes, which are also often convex-concave or partially rectilinear. In general, they are bare, locally cleaned from the alterations, and the rocky outcrops even form abrupt slopes, with escarpments up to 30 m high and steep pontoons. It is also necessary to highlight the existence of recessed areas with flat relief, sometimes wide alveolus that seem to develop at the intersections of some of the adapted valleys, especially in the center-south of the unit. These alveolar planes are surrounded by the structural hills, which sometimes rise like rocky pontoons dominating the ramps. The alveolus where the town of Manuel Vitorino is situated is the one located about five kilometers east of the stone dam, crossed by the Contas River.

Marginal Mountains—are located immediately east of the central massif. It is placed as an extensive longitudinal direction strip, about 30 km wide, on average, with the approximate shape of a wide arch with the concave part facing west. Its main trunk occupies the central part of the V-D grid and west of Y-B, while southwest of the V-B grid and east of V-A are fragmented and discontinuity sectors, totaling about 8590 km^2 in length, in general, above 400 m altitude, mainly between 600 and 1000 m. However, there are more depressed, localized areas around 100 m, and residual elevations up to 1200 m in altitude. The intense dissection of the relief and its alignment along deep grooves oriented approximately in the SSO-NNE and NO-SE direction give this unit its most peculiar aspect in the set of reliefs of the crystalline plateaus.

Such reliefs are hills, that is, elevations with less than 50 m, with heights of more than 50 and up to 100–200 m difference between the top and base of the slope. There are also isolated elevations and bumps, which are lower and elongated hills, eventually wider. In the places near the top pediplains the shapes have more flattened tops, precisely because they are transition stretches where the flattened relief was not completely reached by dissection as some elevations of the Orobó massif, west of Itaberaba. Other stretches where the relief features mountainous features, with deep drainage slot and high shapes, as the valleys of the Preto Rivers, Preto da Crisciúma e Valentim, all with very embedded and rectilinear bed, adapted to structural lineaments.

In most of the area occur charnockites and hornblende-granites that are lithologically very homogeneous. Locally appear schists, quartzites and amphibolites of the lower Precambrian. The shear to which most of the unit was submitted, given the stiffness of the material, is reflected in the current morphology by the numerous river valleys that visibly adapted to fractures or faults, clearly noticeable in radar image mosaics. These are valleys almost always deep, with rectilinear tracings and preferential meaning NO-SE and SO-NE, between crossing in several directions and resulting in an orthogonal drainage network.

They are usually "V" shaped and sometimes have rugged strands adapted to faults. In its central-north part, the unit is drained by the Jequiriçá River and some of its tributaries, as a dendritic pattern, and in the northernmost sectors, by tributaries of the Paraguaçu River, with intermittent regime.

Processes of the current evolution dynamics of the relief are all represented in the different phases of morphogenesis. They occur in the area of the marginal mountains, in general, rock outcrops on the tops of the reliefs and there are large blocks and boulders fallen on the slopes, lose or buried, where the existence of joints and the development of karren can often be observed. The rock is subjected to fragmentation in blocks and slides and granular disaggregation, and is also, in general, profoundly altered. In most cases, an initial phase of the formation of bonhardts is identified. Erosion acts through diffuse runoff, causing local concentrations of sand, granules and detrital spread, as observed in the Preto River Valley, between the cities of Santa Inês, Cravolândia and Theolândia, and through the concentrated runoff, with the formation of grooves and ravines that carve to the bottom of the forms, in some cases discovering the rock.

Among the tectonic events that gave rise to the lines of weakness used by the watercourses stand out the Itaberaba Fault, of Direction NNO-SSE, and the fault of the plateau escarpment, of direction SSO-NNE. The first extends approximately from the northwest of Itaberaba to the northeast of the city of Jitaúna, being covered in the Geraizinhos Plateaus.

In this unit it is also necessary to point out the presence of a circular geological structure, in Brejões, which conditioned the organization of anomalous drainage that guides the intrusion of granite rocks that constitutes it.

10.5.1 Lowered Plateau Region

It covers large tracts east of Sheet SD.24 Salvador, occupying almost the entire center-east of the Y-B grid and east of v-D and extending to the V-B, Y-D and X-A grids. The sector that is in spatial discontinuity with the rest of the region corresponds to the eastern Horst of the Graben do Recôncavo, while the continuous and longer sector of the region includes the western dorsal of this tectonic trench, limited by the Maragogipe Fault, which divides it from the Recôncavo region and the Coastal plains. His contact with the Raised Plateau is made in large expanses through the fault of the plateau to the west and the Itapebi Fault to the southwest.

It encompasses topographically lowered reliefs in relation to the Plateau Raised, belonging to the same domain, with altitudes also decreasing toward the coast. The arrangement of relief in bands that remain constant in the north–south direction is not as sharp as in that region. It includes alignments of mountains with altitudes above 800 m, and pre-coastal and coastal areas where altitudes occur around 200 m. This is an area of occurrence of pyroxene-granulites, granulites migmatized and charnockites of the lower Precambrian. The geomorphological units that make up the region of the lowered plateau are mountains, pre-coastal massifs and pre-coastal tablelands.

Pre-coastal Mountains and Massifs—It covers the mountainous reliefs entreated by relatively flat areas that take all the central part. It has a large altimetric amplitude, since the lowered stretches reach less than 100 m of altitude, while some residual tops reach dimensions greater than 1000 m. The dissected reliefs of the pre-coastal mountains and massifs corresponding to the higher topography sectors were, for the most part, mapped as landscape of differential dissection.

In other areas they were mapped as landscape of homogeneous dissection, as in the western and northern parts of the unit. The resulting relief forms are generally convexized interludes, ranging from small to large hills, which take on features of mountains. The slopes are generally convex-concave, but locally various combinations can be observed composing from perfectly convex slopes, especially in the lower hills, to mixed slopes in the waviest reliefs, where concavities appear next to inclined planes and rocky outcrops that make up steep rectilinear segments.

In the areas of mountains, the slopes are almost always steep, with differences of more than 200 m between the top and the base, and the tops have sharp appearance. The reliefs of the central part of the unit are arranged in extensive alignments of SSO-NNE direction, such as the Pelada mountain range and the Cafundó mountain range, mapped as ridge lines that, between stretches of lowered relief, give this unit a characteristic arrangement. They correspond, respectively, to the axis of syncline and old slopes, already quite eroded.

The lower stretches present fragments, almost always accompanying valleys such as the rivers Novo, Ouro, Pontal do Sol, Oricó Grande, Salgado, Gongogi and smaller tributaries of the Contas River. They form large alveolus surrounded by higher, usually convexized relief forms whose foothills coalesce with the ramps that converge on the thalweg, sketching almost flat top spines, only slightly bulging.

To the south of the city of Dário Meira and west of the Pelada Mountain Range these gullies are translated into the radar image by a high drainage density and a weak notch of the watercourses. It is still in the valley of the Gongogi River, in the extreme southwest of the unit, extensive field of termites.

The slopes in the area under study are generally taken by diffuse surface runoff and elemental concentrate, which leaves its marks through bedside alveolus, grooves and ravines, some covered by vegetation, and that sometimes-cut terraces, another form of widespread occurrence on the slopes and which denotes mass movements intensified by the trampled of cattle.

We observed outcrops mainly in the form of forests and blocks on the slopes, which show signs of fragmentation. It is an area of occurrence of migmatized granulites, pyroxene-granulites and metatexits of the Lower Precambrian, as well as small, localized intrusions of sienites, gabbros and granites.

The unit of the pre-coastal mountains and massifs is drained by tributaries of the Contas River, mainly from the right bank, through the Gongogi River, and by the Salgado, Almada, Una or Aliança rivers (and some of its tributaries) and by tributaries of the left bank of the lower Pardo River, such as the Panelão River. In the area under study, the Contas River presents, in general, a wide valley, with tumbled and disguised edges, locally asymmetric. The canal shows mobile bed, with lateral and axial sandy banks, and rocky outcrops. It is noted the existence of alluvial plain, constituting floodplains, and two levels of terraces. In some stretches, such as in the vicinity of Ipiaú, Ubatã, mainly in the area of the Funil dam, and ubaitaba and Aurelino Leal, the valley is narrower becoming more embedded and the river is superimposed on the alignments of ridges.

Some have rectilinear tracing, having been mapped as valleys adapted to the structure, as is the case mainly of tributaries of the left bank of the Contas River and in the sector northeast of the Y-D grid, where drainage makes deep incisions, carving closed valleys in "V", differently from what occurs in the most extensive block of the unit.

Pre-coastal Tablelands—comprise dissected reliefs installed on the western dorsal of the Recôncavo Fossa.

This unit is topographically lowered in relation to the Marginal Mountains and coastal boards but is placed at a higher altimeter level than the reliefs located to the east, registering altitudes between 100 and 200 m, in general. It is limited in extensive stretches by escarpments that accompany the Maragojipe fault, seizing it from the reliefs of the coastal lowland and the marine plains and fluviomarines, or reaches, other times, to the sea, as south of the city of Itacaré and the site of Salvador.

The relief of the area is quite uniform, with erosion dissected intensely and indifferently the granulites and charnockites much altered from the lower Precambrian. The resulting forms were mapped as landscape of homogeneous dissection, regardless of structural control. The forms usually correspond to hills of convex and convex-concave slopes and bulging tops which in certain stretches make up a landscape of "seas of hills", besides encompassing the hills of the site of the city of Salvador.

The thick mantle of alteration in the entire area of this unit includes, on the rock saprolite, a cover of latosols corresponding to yellowish red colluvial material usually one to two meters thick. The dissected reliefs were primitively covered by ombrophiles forest vegetation, whose remnants are found only on the tops of the slopes or are related to cocoa plantations, being used for shading these, present throughout the southern and southeastern part of the unit. It is an area of ancient occupation where deforestation and replacement of natural vegetation by pastures, another important activity in the area, leads to the acceleration of erosive processes. The mass movements intensify with the trampling of cattle, evidencing itself through entire slopes carved in terraces, while the diffuse surface flow and elemental concentrate deepen the incision of grooves and ravines, even developing bedside alveolus.

The extensive ridge line known as Jiboia Mountain range stands out in the landscape of this unit. These are large elevations, hills deeply excavated by drainage, with steep and convexized slopes where rock outcrop scans are observed. Other residual reliefs correspond to quartzite veins and have been mapped as ridge lines.

This unit is drained mainly by basins of smaller rivers that flow directly into the ocean, such as the Rivers Sauípe, Pojuca, Joanes, Jaguaripe, Dona and Almas, as well as the Paraguaçu, Jequiriçá, Contas Rivers and some of their tributaries. These rivers all perennial, due to the great humidity in these coastal tablelands, guaranteed by the humid climate to which they are subjected. The influence of tectonics on this unit is manifested through these watercourses, mapped as valleys adapted to the structure, which also occur in localized stretches, mainly in the SO-NE direction.

10.5.2 Rejuvenated Plateau Region

Comprises the reliefs located immediately to the east, southeast and south of the southern sector of the top plateaus, approximately between the coordinates of 14° 40′ and 15° S 40° and 41° 30′ W.

The altimetry in the area tends to increase from east to west as it reaches the top of the plateau from depression, ranging from 200 to 800 m. It represents the dissected compartments of the edge of the Vitória da Conquista Plateau, corresponding to the southern sector of the Geraizinhos Plateau, including the Residual Massifs located to the south. It is individualized, in the domain of the crystalline plateaus, by the great energy manifested in the excavation of the valleys, controlled by tectonics. It includes small flat tops, remnants of the conserved surface of the top pediplains.

The advance of dissection on the cover that maintains the plateau reveals rock formations of the undifferentiated Precambrian, Middle and Upper, the latter belonging, respectively, to the Supergroup Espinhaço Indiviso and the Supergroup São Francisco, represented by the Macaúbas Group. There are also granite intrusions. It has a very irregular contour, especially to the west, where it is marked by the notch of the ravines on the plateau. It is subdivided into two geomorphological units: Eastern Piedmont of the Vitória da Conquista Plateau and Residual Massifs.

Eastern Piedmont of the Vitória da Conquista Plateau—extends to the east and southeast of the southern sector of the Geraizinhos Plateau. It represents an intermediate level between those plateaus and the Itabuna-Itapetinga Depression, extending to the south to Sheet SE.24 Rio Doce.

The altimetric dimensions, from 200 to 800 m altitude, vary considerably in reduced stretches, from residual correlated to the top of the plateaus to valley bottoms that are almost at the same level as the depression. The Piedmont area has very busy relief, entirely consisting of hills and hills, resulting from the intense dissection that sculpts the eastern slope of the plateau.

The slopes are steep, with slope between 30 and 45°. The current morpho dynamics in certain locations is so important. At some points there are rocky outcrops at the base of the slope or at the top, showing a localized tendency for the denudation of the slopes, as in the Marçal mountain range and stretches in the southeast corner of the Y-C square. The drainage in this unit is very dense, in dendritic pattern, but composed of small watercourses, mostly intermittent, directed to the Rivers Pardo or Jequitinhonha. The valleys are deep, V-shaped and with a flat bottom, filled by the material of the slopes, composed mainly of rock alteration.

Residual Massifs—It covers the dissected and interspersed reliefs of rocky pontoons south of the Geraizinhos Plateau. The altimetry of the area is around 600–800 m, with residuals reaching 1000 m of altitude.

It is an area of gneiss of the undifferentiated Precambrian, subjected to intense dissection. Some quartzite shafts correspond to lineations that stand out among the dissected reliefs and have been marked as ridge lines, while granite intrusions, represented by large rocky outcrops, were mapped as pontoons, like the area around

Pedra Azul. These steep and bare slopes often appear dominating areas of lowered convexized reliefs, giving the unit a peculiar feature.

The lower choline sections seem to be covered by material carried because of the erosion of the pontoons. They also occur in the Residual Massifs tabular top reliefs corresponding to conserved fragments of the plateau. The sector that emerges southwest of the Gang Y-C, within the plateaus of Geraizinhos, represents an area dissected in granite rocks and is in a higher altimetric position than these plateaus.

This unit is drained by tributaries of the left bank of the Jequitinhonha River, mainly through the São Francisco River and its tributaries. The valleys are deep and have a flat bottom, with flooded areas forming lagoons. The drainage network presents a dendritic pattern, in general, being conducted in many cases by faults, which result in valleys adapted to the structure, composing a landscape differential dissection.

10.6 Final Considerations

A cut out of the thick traces of the studies until the 1980s was made, which will certainly present some gaps, because the objective is not to produce a text on the history of geomorphology of Bahia, but rather present part of the context and the diversity of studies conducted that contribute to the knowledge of the relief and geography of Bahia. However, after that decade, passing through the twenty-first century, XX until reaching the twenty-first century. XXI, there is an increase in the scientific production of Bahian geomorphology.

The size and geographical location of Bahia mark that the physical-natural conditions of its space-territory are heterogeneous. And the main dissimilarities can be observed through the general geomorphological framework of Bahia (a synthesis), represented by eight main units: coastal plain, plateaus, sedimentary basins, recessed surface, Diamantina Plateau, residual mountainous system Espinhaço, lowered surface sanfranciscana and western (Fig. 10.11) (3).

From the highest areas of large tablelands in west Bahian, through Diamantina Tableland (central sector) to reaching the lower areas of the coast (east) there is a unique and diverse geomorphological set, landscapes, of significant scenic beauty, rarity, and socio-environmental importance.

They are extensive or short scenarios, flattened, or mountainous. Terrain shapes, geomorphological systems that are products of geological and historical times also dynamized by work, by culture that urgently demand consistent, planned public or private actions that do not only aim to take advantage of natural potential.

But rather the adoption of scientifically marked socioeconomic interventions, which make it possible to reverse the worrying situation of degradation and environmental conflicts in Bahia (especially, for the greater morpho-climatic domain—the caatingas, not forgetting the Atlantic forest and the Savannah), including: deforestation, transport and illegal timber trade; hunting and illegal sale of birds and mammals, burning, mining, predatory agriculture, accelerated erosion of soils, irregular water

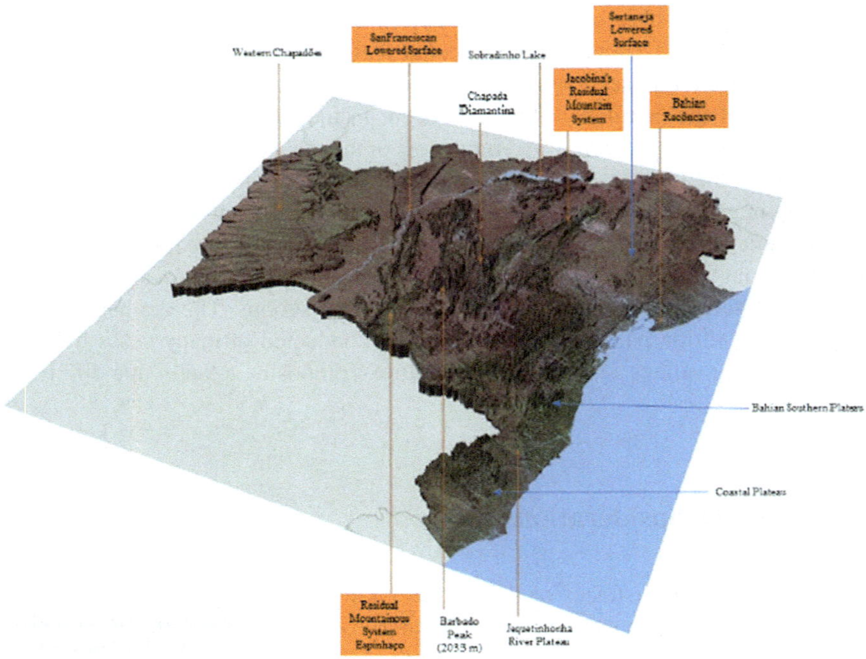

Fig. 10.11 Main geomorphological units of the State of Bahia—2022. Figure adapted by SANTOS, J. M. dos. *Source* Wind Atlas of the State of Bahia—Camargo Schubert, 2013

catchment in rivers, occupation of swampy areas, river banks and steep slopes, pollution, soil contamination and watercourses, release of domestic and industrial effluents into freshwater bodies, disposal and irregular burning of solid waste.

It should be said that the geomorphology of Bahia is also constituted as material and immaterial heritage, which since the sixteenth century, undergoes pressure, alteration, and environmental degradation, sometimes irreversible. Therefore, the relief should be preserved (when exhibiting some significant, rare environmental attribute) and preserved, while it is still possible to ensure the provision and maintenance of environmental services, as well as ensuring the right of populations (rural and urban communities) to live the landscapes and not just survive (at any cost and anyway).

Notes

(1) Text extracted from the work of SEPLANTEC—1976.
(2) Information extracted and synthesized from the Atlas of BAHIA-SEPLANTEC (1976).
(3) In the map in Fig. 10.11 highlight the new terms employed—recessed surfaces and residual mountainous system, in colored square boxes.

References

Ab'sáber AN (1998) Participation of peripheral depressions and flattened surfaces in the compartmentalization of the Brazilian plateau–final considerations and conclusions. J Geol Inst 19(1/2):51–69

Ab'sáber AN (1977) Os domínios morfoclimáticos na América do Sul. Geomorfologia 52:1–22

Alcoforado F (2003) Os Condicionantes do Desenvolvimento do Estado da Bahia. Tese (Doutorado em Planificacíon Territorial y Desarrollo Regional) Universidade de Barcelona. Disponível em: http://www.tdx.cat/bitstream/handle/10803/1944/6.TERRITORIO_SOCIEDADE_ESTADO_BAHIA.pdf?sequence=1

Alcoforado F (2021) The constraints of the development of the State of Bahia. Thesis (Doctorate). University of Barcelona. Available at https://www.tdx.cat/bitstream/handle/10803/1944/0.PREVIO.pdf?sequence=1&isAllowed=y. Accessed 7 Dec 2021

Almeida FFM (1969) Tectonic differentiation of the Brazilian platform. In: SBG, Brazilian Congress of Geology, 23, Salvador. Annals, pp 29–46

Almeida FFM (1977) The craton of San Francisco. Braz J Geo 7:349–364

Almeida FMM, Hasui Y, Brito NBB, Fuck RA (1981) Brazilian structural provinces: an introduction. Earth Sci Rev 17:1–29

Almeida JAP (1996) Le SertãoSisalier de Bahia - BrésilAnalyseD'images Spot Pour Une Typologie Dynamique de L'utilisationdes Terres (1986–1993). Tese de Doutorado. Université de Toulouse - Le Mirail. Toulouse

Angelim LAdA, Kosin M, Aracaju NW (2001) Folha SC.20-V, Bahia, Pernambuco e Piauí. Rio de Janeiro: CPRM, 2001. 1 CD-ROM. Scale 1:500,000. Basic Geological Surveys Program of Brazil. PLGB

Azevedo HCA, Costa PHO (Coord.) (1994) Catálogo de rochas ornamentais da Bahia-Brasil. Salvador: SGM, il. color. Texto em português e inglês, 146p

Bahia (1990) Superintendência de Geologia e Recursos Minerais.Uma Síntese da Geologia da Bahia = A synthesis draftof Bahia StateGeological Evolution, Juracy de Freitas Mascarenhas, Salvador-BA

BAHIA-SEPLANTEC (1976) Atlas do Estado da Bahia. Centro de Planejamento da Bahia/CEPLAB e Secretaria do Planejamento, Ciência e Tecnologia/SEPLANTEC, Salvador-Bahia, vol 1

Barbosa JSF, Sabaté P (2002) Geological features and the Paleoproterozoic collision of four Archean crustal segments of the São Francisco Craton, Bahia, Brazil: a synthesis. Anais da Academia Brasileira de Ciências 74:343–359

Barbosa JSF, Cruz SCP, Souza JSDE (2012) Terrenos Metamórficos do Embasamento. In: Barbosa JSF, Mascarenhas JF, Correa-Gomes LC, Dominguez JML, de Souza JS (Org.). Geologia da Bahia, pesquisa e atualização. 1ed. Salvador: Companhia Baiana de Pesquisa Mineral, vol 1, pp 101–201

Barros O (2020) Cumes da Bahia: As montanhas do 5º maior estado do Brasil. Disponível em: https://blogdescalada.com/cumes-da-bahia-as-montanhas-do-5o-maior-estado-do-brasil/. Acesso em 27 de novembro de 2020

Brasil (1980) Ministério das Minas e Energia. Secretaria Geral. Projeto RADAMBRASIL: Folha Salvador, SD.24. Geologia, geomorfologia, pedologia, vegetação e uso potencial da terra. Rio de Janeiro

Branner JC (1911) Aggraded limestone plains of the interior of Bahia and the climatic changes suggested by them. Geol Soc Am Bull 22(1):187–206

Branner JC (1977) Planície calcária de agradação no interior da Bahia e as mudanças climáticas por elas sugeridas. Sociedade brasileira de geologia. Núcleo da Bahia. Pub. Espec. n.1, 31:50

Barreto AMF (1996) Interpretação paleoambiental do Sistema de Dunas Fixadas do Médio Rio São Francisco, Bahia. Tese de Doutoramento. Instituto de Geociências. Universidade de São Paulo, 194p

Branner JC (1910a) The Geology and topography of the Serra de Jacobina, State of Bahia, Brazil. Am J Sci 30:385–392
Branner JC (1910b) The Tombador escarpment in the State of Bahia, Brazil. Am J Sci 30:335–343
Brito MCRO (1991) Aspectos geomorfológicos do contato inselbergs/pedimento na bacia do riacho seco, Milagres – Bahia. 122f. Dissertação (Mestrado) – Curso de Pós-Graduação em Geociências, Instituto de Geociências, Universidade Federal da Bahia, Salvador, BA
Carvalho LM (2010) Geodiversidade do Estado da Bahia. Organização Luiz M. de Carvalho e Maria A. Barreto Ramos-Salvador: CPRM, DVD, 184p
De Brito NB et al (1995) O evento Cariris Velhos na Província Borborema: integração de dados, implicações e perspectivas. Braz J Geol 25(4):279–296
De Oliveira RG, De Medeiros WE (2012) Evidence of buried loads in the base of the crust of Borborema Plateau (NE Brazil) from Bouguer admittance estimates. J South Am Earth Sci 37:60–76
De Oliveira, Paulo E, Barreto AMF, Suguio K (1999) Late Pleistocene/Holocene climatic and vegetational history of the Brazilian caatinga: the fossildunes of the middle São Francisco River. Palaeogeogr Palaeoclimatol Palaeoecol 152(3–4):319–337
de Souza JS (2013) Geologia, Metamorfismo e Geocronologia de Litotipos de Salvador-Bahia.Tese (Doutorado em Geologia, Universidade Federal da Bahia)
de Sousa AN (2015) Percurso historiográfico do campo disciplinar geográfico na Bahia e em São Paulo: contribuições da Universidade Federal da Bahia e da Universidade de São Paulo. Tese (Doutorado em geografia, Universidade Federal da Bahia)
Falk FH (1977) Esboço geomorfológico da área de Milagres-Bahia. Notícia Geomorfológica, Campinas 17(34):104–111
Fiori SR (2012) Mapas e afins para editoras. Disponível em: https://sergiofiori.carbonmade.com/projects/4379677. Acesso em: 10 de setembro, 2021
Góes LM (2019) Geocronologia das coberturas superficiais nos baixos terraços fluviais e aluviões recentes no médio curso da bacia hidrográfica do Rio Itapicuru (Bahia-Brasil). Tese (Doutorado) – Universidade Estadual de Campinas, (UNICAMP), Campinas, 2019. Disponível: https://repositorio.unicamp.br/Acervo/Detalhe/1091292. Acesso em: 14 dezembro de 2022
Halla MMS, Dominguez JML, Luiz César, Corrêa-Gomes LC (2020) Structural controls on the morphology of an extremely narrow low-accommodation passive margin shelf (Eastern Brazil). Geo-Marine Lett 40(6):865–878. https://doi.org/10.1007/s00367-019-00605-y
IBGE. Instituto Brasileiro de Geografia e Estatística. Cidades e Estados. Disponível em: https://www.ibge.gov.br/cidades-e-estados/ba.html. Acesso em: 03 mar. de 2021
Japsen P et al (2012) Episodic burial and exhumation in NE Brazil after opening of the South Atlantic. Geol Soc Am Bull 24(5–6):800–814
Jesus EFR (1981) Mapeamento geomorfológico da alta bacia do riacho Curimataí (Município de Serra Preta – BA) escala – 1: 30.000. Dissertação (Mestrado) – Curso de Pós-Graduação em Geociências, Instituto de Geociências, Universidade Federal da Bahia, Salvador, BA
King LC (1956) Geomorfologia do Brasil Oriental. Rev Bras Geografia 18(2):147–265. Disponível em: http://biblioteca.ibge.gov.br/visualizacao/periodicos/115/rbg_1956_v18_n2.pdf
Lage CS (1986) Lesmilieuxnaturales de la haute valleeduVaza Barris (Bahia-Bresil): et leuraménagement. Thèse de doctorat – 3 èmecycle em Géographie de L' aménagement. Université de Bordeaux III, Bordeaux
Lima KCO (2017) Holoceno superior na bacia do rio Itapicuru (Bahia/Brasil): proposição de cenário para o baixo curso e planície costeira. 2017. 1 recurso online (127 p.). Tese (doutorado) - Universidade Estadual de Campinas, Instituto de Geociências, Campinas, SP
Lima KC, Lupinacci CM (2019) Fragilidades e potencialidades dos compartimentos geomorfológicos da bacia hidrográfica do Rio Bom Sucesso–semiárido da Bahia/Brasil. Revista Equador vol 8, no 2, pp 503–520
Lima KC, Lupinacci CM (2021) Geomorfologia do semiárido: proposta metodológica de representação cartográfica e interpretação do relevo em escala de detalhe. Revista Brasileira de Geomorfologia 22(2)

Maia RP, Bezerra HR, Sales VC (2010) Geomorfologia do Nordeste: concepções clássicas e atuais acerca das superfícies de aplainamento nordestinas. Disponível em: file:///C:/Users/profe/Downloads/228857-64471-1-PB.pdf. Acesso em: 24 de novembro, 2021

Mendes IA (1978) Estudo Geomorfológico na área dos divisores entre as bacias dos rios Paraguaçu e Jequiriçá. Dissertação (Mestrado) – Curso de Pós-graduação em Geociências, Instituto de Geociências, Universidade Federal da Bahia, Salvador, BA

MME (1981) PROJETO RADAMBRASIL Programa de Integração Nacional (Folha Salvador, V. 23), Rio de Janeiro

Mescolotti PC, do Pupim FN, Sawakuchi AO, Ladeira FSB, Assine ML (2021) Fluvial aggradation and incision in the Brazilian Tropical Semi-arid: climate-controlled landscapes evolution of the São Francisco River. Quatern Sci Rev 263:106977

Neves B, de Bley B, dos Santos EJ, Schmus V, Randall W (2000) Tectonichistory of the Borborema Province, northeastern Brazil. Tectonic Evolution of South America, rio de Janeiro, pp 151–182

Oliveira JH (2020) Influências do quadro litoestrutural e dos processos fluviais na configuração do modelado do relevo no Médio Paraguaçú/Jiquiriçá, Região de Milagres/Itatim-BA. 2020, 191f. Tese (Doutorado em Geografia) – Universidade Federal de Minas Gerais – UFMG

RADAMBRASIL (1983) Levantamento de recursos naturais, folhas SF. 23/24, Rio de Janeiro/Vitória. Projeto RADAMBRASIL, vol 32. Rio de Janeiro, 775p

Santos JM (2016) Indicadores de desertificação no semiárido brasileiro. O caso de Campo Formoso-Ba. 471 f. Tese (Doutorado) – Universidade Federal Fluminense, Niterói

Santos LA (2020) Geomorfologia e geologia do quaternário da planície flúvio-eólica do Médio rio São Francisco, NE do Brasil. Tese (Doutorado) – Universidade Estadual Paulista, (UNESP), Rio Claro

Seabra GF (1991) Estudo geomorfológico da Região Cárstica de Andaraí: uma contribuição à conservação de cavernas. Dissertação de Mestrado. Recife, CFCH/UFPE

Seabra GF (2017) Chapada Diamantina: o falso brilhante. Ituiutaba, Barlavento 176p

Santos EJ, Medeiros VC (1999) Constraints from granitic plutonism on proterozoic crustal growth of the Transverse Zone, Borborema Province, NE-Brazil. Revista Brasileira de Geociências 29:73–84

SEI (2004) Superintendência de Estudos Econômicos e Sociais da Bahia. Mapas Digitais do Estado da Bahia: base de dados. SEI, CD-ROM, Salvador

SEI–SUPERINTENDÊNCIA DE ESTUDOS ECONÔMICOS E SOCIAIS DA BAHIA (1991) Bahia: reconstrução e integração dinâmica. Salvador:SEI

Silva Filho RA (2010) História Geológica da Bahia. CBPM, Salvador

Silva TC (1959) Problemes Geomorphologiques et Paleogeographiques duNord-EstduBrésil. França (Tese de Doutorado) Universidade de Strasbourg

Silva TC (1960) Repercussões da sedimentação Cretácica sobre o relevo baiano. Boletim Baiano de Geografia 1

Stríkis NM, Cruz FW, Barreto EA, Naughton F, Vuille M, Cheng H, Voelker AH, Zhang H, Karmann I, Edwards RL, Auler AS, Santos RV, Sales HR (2018) South American monsoon response to iceberg discharge in the North Atlantic. Proc Nat Acad Sci 115(15):3788–3793. https://doi.org/10.1073/pnas.1717784115

Stríkis NM, Chiessi CM, Cruz FW, Vuille M, Cheng H, de Souza Barreto EA, Mollenhauer G, Kasten S, Karmann I, Edwards RL, Bernal JP, Sales HR (2015) Timing and structure of Mega-SACZ events during Heinrich Stadial 1. Geophys Res Lett 42(13):5477. https://doi.org/10.1002/grl.v42.1310.1002/2015GL064048

Tricart J (1958) Alguns problemas geomorfológicos da Bahia. Notícia geomorfológica, Campinas 1(2):27–32

Tricart J, Silva TC (1959) Programa de Estudos Geomorfológicos para o Vale do Paraguaçu. Universidade da Bahia. Editora: Universidade da Bahia

Tricart J, Silva TC (1969) Estudos de geomorfologia da Bahia e Sergipe. Editora Imprensa Oficial da Bahia

Tricart J, Silva TC (1961) Um exemplo de evolução cárstica em meio tropical sêco: o môrro de Bom Jesus da Lapa (Bahia). Boletim Baiano de Geografia, Salvador 5/6:3–19

Valadão RC (1998) Evolução de Longo-Termo do Relevo do Brasil Oriental (Desnudação, Superfícies de Aplanamento e Soerguimentos Crustais). 243 p. Tese (Doutorado) – Departamento de Geologia, UFBA, Salvador

Wang X, Auler AS, Edwards RL, Cheng H, Cristalli PS, Smart PL, Richards DA, Shen CC (2004) Wet periods in northeastern Brazil over the past 210 kyr linked to distant climate anomalies Nature 432(7018):740–743. https://doi.org/10.1038/nature03067

Jémison Mattos dos Santos He is a Geographer. PhD in Geography from the University Federal Fluminense/UFF and University of Coimbra/UC; Master in Geochemistry and Environment, Federal University of Bahia; Especialization in Management Environmental, by the triad Environmental Resources Center—CRA/ Department of Forest Defense—DDF and Catholic University of Salvador—UCSAL. He is currently Adjunct Professor at the State University of Feira de Santana (UEFS), Coordinator of the Research Group—Science, Technology in Landscape Evolution, Soils and Planning (GEOLANDS-UEFS). Coordinator of the Geography Commission of the Geographic and Historical Institute of Bahia—IGHB. He has experience in studies of Quaternary Geomorphology, Land Degradation and Desertification. International Research Networks: WATERLAT-GOBACIT (Member) and GLP—International Program—Working Group—Coproduction of Sustainable Earth Systems (Member).

SPRINGER NATURE

GPSR Compliance

The European Union's (EU) General Product Safety Regulation (GPSR) is a set of rules that requires consumer products to be safe and our obligations to ensure this.

If you have any concerns about our products, you can contact us on ProductSafety@springernature.com

In case Publisher is established outside the EU, the EU authorized representative is:

Springer Nature Customer Service Center GmbH
Europaplatz 3
69115 Heidelberg, Germany

The manufacturer's authorised representative in the EU is Springer
Nature Customer Service Centre GmbH, Europaplatz 3, 69115 Heidelberg,
Germany. If you have any concerns regarding our products, please
contact ProductSafety@springernature.com

Printed and bound by CPI Group (UK) Ltd, Croydon, CR0 4YY
30/09/2026
02225586-0009